The Richardson Light Guard
of Wakefield, Massachusetts

The Richardson Light Guard of Wakefield, Massachusetts

A Town Militia in War and Peace, 1851–1975

BARRY M. STENTIFORD

McFarland & Company, Inc., Publishers
Jefferson, North Carolina, and London

LIBRARY OF CONGRESS CATALOGUING-IN-PUBLICATION DATA

Stentiford, Barry M.
The Richardson Light Guard of Wakefield, Massachusetts : a town militia in war and peace, 1851–1975 / Barry M. Stentiford.
    p.   cm.
Includes bibliographical references and index.

ISBN 978-0-7864-7348-9
softcover : 50# alkaline paper ∞

1. Massachusetts. Militia. Richardson Light Guard.
2. Soldiers—Massachusetts—Wakefield—History.
3. Wakefield (Mass.)—History, Military.
I. Title.
UA258.R5S74   2013   355.3'7097444—dc23   2013011163

British Library cataloguing data are available

©2013 Barry M. Stentiford. All rights reserved

No part of this book may be reproduced or transmitted in any form or by any means, electronic or mechanical, including photocopying or recording, or by any information storage and retrieval system, without permission in writing from the publisher.

On the cover: Rifle Team, Co. A, 6th Regt M.V.M., Richardson Light Guard 1895 (courtesy Wakefield Historical Society); background (iStockphoto/Thinkstock)

Manufactured in the United States of America

McFarland & Company, Inc., Publishers
Box 611, Jefferson, North Carolina 28640
www.mcfarlandpub.com

For the thousands of men who gave their money, time,
blood, and sometimes lives to serve
in the Richardson Light Guard in all its incarnations
from 1851 through 1975. They showed
the true meaning of community and service.

Acknowledgments

This book has been a work in progress for several years and along the way I have benefitted from the assistance of several people. I owe a great deal of gratitude to Mr. James Buckle, who served in Co. E of the 182d in World War II, earning a Silver Star, and who spent many years afterward compiling records of the men with whom he served. Much of the material for chapters VIII and IX came from his collection. I also need to thank Drs. Sally J. Southwick, William M. Donnelly, and Eleanor L. Hannah who carefully read early versions of the manuscript and offered valuable suggestions and corrections. Colonel (Ret.) Leonid Kondratiuk, the director of the Massachusetts Army National Guard Museum and Archives, gave consistent support both through the collections he oversees and through his own vast knowledge of the military of Massachusetts. Dr. Robert McFarland made a detailed critique of a paper extracted from this and made numerous useful suggestions that greatly enhanced the finished product. The *Journal of Military History* published an article drawn from the manuscript in July 2008 and has graciously allowed the reprinting of the material here. My friend Mr. Michael Boucher and Ms. Nancy Bertrand were very helpful in getting photographs for me. I must also thank the people of Wakefield, who support the Lucius Beebe Memorial Library and its fine local history room, and Mr. Mark Sardella, who graciously allowed me to use images from the library's collection.

And lastly, I must thank my wife, Vitida, and son, Geoffrey, for patiently allowing me to spend too much time engaged in this project over the past few years.

Table of Contents

Acknowledgments vi

Lineage ix

Introduction: A Town Goes to War 1

I. Frontier Town 13

II. New Militia Company 30

III. Volunteers 43

IV. Gilded Age Militia 62

V. A Militia for Empire 81

VI. National Institution 101

VII. In the Army Now 117

VIII. The Oldest Regiment 134

IX. In the Americal Division 149

X. Bedroom Suburb 168

Chapter Notes 183

Bibliography 203

Index 211

Lineage

UNITS BEARING THE HERITAGE OF THE RICHARDSON LIGHT GUARD

Co. D, 7th Reg. Mass. Volunteer Militia	1851–1855
Co. E, 7th Reg. Mass. Volunteer Militia	1855–1861
Co. B, 5th Reg. Mass. U.S. Volunteers (3 months)	1861
Co. E, 7th Reg. Mass. Volunteer Militia	1861
Co. E, 50th Reg. Mass. U.S. Volunteers (9 months)	1862–1863
Co. E, 7th Reg. Mass. Volunteer Militia	1863
Co. E, 8th Reg. Mass. U.S. Volunteers (100 days)	1863–1864
Co. E, 8th Reg. Mass. Volunteer Militia	1864–1866
Co. A, 6th Reg. Mass. Volunteer Militia	1866–1898
Co. A, 6th Reg. Mass. Infantry U.S. Volunteers	1898–1899
Co. A, 6th Reg. Mass. Volunteer Militia	1899–1901
Co. A, 6th Reg. Infantry Mass. Volunteer Militia	1901–1913
Co. A, 6th Reg. Infantry Mass. National Guard	1917–1919
Co. H, 12th Reg. Mass State Guard	1917–1921
Co. A, 6th Reg. Regiment (Provisional)	1919
Co. K, 9th Reg. Infantry Mass. National Guard	1919–1923
Co. E, 182nd Reg. Infantry Mass. National Guard	1923–1945
Co. H, 23rd Reg. Infantry Mass. State Guard	1940–1943
6th Company, 23rd Reg. Infantry Mass. State Guard	1943–1946

After the disbandment of the World War II State Guard, no company officially designated as the RLG.

Co. E, 182nd Regimental Combat Team	1946–1959
Combat Support Company, First Battle Group, 26th ID	1959–1963
Co. C, 1–182d Infantry Mass. National Guard	1963–1975

OTHER READING FIRST PARISH/SOUTH READING/WAKEFIELD UNITS

Reading Company, North Regiment	1644–1680
Reading Company, Lower Regiment of Middlesex	1680–1773
Reading Company, First Regiment of Militia	1732–1812
South Reading Company, First Regiment of Militia	1812–1840

Cavalry Troop	1658
Train Band (Minutemen)	1770–1775
Cavalry Troop	1793–1828
Washington Rifle Grays (aka Greens)	1811–1846
Co. E, 16th Reg. Mass. U.S. Volunteers (3 years)	1861–1864
Med. Det. 182d Infantry Mass. National Guard	1926–1946
HQ Co., 2nd BN 23rd Reg. Infantry Mass. State Guard	1940–1946
2d Batt. Sec., Med Det. 182d RCT, Mass. National Guard	1946–1948
Med. Company, 182d RCT, Mass. National Guard	1948–1950
Tank Company, 182d RCT, Mass. National Guard	1950–1955

READING THIRD PARISH/READING (POST-1812)

Co. B, 7th Reg. Mass. Volunteer Militia	1812(?)–1854
Co. D, 50th Reg. Mass. U.S. Volunteers (9 months)	1862–1863
Reading Home Defense Corps	1941–1943?

Introduction

A Town Goes to War

The declaration of war against Spain on 26 April 1898 sent waves of excitement through cities, towns, and villages across the United States. Although some apprehension must have existed beneath the surface, most Americans responded enthusiastically to the new war. The Civil War had ended thirty-three years earlier, and the public memory of that conflict had evolved from horror into a kind of nostalgia, where the brutality of the battlefield had been romanticized into a noble crusade and test of manhood.[1] A new generation of young men longed for its own test of manhood on distant battlefields. The last Indian Wars had largely been affairs of the Regular Army in remote parts of the West, and had offered no opportunities for civilian men in the East to quench their thirst for adventure.

For the men of the Richardson Light Guard—the local militia company of Wakefield, Massachusetts—news of the declaration of war brought a sense of relief after weeks of uncertainty.[2] The war with Spain would be their chance to show their fellow townspeople that they were worthy successors to earlier generations of men from the town who had marched off to war. Militia companies from the town fought in the colonial wars against the Indians and later the French. Companies from the town fought the British from the opening battles of the Revolutionary War. The town had even supported the War of 1812, which most of New England bitterly opposed. Only during the Mexican War, for which the town shared the distaste most New Englanders felt for that struggle, did no units march from the town to war.

Four times during the Civil War the Richardson Light Guard left the town to enter federal service. Following the end of the Civil War in 1865, the RLG remained the town's militia. For the next thirty-three years, the RLG led an existence typical for many militia units in the Northeast; they competed in rifle matches, fought sham battles, and held numerous banquets and balls. They paraded for their fellow townspeople and participated in most public events.[3] Officers were selected through the time-honored method of company-wide elections. The hometown consistently supported its militia company, building it one of the best rifle ranges in the state in 1891 and a fine new armory in 1895. The RLG had never performed service during labor unrest, and so remained untainted by that onerous duty. Instead the RLG gave men of Wakefield an institution through which they could find camaraderie, athletics, prestige, and a sense of community. But the RLG had always seen itself as a military organization above all, and now that self-image was again to be tested. After years of newspaper stories of Spanish outrages in nearby Cuba, war came to Wakefield. The weeks between the sinking of the U.S.S. *Maine* on 15 February and the declaration of war by the United States on Tuesday, 26 April, had been a time of uncertainty. The declaration ended that uncertainty. Governor Roger Wolcott, who had served as governor since 1896, was eager to get Massachusetts soldiers into combat. Since the Sixth Regiment of Massachusetts was regularly rated as one of the best militia regiments in the nation, and the RLG was Company A of the Sixth, everyone understood that the RLG would soon be called to federal service.

On Thursday, 28 April, the RLG held a special meeting at their armory on Main Street, at the southern end of Wakefield Center. The commander of the RLG, Captain Edward Gihon, asked each man to volunteer for federal service. A total of sixty-nine members did, with another fifteen taking a day's reflection to make the commitment. Many local men attempted to join the company to fill the "two or three" vacancies that had arisen due to members who were not able to go.[4] Most of the aspiring recruits had to be turned away — not because of physical shortcomings but simply because too many sought positions in the RLG. However, their names were taken in the event that the town needed to raise another company for the war, as had happened during the Civil War.[5] Few men if any in the RLG were in any way "strangers." Even the new members had community or even kinship ties to the company. If the soldiers in the ranks felt any trepidation at leaving their pleasant town for distant battlefields, none expressed it openly. Instead, the members of the RLG allowed themselves to be swept up in the carnival atmosphere of the town in the few days before the RLG entrained for their great adventure.

Although most of the men in the RLG were residents of Wakefield, the surrounding towns also contributed men to the ranks of the RLG. The neighboring town of Reading, whose history had overlapped with that of Wakefield for almost two centuries, contributed the most outsiders, but men from Reading were more like cousins than strangers to Wakefield men. Although all the officers were from Wakefield, most of the noncommissioned officers were from Reading. A few days after the special meeting was held, the RLG marched through the town of Reading, with the procession ending at that town's Masonic Lodge, where the militiamen and their friends listened to speeches and enjoyed a banquet provided by the town. One of the speakers, James H. Griggs, was a former resident of Reading who was now an alderman from Charlestown. Griggs had been captured at the First Battle of Bull Run during the Civil War while serving in the RLG. The church bells in Reading Square were rung in honor of the RLG, bells which could clearly be heard in Wakefield, a short three miles away.[6]

At the regular Monday evening drill, held on 2 May, a large crowd came to the armory in Wakefield to watch the whole company go through its training. All of the seats around the drill hall quickly filled and spectators stood ten or twelve deep.[7] The crowds reflected the ethnic and religious base of the town, with a mixture of old-stock Protestants, Swedes, and Irish Catholics whose parents or grandparents had immigrated to the United States, and Italians who were then settling in the town. James H. Carter, who had served in the company before the Civil War, came to the armory to present each married member of the RLG with five dollars in gold from the Wakefield Citizens' War Committee. Local businesses promised to hold jobs for members going to war. Second Lieutenant Frank Gray was given a "handsomely engraved sword" by his employers.[8] The next day, at a home on West Chestnut Street, just off of Main Street, a large bonfire attracted lots of younger boys caught up in the excitement of the evening. Effigies of Spaniards went up in flames to the cheers of the crowds.[9]

Dr. Solon O. Richardson, son of the man for whom the company was named, arrived back in Wakefield from his son's home in Ohio, during the first week of May and immediately went to Boston to buy a new American flag to fly above his family's estate on Main Street. This younger Solon O. Richardson had served in the company for nine years prior to the Civil War, being elected lieutenant shortly before he left the company due to increased business and family obligations. The flag was to remain flying until the company returned from its federal war service, as similar flags had flown over the estate three times during the Civil War.[10] According to the local newspaper, the governor issued orders for the 6th Volunteer Regiment to arrive at the state camp grounds in South Framingham on Friday, 6 May.[11] With the RLG due to leave town early that Friday, the big public send off was held on Thursday. That evening, the residents of Wakefield held a "grand farewell reception" for the RLG at the Town Hall.[12] For so peaceful and pleasant a town, Wakefield nevertheless suddenly seemed to brim with martial organizations. Escorting the RLG up the block to the Town Hall were the posts of the Grand Army of the Republic (GAR) from both Reading and Wakefield. The local posts of the GAR, a politically powerful organization of Union veterans of the Civil War, contained many men whom themselves had left the town for war as members of the RLG decades earlier. Joining the older veterans were members of the Sons of Union Veterans, of which Captain Gihon was a member. Also on hand was the organization for former members of the RLG, dubbed the Fine Members Association. Many marchers belonged to both the Fine Members and the GAR. Another group marching was composed of honorary members of the RLG. Like similar militia organizations, the RLG had long admitted prominent local men as honorary members, meaning that such men were invited to company social functions and could link their prestigious names with the company, and in return were expected to provide financial and personal support. The youngest men participating in the parade belonged to the high school company of cadets. The cadets turned out in their stylish uniforms, under the watchful eye of their drill master, himself a former longtime member of the RLG. Many members of the high school cadet company wanted to go with the RLG, but most were too young to join.[13] Also joining in the celebration were the members of Wakefield's fire departments—mostly volunteer firemen—looking in their uniforms and mustaches as martial as any group.

Main Street in Wakefield Center was predominantly filled with businesses and civic buildings, but a few private residences still fronted the street. At the northern end of the Center, surrounding the Common, stood four Protestant churches—the Congregational facing south at the southern end of a small lake, the Baptist, facing east, and the Unitarian and Episcopal, facing west. Saint Joseph's church, which served the town's growing Catholic population, was a few blocks west of the Common, on the other side of the tracks, both literally and figuratively. The marchers paraded the short distance to the Town Hall, up Main Street, in which every building stood heavily decorated with red white and blue bunting, American flags, and a few Cuban flags.[14] Once at the Town Hall, the celebrants packed into the building. The stage at the front of the auditorium displayed a large American flag, with a banner hanging from the center that read, "Co. A, Sixth Regiment—Onward to Victory—Richardson Light Guard." Inside the auditorium, the militiamen and their escorts listened to a series of patriotic speeches by a host of well-wishers from the town's military, civic, and religious spheres.[15]

At 7:30 in the morning of 6 May, the day after the parade and reception at the Town Hall, the town's fire whistle blew the 12–12 signal, calling the militia to assemble. This was a formality, more for the sake of tradition and to let the residents of the town know the time had come, rather than an actual call for the men of the RLG. The company assembled at the armory and marched west from Main Street the short distance to the railroad station, accompanied by a large crowd. The men were clothed for field service and not show, wearing their "fatigue uniforms, with leggings, canteens, haversacks, overcoats, and car-

The scene at Wakefield's Upper Station as the train carrying the Richardson Light Guard was about to depart for Boston for the start of its service in the Spanish American War, May 6, 1898 (courtesy Wakefield Historical Society).

tridge belts."[16] Across Railroad Street from the train station, each building, mostly two and three story residences with storefronts on their first floors, were also bedecked with flags. One small store had its display windows almost completely obscured by a large American flag that draped over the front, below a sign urging all to "Remember the *Maine*." Atop the same building flew small flags on the corners and another large flag on a pole in the center. Apparently, before the flag went up the pole, an effigy of the Spanish general Valeriano Weyler y Nicolau was hauled up the pole, where locals blasted it to pieces with guns. Schools were closed for the day in recognition of the importance of the event, and the scene at the train station was one few present would forget, being larger than similar events during the Civil War. Large groups of formally dressed men, women, and children crowded around the train cars as the men boarded shortly after nine o'clock in the morning. The windows of the train were open, allowing members of the RLG to wave once last time to family and friends. The train blocked Albion Street, which ran from Stoneham to Wakefield Center, and so passengers on street cars and men driving delivery wagons added to the throngs of spectators for the emotional event.[17] A few minutes after the train had arrived in the station, it departed for Boston, carrying the RLG away from Wakefield to war for its fourth time.

Through the Spanish American War, the bond between the militia company and the town of Wakefield remained strong. The town had supported the company morally and financially during peacetime, and eagerly followed the experiences of its company during wartime. The town created committees and organizations to provide moral and economic support to the families of soldiers who were away, and to the soldiers themselves. In short, the Richardson Light Guard was the vehicle through which the town of Wakefield, Massachusetts, participated in war, and remembered war during times of peace. A few residents and natives of the town would serve in other branches of the military during the war, or in other companies or regiments of Volunteers that Massachusetts contributed for the war with Spain, but the vast majority of Wakefield men who fought in the war did so as members of the Richardson Light Guard. After the war, the RLG would be the main vehicle through which the town memorialized the war. As the schools of Wakefield prepared youngsters to assume their adult roles in life, and the churches prepared souls for the afterlife, the Richardson Light Guard prepared the men of Wakefield to show the current residents of the town that they were worthy heirs to earlier generations who had marched from the town to battlefields near and distant. The return of Wakefield's militia company, to its hometown following the war, rather than the surrender of Spanish troops in Cuba or the acceptance of a peace treaty by Spain, would mark the end of the war for Wakefield.

* * *

A militia company departs its hometown, fights the enemies of colony, state, or nation, and after days, weeks, months, or years, returns to its hometown. The town celebrates the return of its sons. The scene is a nostalgic image of American communities before the twentieth century, existing in a timeless flow of Americana, harkening back to an age before world wars, Cold War, and urban sprawl. The image became part of the national myth but was based on reality. The pattern of the locally organized militia company marching away and returning as a group was very real for many towns before the twentieth century, and was at least as old as the town of Wakefield itself. Although seldom repeated more than once in a generation or two, it had nonetheless embedded itself in the self-identity of the town, and in other towns throughout the nation, making war real and present in the town, even if the scene of fighting was miles or even continents away. Numerous towns and cities, especially in the longer settled areas of the nation, had similar militia companies, although few had militia traditions as long as Wakefield's. The United States traditionally kept a very small standing army during peacetime, and relied on Volunteer regiments drawn from and organized through the state militia for expansion during wartime. Although states and towns often raised new companies specifically for war, ideally Volunteer regiments contained companies that were already in existence, like the Richardson Light Guard. Wakefield's militia company was the ideal and the stereotype of the American militia prior to the world wars.

Changes in the way the United States raised armies for war in the twentieth century destroyed this pattern, this intimate linkage between town and war. The increasing complexity of war and society would make the experience of a town-based militia company marching off to war, and returning as a unit after the war, increasingly rare. Reforms in the Army following the Civil War led to more emphasis on professionalism, and called into doubt the future of the state-raised Volunteer regiments during wartime. The rise of the National Guard in the late nineteenth and early twentieth centuries preserved a role during wartime for military units organized through the states, but also began the slow and largely unintended process of divorcing militia companies from their hometowns. Town-based militia companies across the nation had either to become part of the National Guard, and as such surrender much of their independence, or remain outside of the National Guard and become relics existing for tradition rather than utility. At the same time, towns such as Wakefield, lying within commuting distance of a major city, underwent profound demographic and economic changes as the railroad and later the automobile turned Wakefield and other similarly situated towns into bedroom communities.[18] The domination of such towns by a handful of economic and social elites gave way to more professional and impersonal forms of town government. In the suburbs after World War II, local government was more professional and less likely to be dominated by the same prominent residents who also controlled local business, fraternal organizations, churches, and the militia company.

The more centralized control represented by inclusion of local militia companies into the National Guard was only part of reason for the separation of towns from national war efforts. The episodic nature of Volunteer wartime service, with regiments drawn from the militia usually serving for specific periods of months, gave way to service "for the duration." During the twentieth century, few National Guard companies returned to their hometowns as units after fighting in wars.[19] The world wars were too long and too complex, and the Korean War and especially the Vietnam War saw little involvement by the National Guard.[20] By the late twentieth century, few National Guard companies in more heavily populated areas would draw predominantly from a single community, and Wakefield no longer had a National Guard unit based in it. War in the United States became an individual and federal matter; states and even more so towns became increasingly less relevant to fighting wars.

The aim of the National Guard movement in the late nineteenth century was to bring state militia across the nation up to levels of funding and professionalism more common in the wealthier states of the northeast, especially Massachusetts, Connecticut, and New York, which would allow the continued participation of state-based military units in war. The enormous casualties of the Civil War had led to a period of reflection and military reform, although the lessons to be drawn from the war remained in flux for a generation or more, with the extremes of the debate over the composition of wartime armies centering on the supremacy of either the wartime Volunteer or the professional Regular. One of the strongest salvos for depending on state-raised Volunteers came from John Logan, Congressman from Illinois. In his tome *The Volunteer Soldier of America*, published posthumously in 1887, he celebrated the civilians who temporarily take up arms to defend the nation while he chastised the Regular Army and Navy and their respective academies.[21] For Logan, the large numbers of academy-trained officers who served the Confederacy was proof enough that the academies did little to instill patriotism or duty to the nation in professional officers. Instead, Logan argued, the volunteer militiaman was the true savior of the nation, from the colonial era to the present. While Logan's work was largely ignored by the regular military establishment, it did help fuel the growing interest in the organized militia in the decades after the Civil War. The champion of the Regulars was Emory Upton, a Civil War hero who argued that state influence on militia would always make it unreliable as a reserve for the federal army.[22] Upton advocated the creation of a wholly federally controlled reserve force to replace the militia. From the two extremes arose the modern National Guard, with increased federal funding and control but retaining its links to the states. Both Logan and Upton used American military history to make their points, but both treated the organized militia as a whole, and focused on its wartime employment, although they came to polar opposite conclusions as to its effectiveness as a fighting force. Their works would stand as the main book-length treatments of the organized militia until after World War II. Aside from a multitude of unit histories, mostly of regiments, written in the late nineteenth century documenting the experiences of state-raised Volunteer units during the Civil War, and a small coda of such publications following the Spanish-American War, few histories of militia units existed, and very few explored the life of a unit during peacetime.[23]

Academic historians had some interest in the militia of the colonial and early republic periods; indeed, most military histories of the colonial period were by default histories of the militia, at least of militia units during wartime. Probably because of the iconic role played by the settlers of Plymouth and later Massachusetts Bay in American history, as well as the role of the militia during the outbreak of the Revolutionary War, the colonial New England militia in general, and the Massachusetts militia in particular, has garnered the most interest by historians.[24] Steven C. Eames's *Rustic Warriors: Warfare and the Provincial Soldier on the New England Frontier, 1689–1748* argued that Massachusetts Bay and New Hampshire created wartime armies that reflected their society and performed competently, but that later historians accepted the negative judgments made by class-conscious British officers.[25] Howard H. Peckham's *The Colonial Wars 1689–1762*, published in 1965, explored the role colonial governments played, with only limited assistance from Great Britain, in raising forces and fighting Indians and French,[26] while Harold E. Selesky took a more focused view in his *War and Society in Colonial Connecticut*, which demonstrated the growth in sophistication Connecticut employed in raising armed expeditions as the colony grew and matured.[27] Fred Anderson's *A People's Army: Massachusetts Soldiers and Society in the Seven Years War* demonstrated the leading role colonial governments took not only in raising forces for the war, but in formulating strategy for the North American theater of the war.[28] The fabled minutemen of Concord, Massachusetts, and their town were studied in depth by Robert A. Gross in his *The Minutemen and Their World*. Gross examined the town's politics and social order in the years leading up to the historic fight in the town in 1775. Gross found a town often divided on economic and religious issues. Although the opening battle of the Revolution was fought mainly by the yeoman farmers, as the war wore on, the town increasingly turned to using economic incentives to entice men from the lower strata of society to serve in the town's militia quotas. The persistence of theological controversies in a period in which most historians had assumed the Puritans had all transformed into Yankees was one of the book's major surprises.[29] More focused on the ideology of Americans regarding a standing army in the early republic was Lawrence Dilbert Cress's *Citizens in Arms: The Army and Militia in American Society to the War of 1812*.[30] Cress traced the struggle between radical and moderate Whigs over the nature of the militia after independence, and the debate over whether the nation should have a standing army in peacetime. Cress believed the opposition to a peacetime standing army or federally controlled militia created an ineffective military policy through the War of 1812.

Few works examine the militia in its fallow years, from the end of the War of 1812 until the rise of the National Guard after the Civil War, in part because there was little to write about. A few of the more urbanized states of the northeast occasionally maintained credible organized militias, but in general the peacetime militia was a moribund institution. The only real development regarding militia in the 1820s to 1840s was the abandonment of even the pretense of a general militia and regular musters in most states. The only book-length academic study of the militia in the period was Marcus Cunliffe's *Soldiers and Civilians: The Martial Spirit in America, 1775–1865*. Cunliffe argued that Americans embraced the myth of the militia, but had little enthusiasm to bear the burdens of serving in the militia. Americans grudgingly accepted the necessity of the regular military establishment as an unpleasant but acceptable alternative to participating in a functional militia system.[31] Cunliffe documents well the decline of the militia in most of the nation as an active component of order and national defense, into a mythologized but largely neglected institution.

The National Guard that arose after the Civil War was overlooked as a subject for academic historians until the latter half of the twentieth century. Early histories of the Guard were largely celebratory, such as *The Nation's National Guard*, which was published by the National Guard Association in 1954.[32] In a similar vein was R. Ernest Dupuy's *The National Guard: A Compact History,* published near the end of the Vietnam War, a war in which the National Guard's involvement was conspicuously small.[33] This trend reached its highest form in Jim Dan Hill's *The Minuteman in War and Peace: A History of the National Guard*, published in 1964.[34] Hill, like Logan before him, saw the modern National Guard as the inheritor of both the militia and Volunteer traditions, and a bulwark of the nation. Hill's book served as the only in-depth study of the militia and National Guard, and he argued that whatever shortcomings the National Guard had were mainly the fault of the petty jealousies and slights of the Regular Army. If Hill's work was overly defensive on the prowess of the National Guard during war, then it was an appropriate response to William Riker's *Soldiers of the States*, published in 1957.[35] Riker's book, one of the first scholarly works on the National Guard, argued that the National Guard, especially its officers, was severely lacking in competence in performing its federal war-fighting missions, largely as a result of state interference—an echo of Upton's opposition to any form of state militia involvement in national defense. For Riker, and many others, the National Guard was better than its nineteenth century forbears, but the remaining state controls over the National Guard were anachronistic. The states wanted federal funds for the National Guard but opposed most federal controls. Riker was a political scientist, and not a military historian, and looked at the National Guard as a case study in American federalism. He was joined by Martha Derthick, whose *The National Guard in Politics*, pointedly ignored the military effectiveness of the National Guard as a fighting force, and instead focused on the lobbying efforts of the National Guard Association in Congress.[36]

The historiography of the militia and National Guard tends to be whiggish in that the rise of the National Guard is viewed as wholly progressive. Locally created and supported nineteenth century militia companies are criticized for such archaic practices as election of officers and overt political involvement of such companies. The men in them are generally seen as good material from which to build the National Guard, but the informal relations between Guardsmen, the influence of local and state politics, and the lack of consistency with the Regular Army are judged to make the militia as it existed in the nineteenth century unsuitable as a reserve for the federal military. Increased federal control over the National Guard in the twentieth century, in function of units, training, equipment, and especially selection and education of officers, with the concurrent weakening of state control over these matters, were all positive developments. Little attention is given to the downside of the absorption into the National Guard of militia companies with deep community connections, generous funding, and well-earned reputations for professionalism. John K. Mahon's *History of the Militia and National Guard* broadly traced the colonial militia's evolution, after a few missteps and a lot of political wrangling, into the modern National Guard.[37] Cooper's *The Rise of the National Guard* focused on the struggle between the Regular Army and the National Guard between the 1880s and the adoption of the "National Defense Act of 1916." Cooper showed the wide disparity in support given by the states to their organized militia in the late nineteenth and early twentieth centuries. States such as Massachusetts were on the forefront of militia reform, with state bureaucracies and dependable budgets, while the militia of states such as Arkansas received no state financial support. Increased federal funds and involvement brought higher standards of training and uniformity in the twentieth century, but at the cost of federal oversight and control.[38] The current standard history of the Army National Guard, Michael Doubler's *I Am the Guard: A History of the Army National Guard, 1636–2000*, followed the progressive trend, showing the long and sometimes bitter struggle waged by National Guard leaders in the early twentieth century to ensure a continued role for the forces organized at the state level in fighting wars.[39] The result of that struggle is the modern National Guard — well equipped, professional, and fully integrated into the federal military. Most general works on the National Guard as a whole tend to emphasize the militia and National Guard only during wartime, or other periods of strife, to the exclusion of most of the peacetime existence of military units. Two exceptions to this trend look at the National Guard of a single state over an extended period. Jerry Cooper's *Citizens as Soldiers: A History of the North Dakota National Guard* underscored the positive influence that additional federal involvement in militia had in a poor and sparsely-populated state.[40] Eleanor L. Hannah looked at the growth of the National Guard in Illinois between the Civil War and the entry of the United States into World War I. She showed that participation in the organized militia was for many members a way to define themselves as men and as citizens.[41] Both Cooper and Hannah looked at the development of the National Guard during the crucial years when militia service evolved from grassroots organizations with members bearing the financial burden of maintaining the militia, to a more centrally controlled and financed organization. When most works mention the militia or National Guard performing local service at all, they tend to focus on the employment of the militia during labor unrest, often seeing the militia as a reactionary force, or in providing disaster relief. The Railroad Strike of 1877 is often seen as a catalyst for renewed interest in the organized militia, and many historians have overemphasized that role for the militia.[42] The employment of militia companies for strike duty was not even across the nation, as some state militias, such as Illinois, called out their militia often in the last half of the nineteenth century to deal with labor problems, while other industrial states did so less. The theory that labor unrest led to the development of the National Guard fails to account for Massachusetts, which did not use its militia at all in response to labor unrest in the second half of the nineteenth century, yet was at the forefront of the National Guard movement.[43] Even in states where militia was often called out for strike duty, militia companies also often filled a myriad of roles in a community, such as forming business and social connections, incorporating newcomers into a community, and celebrating past martial achievement. George N. Vourlojianis's *The Cleveland Grays: An Urban Militia Company, 1837–1919*, and Hannah's study of the National Guard in Illinois are two of the very few works on the organized militia and National Guard that explore their social function. No study adequately addresses the gradual breakdown of local connections between some National Guard companies and their home communities that occurred, particularly in more heavily populated areas.

The current study is at heart a traditional unit history, but whereas most unit histories cover a regiment during war, this one covers a single militia company for over a century, following the company in war and in peace. The company was in some ways an average company in the organized militia, but in other ways it was an extraordinary company. The company, the Richardson Light Guard, was in one of the states that stood at the forefront of militia reform by creating bureaucracy within the state government that would eventually lead the way toward the creation of the National Guard. Within the Massachusetts Volunteer Militia the RLG consistently ranked as one of the top companies, scoring high in state evaluations in peacetime, and proving its worth in war, from the Civil War through World War II. Over the course of the century, the RLG constantly adapted to changes in state and federal laws on the organized militia, riding the crest of the wave that was the rise of the National Guard. During the first decades of its existence, the RLG enjoyed the support of generous local benefactors. Membership in the company carried with it prestige. By the Spanish American War, the RLG benefitted from being from a prosperous town in a relatively wealthy state. It was fully intertwined politically, socially, and economically into the fabric of the town. Into the twentieth century, the RLG remained respected both as a military institution and as a social institution. The town produced handsome volumes on it in 1901 and 1926, it was selected as the best rifle company in Massachusetts in 1917, the Commonwealth featured it in a mural on a wall in the State House in 1931, it maintained unique ceremonial uniforms until 1941, and during World War II, it was the only unit in the Americal Division to be awarded the Distinguished Unit Citation. Despite the continued high status of the company, the increased federal involvement in state militia continually weakened the once deep bonds between the Richardson Light Guard and its hometown. After the Second World War, the bonds completely broke. Despite the fame of the company, the story of the rise and fall of the Richardson Light Guard was similar to the experience of many of the better militia companies across the nation as they attempted to adapt to fundamental changes in the way the land forces of the United States were organized for war during the early twentieth century. While some National Guard companies in some less urbanized parts of the nation continue to maintain strong links with communities, the history of the RLG and Wakefield is more typical of the changing interaction between National Guard companies and their hometowns.

The current study is also in part a history of a town, South Reading/Wakefield, with an emphasis on the military history of the town. The story of the militia company is set on top of the development of the town from a small, mostly agricultural community not far from the frontier, into a factory town, and finally into a modern bedroom suburb of Boston. As the town developed, the militia company became less based on the town, and more entwined with, and dependent on, the federal military establishment. The military history of the town underscores the dramatic changes in how Americans interact with the military and war. As the RLG became more professional and less tied to its hometown, so too did the town of Wakefield evolve during the twentieth century. During the nineteenth century, local elites, usually from the oldest families in town and who drew their wealth from the community, ruled the town largely unchallenged. They served on numerous town committees, established banks and volunteer fire departments, and gave generously to local causes. Elites linked themselves with the RLG as active, former, or honorary members. Modernization and the absorption of the town of Wakefield into Greater Boston ended the long reign of local elites, and instead professionals began to run the town and its institutions. The deeply intertwined fabric of community and militia company unraveled as both evolved into something more professional but less personal in the twentieth century.[44]

The Spanish American War occurred roughly at the midpoint in the life of the Richardson Light Guard. As the men of the company boarded their train to leave Wakefield that spring day in 1898, the institution of the local militia company seemed as enduring as ever, supported by the solid bonds between militia company and hometown. But changes underway in domestic politics and in the science of war slowly but certainly weakened and eventually broke the critical bonds between militia company and town. While the rise of the

National Guard and eventual incorporation of most companies such as the RLG into it brought increased uniformity and professionalism to the militia as a whole, the leveling effect of the National Guard also meant that the better militia companies became average National Guard companies. Increased state and later federal involvement in militia in the name of efficiency and professionalism brought an unprecedented level of preparedness, but at the expense of close ties to the community.

By the Vietnam War, a conflict in which all Wakefield participants entered the federal military as individuals, mostly through the draft, the Richardson Light Guard no longer existed. A National Guard company occupied the state armory that had been built for the RLG, and carried the lineage of the RLG, but it never received the official designation as the RLG. The unpopularity of the war in Vietnam, and the use of the draft to fill the Army during that war, altered the image of the soldier and veteran in American society from hero, or at least solid citizen, to victim or worse. During the Bicentennial celebration of the American Revolution in the mid-1970s, the very concept of the town militia became a nostalgic icon from an imagined static distant past. "Militia" had come to mean in East Coast states small groups of re-enactors marching in parades attempting to represent the town's contribution to the opening battles of the War of Independence, complete with tri-cornered hats, smoothbore muskets, and rope-tensioned snare drums. While the local militia re-enactor company in Wakefield participated in celebrations for the two-hundredth anniversary of Independence, the local National Guard company — the inheritor of the lineage of the RLG — was quietly reorganized out of existence by forces outside of the town. The town put up no protests at the ending of over three centuries of its town militia. Wakefield residents interested in local history might take a passing interest in the former Richardson Light Guard, and even point with pride to its past accomplishments, but these were all safely in the past, and not part of the current life of the town.

The current work does not dispute the generally positive impact the rise of the National Guard had on the organized militia across the nation, but intends to show that for some units and communities, that positive impact came at a heavy price. The history of a single militia company, the Richardson Light Guard, demonstrates that increased federal involvement created an average National Guard company out of what had been one of the best militia companies in the nineteenth century, and undermined the bond with its hometown. The National Guard as a whole is today more professional, better equipped, and better trained than at any time in the past, yet its links to communities in most areas are much weaker. The armory in a town, where it still exists, is more like a post office — a local representative of the federal government — rather than an institution of the town such as the library or a school. National Guardsmen drilling at an armory are likely to come from another town, while town and community elites have little direct involvement with the local National Guard unit. As a result, towns and states in the twenty-first century play a much diminished role in fighting wars, in contrast to before the rise of the National Guard. For towns as corporate bodies, war in the late twentieth century and beyond was simply not a town matter. Residents or relatives of residents might serve in the military during wartime, but they served directly under the federal government and not as representatives of their hometown at war. The history of the Richardson Light Guard, one of the better militia companies from the nineteenth century that maintained a reputation for military prowess through the Second World War demonstrates well the gradual separation of the militia company from its hometown.

The history of the Richardson Light Guard and its hometown is also a story of names and what they represent and how they change through time: the names of prominent residents, the names of a town, the names of institutions, and the names of a militia company. The changes were not arbitrary; they reflected changes in both reality and perceptions. The story of the RLG demonstrates what has been lost with the disappearance of the town militia. During the nineteenth century, the RLG was as much an institution of the town of Wakefield as were the schools, library, and fire departments. The RLG was created by the town, maintained by the town, later with the assistance of the commonwealth, and served as an institution of the town in war and peace. As the local National Guard company, which bore the heritage of the Richardson Light Guard, became more of an institution of the state and later the federal government, the bond with the town was broken, and the Richardson Light Guard ceased to exist.

Chapter I

Frontier Town

The modern town of Wakefield, called Reading until 1812, and South Reading until 1868, traces its origins to 1639. In that year, a small group of residents of the town of Linn, a few miles up the coast from Boston in Massachusetts Bay Colony, petitioned the General Court in Boston for the right to settle about eight miles inland, at the southern end of a large pond at the headwaters of the Saugus River. The area had previously been controlled by the Saugus Indians, Algonquin speakers, who hunted in the woods, fished in the ponds and brooks, and grew crops. Although Linn, or *Lynn*, had been settled by Englishmen only in 1629, most desirable land in the town had already been claimed, and some men sought to settle farther inland to establish their holdings in the colony. The General Court, as the Massachusetts Bay legislature was called, valued the expansion of lands under English cultivation and granted their request,[1] provided that the petitioners established a village in the new grant within two years. Technically, Massachusetts Bay, as a chartered company itself, had no legal authority to charter other towns, but did so anyway, chartering most of the towns in Massachusetts Bay Colony and other New England colonies. The General Court further ruled in 1640 that the new settlement, to comprise an area of "four miles square," (sixteen square miles)[2] would be called "Lynn Village" until the required seven men who were church members, and their families, established households in the new settlement. In 1643, when Massachusetts Bay established counties, Lynn became part of Essex County, while Lynn Village became part of Middlesex County. By 1644, after the establishment of seven households, the new settlement petitioned to become a town and chose a new name, "Redding."[3] When the General Court actually incorporated the town, the name of the town was spelled as "Reading," in the manner as the town of the same name in Berks Country, England, although the correct spelling of the name remained in flux for decades.[4]

The inspiration for the name remains uncertain. Some of the original settlers, specifically John Poole and Richard Parker, might have come from Reading, England, and thus named the new town after their hometown in the old country. This theory comes from the extensive comparative research by Loea Parker Howard of family names among early settlers of Lynn Village and Reading, England, in the 1630s.[5] However, the first English Civil War[6] was ravaging England at that time, and Reading, England, an inland town, had refused to pay the tax King Charles I sought to levy to avoid dependency on Parliament for funds. In 1644, the year the name "Redding" was chosen for what had previously been "Lynn Village," a Parliamentary army defeated a royalist army at Reading, England. Massachusetts Bay, as a Puritan colony, sympathized with the Puritan–dominated Parliament, and the name might have been chosen to commemorate the battle in England.[7]

As with most New England towns, Reading in Massachusetts Bay Colony included a main village and several square miles of meadows, hills, swamps, and forests. The lack of consistency with the rest of the United States over what constitutes a "town" causes some confusion. In most of New England, the political division known as a "town" includes sometimes farms, woods, swamps, as well as a main village and maybe other villages. Except for some areas of northern Maine, all territory in modern New England belongs to some town or city. The original charter from the General Court gave the original free adult male settlers collective ownership over the territorial expanse of the grant, which they divided among themselves as they began to cultivate the land. Most of the grazing land was designated as "common" land, and all residents had the right to set their animals to roam it, while cultivated lands had to be enclosed for protection from the free-range animals.[8] Part of the land owned in common formed a grassy pasture area in the middle of the original village. As more of the lands of the grant were divided up and became privately owned, the grassy area in the middle of the village became known simply as the Common.

The first settlers built their village at the southern end of a 247-acre pond, originally referred to by the English as the "Greate Pond," situated about 90 feet above sea level at the head of the Saugus River. The Saugus River, called *Aboutsett*, meaning "(winding stream" by the Natives, was an insignificant stream for most of its 13-mile course to where it emptied into the sea at Broad Sound, at Lynn. It traveled from the Greate Pond due east, then turned south and defined much of the border between Reading and Lynn. Through its course in Reading and on the border, the "river" was only a few feet wide and shallow. In the neighboring village of Saugus, part of Lynn until 1815, it became large enough to power an early mill and iron works, but remained negligible as a means of shipping. The Greate Pond became so entwined with the identity of the town, that residents soon began calling it "Reading Pond," a name it would keep until 1847. A smaller pond, Wahpatuck Pond, later called *Smith Pond* after a settler who established his holding on the northern end of it, covered about sixty-four acres and sat less than a mile to the south. Smith pond had its outlet in an inconsequential brook, later known as Mill River, which flowed east into the Saugus River. Most of the land in the original grant lay to the north of the main village. The Ipswich River, running east about five miles north of the main village, formed the original northern boundary of the grant. The Ipswich River is only slightly larger than the Saugus River, and in the section that ran through the northern reaches of Reading, it too seldom reached a breadth greater than ten feet or depth greater than four feet. The Ipswich River eventually empties into the Atlantic Ocean north of Cape Ann, at the town of Ipswich. Aside from canoes or other small boats, the river was useless as a communication or trade link. In 1651, the boundaries of the grant were enlarged, with a section stretching two miles north from the Ipswich River added to Reading. In 1727, a small village of about ten families, known as Greenwood, laying about a mile south of a narrow pass between Smith Pond and a steep hill, was annexed to Reading from Malden. At its largest, the grant totaled about thirty square miles.

As with the rest of New England, no separation between church and town existed in Reading until after the Revolutionary era: church and town were essentially the same. The town outgrew its initial meeting house by the 1670s, but not until 1689 did residents erect a larger meeting house, at the northern end of the main village on the southern shores of Reading Pond. The meeting house served as both a place for Sunday worship and as a place to conduct town business. In the ensuing years other villages were established within the boundaries of the grant. The settlers from a small village just north of the bridge over the Ipswich River found the roughly six-mile journey to the main village for Sunday services an arduous task, and in 1713 petitioned the General Court for the right to build their own meeting house for Sunday services, which was granted, and these northern settlers established the Second Parish of Reading, the current town of North Reading. The warrant, dated 27 March 1713, required a town meeting of residents who lived north of the Ipswich River to be held on 7 April 1713 at the meeting house in Reading to vote on the matter.[9] The main village in the Second Parish was less than a quarter-mile north of the bridge, but the southern border of the Second Parish was placed about two miles south of the bridge, at Bear ("Bare") Meadow, a small seasonal stream that flowed west-by-northwest to join the Ipswich. The remaining area of Reading became the First Parish. The residents of a village in an area known as "Woodend," about three miles by foot northwest from the main village, resented having to share the cost for a new meeting house in the main village in 1767. The General Court eventually required the main village to refund the money spent by Woodend on the new meeting house, and Woodend established its own meeting house and became the Third Parish of Reading. Other villages nearer to the main village, such as the East Ward, later known as "Montrose," which was about one mile northeast of the main village, and Greenwood, remained in the First Parish.

From earliest settlement, the town of Reading defined itself in part by martial prowess. It was a Puritan town, settled as part of the great Puritan experiment to build a community where the church recognized by men and the church recognized by God came closest.[10] Puritans in New England still claimed to be loyal members of the Church of England, but wanted to reform it by example. True members of the church, the ones recognized by God, had "saving grace." The Church of England admitted all residents within a geographic area, regardless of whether the person showed any signs of having received saving grace, while the Puritans in New England believed the church should admit as members only those showing signs of having received saving grace. However, all residents of the town were required to attend Sunday service, whether members of the church or not. In 1645, the twelfth church — Congregational, that is to say, Puritan — in Massachusetts Bay, was organized in Reading and all residents were required to attend weekly divine service. But for several decades Reading was also a frontier town, existing on the border between areas settled by the English and lands still under Native control.[11] The General Court ruled in that same year that every town in the colony should keep a guard of militiamen to prevent surprise attacks from the Indians.

Although relations with the Indians were generally peaceful in the early decades of the town, a training band was organized in the same year the Reading church was organized, with Richard Walker, who also served as the town's representative in the colonial legislature, elected captain. The second captain was John Poole, originally from Reading, England, and who was probably the wealthiest man in Reading, Massachusetts Bay. During a division of land near the southern end of the grant among the thirty-four men in the town in 1652, only six, including Poole, got twenty acres, the largest amount granted. The amount granted to each male resident was based on their livestock holdings, although no man was to receive below the minimum of two acres. Half of the grants were for ten acres.[12] As one of the first investors in the establishment of the town, he had an exclusive right to maintain a grist mill. Colonial law did not consider a militia company to be at full strength until it had sixty-four men, although that number was soon raised to one hundred. When a company contained two hundred men, it was to divide and make two companies.[13] The existence of the training band — a militia company organized for military training — suggests unease over relations with the Natives. Almost all adult, free Englishmen in the town belonged to the militia company, which was included in the Middlesex Regiment. The Middlesex Regiment had been created as the North Regiment in 1636, and originally included the militia companies from Charleston, New Town (Cambridge), Watertown, Concord, and Dedham. In 1643, concurrent with the division of Massachusetts Bay Colony into counties, the North Regiment was redesignated as the Middlesex Regiment. The militia tradition brought by the English colonists to the New World was several hundred years old.[14] Although in decline in England, the institution was revived by colonists. Law and tradition required most adult men in the town to attend training. The colonial government also required that a militia officer from each town train boys between ten and sixteen years — those too young to serve in the militia — in the use of firearms, half-pikes, and bows and arrows, during muster days. Massachusetts Bay also required that "children, Negros, Scotsmen, and Indians" be trained, although the requirement, and permission, to train Indians and blacks was soon eliminated.[15]

Under Massachusetts Bay laws, almost all adult, free, English men belonged to the militia. Militia was the obligation to bear arms for training or fighting when directed by town or colonial government. Militia was an obligation of belonging to the society, similar to the obligations to pay taxes, serve on juries, attend town meetings and divine services, and serve in posses. The members of the militia elected their own officers until the General Court assumed the right to appoint officers in 1668,[16] and captains, usually the highest ranking man in a town's militia, had the authority to appoint sergeants. Musters for training usually were held on Saturdays, four times a year during peace, and monthly during war. During musters, the men practiced a simplified system of loading and firing in volleys. Musters during peacetime were more than just an obligation for military training — they were also a sort of holiday, a break from routine. They often ended with eating and drinking. Fines were levied on residents who failed to attend musters, and the funds thus raised were spent purchasing arms and equipment.[17] Over time, as communities grew larger, fines for not attending became more of a tax on those, usually the wealthier members of the community, who wished to avoid service. In 1658, the Court granted the towns of Reading, Lynn, and Rumney Marsh the right to raise a troop of cavalry in the militia, although without the normal right of such troops to use the ferry for free or to draw the nominal fee from the county treasury. The cavalry troop would draw from a more select group, those wealthy enough to own a horse, saddle, holster, pistol, and sword, and had more prestige than the normal town-based militia. The law specifically required a potential cavalry man or his parents to have an estate worth at least £100. Perhaps because of the restrictions on the rights of the cavalry troop, or the need to draw militiamen from several towns, it never seems to have become viable and it vanished from the records as quickly as it appeared.[18]

The town grew throughout the period, containing fifty-nine houses by 1667. Although many of these residents were in the main village, others were in the smaller villages and clusters scattered throughout the grant. Occasional scares swept through the town, but not until King Philip's War in 1675–76 did Reading men fight in large numbers. King Philip's War, also called Metacom's Rebellion, is the bloodiest war in American history in percentages of people killed.[19] The name of the violence between the English settlers in New England and their Native allies on one side, and the Wampanoags and their allies on the other, that ravaged New England in 1675–1676 is a matter of some dispute, with the names Metacom's Rebellion, Puritan Conquest, as well as an Indian civil war also proposed and used to some degree. At least twenty-one Reading men served in the war, and several were recorded as receiving property on the frontier from the colony for their services in the war. Captain Jonathan Poole, son of settler John Poole, fought at Hatfield and Hadley. Major Jeremiah

Swain, who had earlier been the fourth commander of the Reading militia company, commanded all colonial forces on the eastern frontier, and led the column to the defense of Deerfield and Hadley in the western part of the colony. Following the shock of King Philip's war, Reading, as with all of Massachusetts Bay, placed greater emphasis on local defense. In 1680, the colonial government divided the Middlesex Regiment of the militia, to which the Reading militia company belonged, to create the Upper Regiment and the Lower Regiment of Middlesex. The Reading militia company joined those of Charlestown, Cambridge, Watertown, Woburn, and Malden in the Lower Middlesex Regiment. Reading was the northernmost town whose militia belonged to the regiment.

Besides organizational changes, Massachusetts Bay contemplated building a line of stockades or a stone wall between the fall line of the Charles River north to the Concord River as a bulwark against Native attacks, which would have put this wall a few miles west of Reading. This scheme does not seem to have gotten beyond the planning stage. However, Reading did build garrison houses, fortified homes in which members of a community could seek shelter during attacks by the Natives. One such home in the more exposed Second Parish, to the north, kept its door with its marks from Indian bullets for many years after as a local curiosity.[20] Not until the frontier moved far enough away a generation later did Reading face no more immediate threats from Indian attacks. The last attack came in 1706, during Queen Anne's war, when five Indians who had earlier raided the town of Dunstable, attacked an isolated home in the far northwest part of Reading, in an area that later became part of the town of Wilmington.[21] The wife and three children of John Harnden were killed, and another five children taken captive. A party of pursuers eventually recaptured the children.[22]

After the Harnden attack, Reading men continued to participate in colonial wars, but the fighting was away from the town. The rising tensions between France and Great Britain, and especially between their respective colonies in the New World, increasingly involved the colonists of Massachusetts Bay during the eighteenth century. Reading men fought in five colonial wars, known collectively as the French and Indian Wars, between 1689 and 1763.[23] In 1690, the Reading militia company served in an expedition to Canada to fight against Indians who were allied with the French.[24] Several Reading militiamen fought at Nova Scotia in 1711, with some dying of illness or being killed in that struggle. In 1745, Reading furnished its quota for another expedition to Nova Scotia. During the last two struggles against the French, Reading contributed about ninety-six foot soldiers and another seventy-five cavalrymen. In addition, several French captives were taken to the town as prisoners, with a few dying while in custody.[25]

As new parishes were carved out of the original grant, additional militia companies were created. The Second Parish created its own company in 1713, followed by the Third Parish in 1770. Sometime before 1733, Massachusetts Bay Colony redesignated the Lower Middlesex Regiment as the 1st Regiment Militia of Middlesex.[26] The charter for Massachusetts Bay granted by King William in 1691, following the Glorious Revolution in England, gave royal governors the authority to appoint militia officers. For practical reasons, company officers remained prominent men from the town. Militiamen elected their captains, whose names were then submitted to the colonial government for appointment. However, in the eighteenth century, the general militia obligation in eastern Massachusetts became an increasingly neglected institution as the immediate threat receded. Towns still held their training days four times a year against the possibility that the colonial government might call for drafts from the militia for expeditions against the distant Indians, or, increasingly, the French. While Reading as a whole was no longer susceptible to attack, other parts of Massachusetts Bay and New England remained vulnerable to raids until the close of the last French and Indian War in 1763, and so like all New England towns, indeed like almost all British colonies in North America, Reading maintained its militia companies throughout the colonial era.[27] But these general militia companies were becoming more a pool of obligated men rather than fighting units. Raising temporary volunteer companies for wartime service, which were technically detachments from the militia, increasingly become the method by which towns in Massachusetts Bay contributed to war efforts. The Reading militia companies, to which almost all men in the town belonged, by this time existed more as an obligation. After the overwhelming defeat of France in 1763, with the resulting withdrawal of French forces from mainland North America,[28] militia training again fell to a low priority. Most adult free men continued to have an obligation to serve in the militia when called, but with no obvious threat, the obligation became increasingly theoretical.

During the years of rising tension between Massachusetts Bay and the British Parliament that began in the 1760s, militia training again took on a new earnestness, especially after 1768 when British soldiers were stationed in Boston. Reading erected a new powder house in 1770, and continued to maintain three militia companies, one from each parish. Royal authority had replaced colonial autonomy in the selection of militia officers, especially in the higher ranks. The presence of Regular British soldiers in Boston, and the implications of a military force in the colony that was not under the control of the local legislature, led to a revival of suspicion of standing armies and a celebration of the militia as the true guardian of a free society.[29] Representatives from towns not in the immediate vicinity of Boston began meeting in the town of Concord, eventually creating the Provincial Congress of Massachusetts Bay, as a rival government to the Royal government under General Thomas Gage in Boston. In October 1774, the Concord government instructed local committees of public safety to assume control over militia companies, and usurped the royal prerogative to appoint militia officers. The Concord government, acting on a motion from Suffolk County, directed the enlisted men of militia companies to once again elect their company officers, with company officers electing regimental officers, and the government in Concord appointing general officers.[30] At the time, the Reading First Parish militia company contained about 81 men, the Second Parish company 71 men, and the Third Parish company had 62. In December of 1774, at a town meeting held in the Second Parish, to the north, Reading voted to send its appropriations to the Concord government rather than to Boston.[31]

That same year Reading formed a train band; a company of self-selected militiamen from all three parishes who wanted to train more and be the first to respond to an attack.[32] These train bands became known as "minutemen," and contained about one-quarter of all men with a militia obligation.[33] The Reading minuteman company had 58 men, of who 21 came from the First Parish, 17 from the Second, and 20 from the Third. They trained at least three times a week, usually practicing aimed fire, and were required to assemble during an emergency within thirty minutes of being notified.[34] One resident of Reading, Dr. John Brooks, who would later serve as governor of Massachusetts from 1816 to 1823, was elected captain of the minuteman company. Being unfamiliar with military drill, he spent many days in Boston watching the British Regulars stationed in that city going through their paces, before returning to Reading to train the minutemen.

Reading, with strong historical and economic ties to Boston, was generally sympathetic to the Patriot cause but, as with most towns in the colonies, sought to regain harmonious relations with the mother country until actual fighting began. The town instructed its representative, Ebenezer Nichols of the First Parish, to work to repeal the Stamp Act in 1765. Although one resident, Andrew Oliver, would long brag that he had participated in the Boston Tea Party in 1773, the same year the town instructed its representative, Daniel Putnam, of the Second Parish, 'to seek redress against "public grievances," and restore that happy harmony between Britain and the colonies.'[35] Daniel Putnam was a son of the first minister of the Second Parish, and a deacon of the church. The town hesitated to create a Committee of Correspondence, and instead attempted to use traditional channels to address grievances with Parliament. The conciliatory stance by Reading ended suddenly in the spring of 1775 when a task force of British Regulars left Boston on a raid on the town of Concord, site of the rival government. When word of the British raid reached Reading on the morning of 19 April 1775, the Reading minutemen rushed toward Concord under the command of one of their lieutenants. Captain Brooks caught up with them in time to take part in the running battle with the British column on its bloody retreat back to Boston. The Reading minutemen arrived with the Billerica company in time to join some of the Concord minutemen at Merriam's Corner, a few miles east of Concord. There, along the Bay Road, the combined minuteman force ambushed the retreating British, while suffering no casualties among the Reading men.[36] The Continental Congress passed legislation on July 18, 1775 for one-fourth of all militia to be enrolled in minutemen companies, but as a practical matter, minutemen companies disappeared shortly after the Battles of Lexington and Concord, mainly because militiamen so inclined for more active service tended to enlist in provisional regiments or the Continental Army.

Reading in 1776 contained a population of around two-thousand, including about five hundred refugees from Boston.[37] Reading remained firmly behind the Patriot cause, with the Town Meeting voting to support the Declaration of Independence after it was read locally. The town's militia companies continued to serve episodically during the war, while other residents joined the Continental Army or the Continental Navy for longer periods of service. The town later estimated that an average of one-hundred men from the town served at any given time during the war. About four hundred to five hundred Reading men served in some capacity during the war, about one resident in four. Although no battles took place within the limits of the town,[38] the town did house some prisoners of war, as it had during the French and Indian War.[39] One prisoner, Lieutenant Colonel Archibald Cambell, M.P., of the Seventy-First Highlanders, soon proved himself to be unpleasant company for the town. Local sources give the image of Reading being completely behind the Patriot cause, with no Loyalist sentiment existing in the town. This image, recorded about a century after the Revolution, probably owes something to selective memory and the creation of national myth. The claim of almost constant employment of the Reading militia during the war, despite the lack of British activity in the area after 1777 suggests that the Reading militia had a role in keeping Loyalist sentiments suppressed

After the Revolution, signs of the changes underway in society began to manifest themselves. Massachusetts Bay Colony became the Commonwealth of Massachusetts. Slavery in Massachusetts was abolished by court decision in 1783. In Reading, more subtle changes occurred. Previously, in theory at least, all members of the community formed a single group. In the years after the Revolution, new associations began forming, increasing the complexity of the town. In 1793, the first non–Congregational religious group formed openly in the town.[40] This group, consisting of Baptists associated with a church in nearby Woburn, began holding meetings and in 1800 they built their first church in Reading. The Baptists were an illegal sect in the seventeenth-century Massachusetts Bay, with practitioners jailed

Colonel James Hartshorne's house, built around 1681, near the southwest corner of the Greate Ponde, later referred to as "Reading Pond," and still later as "Lake Quannapowitt." Colonel Hartshorne long served as a militia officer. His home later hosted the first recorded Masonic Lodge in what was then Reading (courtesy Michael Bouche).

and forced to pay fines or suffer whipping. In 1813 the Universalists began holding meetings in the town, although they did not have a minister until 1833, nor a church building of their own until 1839.[41] To some extent the growth of non–Congregational churches in the town reflected the increasing complexity of society in Reading, as new people and new ideas entered the town. The town, as with all Massachusetts, was taking steps toward separating the Congregational Church and the town government. The Congregational Church would continue to be funded in part through taxes until the final separation of church and state in Massachusetts in 1833. In 1834, the town built the Town House, next to the Congregational Church, moving the functions of town government out of the church.

In the decades after the Revolution, few opportunities for strictly masculine association existed in Reading, aside from taverns. The churches admitted both men and women for attendance and membership. On 12 March 1798 three men from Reading petitioned the Grand Lodge of the Masonic fraternity in Boston to recognize their lodge. The lodge, called "Mount Moriah Lodge," remained in existence perhaps until as late as 1842.[42] The lodge met in two rooms in the upper floor of what eventually became one of the oldest houses in Wakefield, then called the "old Lafayette House," now referred to as the "Colonel James Hartshorne House." James Hartshorn commanded a cavalry troop during the War of 1812, and eventually rose to the rank of colonel in the Massachusetts militia. The house sat on the southwest shore of Reading Pond, just west of the Congregational Church and cemetery. As with over half the Masonic lodges in Massachusetts, Mount Moriah Lodge was mortally wounded by the anti–Masonic movement that swept across the United States following an incident in 1826. In that year, some Masons in north-western New York state were implicated in the kidnapping and presumed murder of William Morgan, a former Mason, for threatening to publish the secrets of the fraternity.[43] Of the 107 lodges registered with the Grand Lodge of Massachusetts in 1826, only fifty-two were still functioning in 1844.[44] In the hostile atmosphere of the times, local lodges became more secretive, and the Grand Lodge in Boston had difficulty maintaining control or even contact with local lodges. When in June, 1848, the Grand Lodge sent its "Grand Sword Bearer" to retrieve the charter and regalia of Mount Moriah Lodge from Joel Winship, apparently one of the last members, Winship was "very abrupt" and refused to surrender the items.[45] Records mention only that the Grand Lodge "took action at a later date,"[46] but for practical purposes, the Masons no longer existed in the town.

By the time the lodge had ceased to function, the town was known as "South Reading." The change in the name of the town came as a result of partisan politics. In the years after the Revolution, two distinct political parties emerged: the Federalists, based largely around those who had supported the adoption of the new national constitution in 1789, and the Democratic Republicans, who had opposed the new constitution, but were somewhat mollified by the adoption of the Bill of Rights as the first ten amendments. In general, the Federalists favored a strong central government, whereas the Democratic Republicans were more suspicious of the centralizing tendencies of the administrations of Presidents George Washington and John Adams. The two parties had disagreed vehemently over the role and status of the militia. The Federalists sought to rely on a standing army in peacetime and war, and desired a federally controlled militia that would augment the federal army during wartime. Most Federalist proposals for reform of the militia included some system to classify the militia by age, with men between 21 and 25 or older, depending on the proposal, bearing the heaviest burden for training and service. Democratic Republicans opposed maintaining anything but the smallest standing army in peacetime, and believed that the militia should be the nation's first line of defense during wartime.[47] With the two parties finding little common ground on the subject of militia, no reform of the militia would come from the federal government as the nation entered the nineteenth century.

The First Parish, containing the original settlement, was a firm supporter of the Democratic Republican party, while the other two parishes were solidly Federalist, as was most of New England. Although the First Parish had the largest population of the three, the combined populations of the Second and Third surpassed that of the First. Due to the balance of the population being in favor of the Federalist sections, most of the representatives Reading sent to the General Court in Boston were Federalists. The First Parish, centered on the original village of Reading, found this situation intolerable. Lingering resentment between the First and Third Parishes over the cost of constructing the meeting house in 1767 created a rift that only got worse over the next generation. The First Parish had originally petitioned the General Court to become a separate town as early as 1785, although that petition was denied.

During passions created as the nation drifted into the War of 1812, the town of Reading formally split. The First Parish strongly supported the war, while the Second and Third Parishes opposed it. A meeting of the First Parish was called for 28 May 1811 to decide on whether or not to petition for separation. At the 11 June meeting, the First Parish "voted unanimously to petition the Hon. General Court to set them off from the North and West Parishes of the town," and that the name for the First Parish as a separate town would be "South Reading." Granting the request submitted by the First Parish in 1811, the General Court granted the First Parish the status of a separate town in 1812 At the time, the First Parish contained about eight hundred residents,[48] the combined Second and Third Parishes slightly more. The former Second and Third Parishes thus kept the name "Reading," while the First Parish became "South Reading." The name "First Parish" would continue to be used by the town for many years afterwards, although the term had become superfluous. The parishes were numbered according to the order of their creation, not geographically. As a result, the Second Parish was geographically separated from the First Parish by the creation of the Third Parish in 1763, and thus had to remain with the Third Parish when it broke from the First in 1812. In 1853, the former Second Parish, then called the North Parish, broke with the town of Reading and became the town of North Reading. The former First Parish contained the oldest and largest village in the old grant, but the new name implied a subordinate relationship to the former Second and Third Parishes.[49] Consequentially, the modern town of Wakefield, as South Reading was renamed in 1868, officially dates only back to 1812, while the modern town of Reading, the former Third Parish of Reading, claims 1644 as its date of incorporation The ambiguity the people of South Reading felt toward their town's new name would bring opportunity for immortality years later to a man born one year before the split.

Whatever misgivings South Reading residents might have had over their town's new name, the Democratic Republicans were firmly in control. Unlike much of New England, South Reading backed the war against Great Britain that began in 1812, and raised militia units for state service in the event that the British raided or invaded the long exposed Massachusetts coast. A militia cavalry troop formed by men from South Reading and towns to the south in 1797 served during the war as a minute man-type organization, alert to the possibility of a landing by British raiders. The troop managed to survive the war without being called to active service, and remained in existence until its final disbandment in 1828.[50]

Commanders of the cavalry troop tended to be prominent men from Reading, such as James Hartshorn, who also served as town treasurer for many years.[51] South Reading also contributed soldiers to a new company organized for the war, the Washington Rifle Greens, organized 11 September 1811.[52] Although many of the members of the Rifle Greens came from South Reading, the company also drew many of its members, and all of its officers, from neighboring Stoneham, to the southwest.[53] The Rifle Greens served on active service during the war, garrisoning Dorchester Heights outside of Boston from 22 September to 31 October 1814. After the war, it became a South Reading company in what was termed in state law as the "active militia." With the color of its uniform and its name altered to gray, the Washington Rifle Grays survived until 1846, occasionally noticed by townspeople as men of the company drilled on the Common, or paraded to the sound of fife and drums in their gray trousers, coats with bright stripes, and feathered caps.[54] In recently settled towns, such militia activities served to bind communities together, but they served similar purposes in longer settled communities such as South Reading. The militia helped give at least the semblance of unity while some of the older institutions such as the Congregational Church were losing their central role in the community.[55] The Revolutionary era had broken so much that was certain, and militia, which built on the mythology of the Revolution, gave certainty not only to men who served, but also to other members of the community who saw them drilling or marching, or even just knew of their existence. While nationally enthusiasm for militia activities declined precipitously after the War of 1812, the existence of voluntary militia units such as the Rifle Grays, which had no explicit standing in federal law and were weakly defined in state laws, allowed states to believe they had militia forces to call during an emergency without the bother of enforcing the obligations of a true enrolled militia. That the Rifle Grays existed for a few decades, although often limping along, bucked a trend for such voluntary companied to last only a few years.[56]

Along with the Washington Rifle Grays and the cavalry troop, organizations in the active militia, the town also maintained, on paper at least, the South Reading militia company in the enrolled militia. The U.S. Constitution recognized the militia in Articles I and II, and in the Second Amendment, while the federal *Militia Act of 1792* required most free white males between the ages of 18 and 45 to arm themselves and attend regular muster, but neither the federal government nor most states enforced the *Act*.[57] Instead Massachusetts, like many states, recognized self-created companies as the "active militia." The active militia was comprised of units whose members volunteered to serve in them, while the enrolled militia was comprised of all male residents liable for service who were not in the active militia. In theory the South Reading militia company existed continuously from settlement of the town in 1644 until it was formally disbanded in 1840.

The regiment to which the South Reading militia company belonged remained relatively consistent, although its designation changed several times since settlement. By the early nineteenth century, the South Reading company belonged to the 1st brigade of the 3rd division of the Massachusetts militia, along with the militia of nearby towns to the south such as Malden, Woburn, and two companies from Charlestown.[58] However, such enrolled militia companies existed more as a theoretical obligation for adult free males to attend muster days or pay a small fine, and by in large the enforcement of this obligation was spotty at best. The town however continued to commission the officers elected for the company and to keep lists of men with an obligation, and the commonwealth granted commissions for regimental officers in this increasingly paper military force.[59] Still, as late as 1823 the adjutant general of Massachusetts, William H. Sumner, was still championing the general militia of almost all adult males as more suited to a republic than any scheme for classifying the militia by age, or other proposed reforms. For all its weaknesses, and men such as Sumner were quite aware of the problems in the militia, Massachusetts was still generally recognized as having the best militia system in the nation, even if it was largely theoretical.[60]

Theoretical or not, the existence of the South Reading militia company served to remind residents that almost every adult free man under forty-five years of age had an obligation to serve if called. For men who wanted active militia service, with regular training and a uniform, the Washington Rifle Grays, and for a couple of decades, the cavalry troop, gave South Reading men a choice. South Reading, which upon its establishment as a separate town in 1812 had a population of only eight hundred or so, supported at the time three militia organizations.[61] Most men attended at least a few training days in their lives, especially if they could not afford the fine for not showing, although several categories of better-off men were exempt, much to the chagrin of those not exempt.[62] During the War of 1812, the newly styled South Reading Infantry Company — the company in the enrolled militia — began more active training, with the increased training days on the Common reminding older residents of the Revolutionary era. The training does not seem to have been too strenuous, and memories of the training days involve sergeants bringing pails of grog and pails of gingerbread, cheese, and crackers from nearby Hale's Tavern to the Common for all to enjoy, more than military drill.[63] The increasing consumption of alcohol became one of the arguments for abolishing the musters of the enrolled militia in states such as a Massachusetts that actually conducted them, as musters were increasingly seen as scenes of drunken debauchery and licentiousness, if not lawlessness, far more than events of a military, patriotic, or republican nature.[64]

Despite the continuation of this martial tradition, militia service was an obligation, not an occupation. South Reading was an almost constantly growing town, but into the nineteenth century most of the residents of South Reading continued to derive their livelihood from small, family farms. South Reading in 1832 had 1,311 residents, living in 163 houses, and tending to 275 cows, 191 hogs, 58 oxen, eighteen steers, and eight sheep. The town had 110 barns, along with three grist mills. Geography played a decisive role in the evolution of the town, as it had for most towns. Prior to the invention of the railroad and the internal combustion engine, farmlands within ten miles or so of a city could sell surplus goods in the city, and so joined the market economy almost from inception. Communities outside of the ten-mile limit, without a navigable river or canal linking them with the city, or along a common seacoast, could not economically ship wheat, milk, eggs, or meat to the city; oxen or horses would eat more than they could carry or pull. Farmers outside of the radius had little incentive to produce more than could be consumed on the farm and in the local village.[65] South Reading, lying roughly ten miles north from Boston, maintained its independence from that city longer than towns closer to Boston, yet was close enough to benefit from the larger markets that proximity to Boston and the coast offered.

With limited territory, and much of the soil rocky and not fertile, industry became necessary for the growth of the town. Most of the early businesses and industries were common to similar New England towns. From the time of settlement, some industry existed in the main village. Most of these early industries were in blacksmithing, with a blacksmith shop long established at the northern end of the village, on Church Street, with Reading Pond behind it, but milling and other agriculture related industries also appeared. Trade with coastal Salem encouraged farmers to raise cattle, pigs, and a few sheep for the market. The presence of cattle gave rise to a leather industry. Two acres at the southern end of Reading Pond, just east of the burial ground, were set aside for a tannery within a few decades of settlement. The availability of leather, and the need for extra income on small family farms, gave rise to shoe making as an adjunct profession. Shoe making took place on the farm, with a businessman who would collect the finished products for shipment to more distant markets. Small cobblers' shops appeared behind homes throughout the town. In 1812, a resident named Bolles Evans expanded his small shoe making operation by purchasing the materials needed to complete shoes, and distributing the materials among residents. He would later return to collect the finished products, paying a fee for each completed pair of shoes.[66] The collective system for shoe manufacturing as practiced by men such as Evans gave way in its turn to factories. By the middle of the nineteenth century, South Reading boasted at least six shoe-making factories, including the large Thomas Emerson and Sons, on Main Street, near the southern end of the Common. Related to the shoe making business was the manufacture of shoe making tools. Thomas Woodward made his mark on American manufacturing when in 1810 he developed a new type of shoe awl, which was an improvement on English designs. His son, James, inherited the family business and under his direction it was soon making tools for the national market. Other factories in the town were producing razor strops—another product made from leather—and rope.

In addition to these industries, several general stores served the town, with the Cutler Brothers running the most prominent one. But the most notable of the local industrialists of the first half of the nineteenth century was Burrage Yale. Yale, born in Meridian, Connecticut, in 1781, traced his ancestry back to the earliest English settlers in New England. He liked to boast that he first entered Reading about 1810, barefoot and pushing a tin cart. He soon established a small tin shop on Main Street, and built his business until he employed a hundred tin peddlers throughout New England. Sometime around 1812 he opened a large general store, which his brothers ran. He later branched out to other businesses, opening the Burrage Yale Inn and Tavern. Yale was noted about town for his careful accounts and his cold and haughty manner. Although he had a reputation for paying his employees fairly, he also expected prompt payment for all debts. Despite his Yankee frugality, his fellow residents of South Reading had little warmth for him. At one point, some of the more rowdy elements of the town hanged him in effigy from an oak tree on the Common, then tossed the likeness onto a large fire to the cheers of the gathered crowd. The next morning, a board nailed to a great oak tree read, "The great and mighty lord, he is no more!"[67]

Despite this animosity, Yale fulfilled his expected role as a prominent man of the town. Due to his financial involvement in the acquisition of the town's first fire engine in 1852, it and the volunteer fire company that ran it were dubbed "Yale No. 1"; the first officially recognized fire department in South Reading.[68] The engine was a William Jeffers hand-drawn tub, which cost $1,506. The town had acquired its first fire fighting apparatus in 1804, a Republican Extinguisher, and its second, a Black Hawk, in 1830, but only with the acquisition of Yale No. 1 did the General Court of Massachusetts approve the establishment of a fire department as an agency of the town government of South Reading. The town built a two-story fire station next to the blacksmith shop on Church Street, which ran east-west between Reading Pond and the Common, to take its place with the Congregational Church and Town House. The new station housed Yale No. 1 and its accompanying hose reel, but ironically the wooden structure burned in 1859. Its brick replacement would last throughout the nineteenth century, until only it and the church stood between the Common and the pond. The volunteer fire company that operated Yale Number 1 remained the only fire company in South Reading until a proliferation of fire companies occurred in the decades after the Civil War.

Yale's interest in creating a fire company may have stemmed from a fire that occurred in 1835, when the Baptist church burned. The Baptists, who had been outlawed during the colonial era, had established themselves in the fabric of the community, and by mid-century residents from some of the oldest and most respected families in town belonged to it and not the Congregational Church. The building committee for the new Baptist Church, erected the following year, included not only Yale, but an Eaton, a Sweetser, and a Wiley, names of some of the first settlers.[69] The new Baptist Church—both in edifice and membership—was an indicator on how much the town had changed in the past two centuries. Burrage Yale was a prominent local man, and he fulfilled his expected obligations, but he left no record of involvement with the local militia companies. The ambivalence of New Englanders to the militia allowed prominent men such as Yale to avoid meaningful connections with the militia. Such aloofness would not be common in the next generation of prominent men in South Reading. Yale had been one of the most prominent citizens of the town in the first half of the nineteenth century, but other men were rising to prominence in the town. A banner carried during the town's 1844 celebration of the Fourth of July listed the manufactured items of the town worthy of "protection," and included razor strops, shoe tools, boots, shoes, tin wares, and bitters.[70] The tin referred to Yale's operations, while the others were related to leather products. All, that is, except the reference to "bitters," which referred to a patent medicine manufactured in the town by the Richardson family.

Dr. Solon O. Richardson came from one of the most respected families in South Reading, with a reputation that stretched far beyond the borders of the town.[71] His grandfather had been a prosperous farmer. His father, Nathan, born around 1781, had been a highly respected physician, first in the Second Parish of Reading, then moving to the Third Parish of Reading, before finally settling in South Reading. According to local stories, the elder Dr. Richardson had originally sold his practice in the former Third Parish, by then the town of Reading, following the death of his wife, with the intention of moving his family to Cambridge, across the Charles River from Boston. Under the terms of the sale of his Reading practice, he posted a bond of $200 to refrain from practicing medicine within ten miles of Reading. His new practice in Cambridge would place him just beyond the stipulated ten mile ban. But so great was his reputation, that the town of South Reading offered him a tract of land worth an estimated $1,500 fronting on Main Street to forfeit his bond and practice medicine in South Reading. The elder Dr. Richardson accepted the offer, and built his new residence and new practice in South Reading. He remarried a year after the death of his first wife, and fathered another seven children. Dr. Nathan Richardson's reputation grew throughout the remainder of his life. The raising of the framework for his house on Main Street long remained in town memory as a grand festival. His support of public causes, and his financial generosity won him many friends. Dr. Nathan Richardson later opened a small hospital at his home in South Reading. His reputation spread throughout New England, so much so that the streets outside his home were often lined with carriages of patients coming to be examined by him. According to later accounts, he accepted whatever payment patients offered, never bothering to present formal bills for his services.

When he moved to South Reading, he brought with him his six children from his first marriage, including two sons, Winslow and Solon. Winslow died at the age of 18, and Solon became the heir apparent. With a father of such standing and reputation, becoming a physician was perhaps inevitable for Solon. Solon Osmond Richardson was born 19 July 1809 in what later became North Reading. He received his education at local academies in North Reading and Reading, before going to Atkinson, New Hampshire, and later to the Pinkerton Academy in Derry, New Hampshire. After writing his thesis on diseases of the heart and lungs, Solon Richardson received his medical degree from Dartmouth College. Townsfolk noted his graceful handwriting and his generally pleasant manner. He took several jobs prior to completing his medical education, after which he joined his father's practice. Married to a woman from North Reading in 1837, he became the father of two children. The elder, a daughter, died in infancy but his son, also named Solon, lived a long life and would carry on the family traditions.

From his childhood, Solon Richardson had assisted his father in mixing medicines. Around 1808 the elder Doctor Richardson had developed a "proprietary medicine" that he sold in powdered form, called "Jaundice Bitters." Eventually he found a much larger market selling it in liquid form, packaged in quart wine bottles. He christened this liquid "Sherry Wine Bitters." The exact ingredients of this medicine are today unknown, but the concoction seems to have been intended as a general elixir, advertised to restore health. Richardson claimed in advertising that his Bitters cured many ailments of digestion. Like most such tonics, it was essentially an alcoholic drink sold as a medicine, later testing out at 95 proof. A June 1887 study by the Massachusetts State Board of Health of such widely available "Temperance Drinks" in Massachusetts found that Richardson's Sherry Wine Bitters contained about 47% alcohol (almost 100 proof), making it the strongest of the nine tonics and thirty-two bitter drinks tested. According to an inventory in the 1860 Census Records for the U.S. Commerce and Industry, it also contained various barks, roots, seeds, herbs, and opium. No doubt it lived up to its advertised claim that "its good effects are immediate."

Following the death of Nathan Richardson in September 1837, Solon O. Richardson assumed control of his father's medical practice and patent medicine business. Although he followed his father into medicine, the calling did not have the hold on him as it did his father. For a few years he continued treating patents, paying weekly visits to branch offices in Lowell, Salem, and Boston, but this exertion took its toll on his health. He eventually retired from the medical practice and instead put his energies into the manufacture, advertising, and sale of the Sherry Wine Bitters.[72] He also tinkered with the recipe, changing its base from sherry to less expensive rum at some point. For the next three decades, his fortune would grow from the sale of this

Bottles that once held Richardson's Sherry Wine Bitters. The near bottle has "S.O. Richardson" in raised letters, while the back one reads, "South Reading." From stout bottles such as these the pious of New England could take their daily dosage of tonic (courtesy Nancy Bertrand).

tonic. The concoction became popular throughout New England, laying some claim to being the first regionally popular patent medicine. Folks who would not dream of visiting a saloon, or being seen buying alcohol, could buy this "medicine" and take their daily dose without damaging their reputations or self-identity as sober and pious members of society. Early problems in finding a suitable bottle that could survive rough handling and shipping yet not be excessive in weight were finally solved by a manufacturer from Philadelphia. Dr. Richardson created his own advertising, and formed a shipping company to move his stock throughout New England. As a businessman, the younger Dr. Richardson found his true calling, and he prospered financially.

The Richardson family personified the elite of New England in the early industrial revolution. They were among the most prominent citizens of South Reading, and their kind enjoyed similar respect and deference in most New England towns. Their standing came from a mixture of education, public service, financial generosity, standards of behavior, and reputation. As wealthy as they appeared to their neighbors, with larger houses, nicer carriages, and sources of income that did not involve manual labor, the economic gap between themselves and all but the poorest of their fellow townsmen could be measured in terms of small multiples. Perhaps Dr. Richardson, like his father before him, was four or five times wealthier than the average resident of South Reading. The vast differences in wealth of the Gilded Age were still decades away from South Reading.

Although life in South Reading seemed relatively static in the 1840s, plans for the construction of the first railroad line, which would link the town to Boston, promised great opportunity. Other changes were underway in military affairs, and laws regarding the militia would remain in flux for a decade. Various proposals to reform the Massachusetts militia had been proposed with regularity in the proceeding decades, but under adjutant general Sumner, all had been defeated. One proposal in 1833 aimed to create battalions of volunteers, but Sumner argued that such a move was unconstitutional, but more fundamentally, would break the connection between the "people" and the "militia." The commonwealth, he argued, could no more depend on a totally voluntary militia system than it could on a voluntary tax system.[73] In 1840, under a new adjutant general, Massachusetts took notice in law of the difference between men who actually served in militia companies and men who did not. Previously the commonwealth had been haphazard in recognizing militia companies, and many recognized companies had not existed in reality for generations.[74] The South Reading Infantry Company, to which in theory most adult men in the town belonged, whether they had ever turned out for muster or not, was eliminated that year when changes in Massachusetts law abolished all such notional companies in the enrolled militia.[75] The new law stated that "[e]very able-bodied white male citizen, resident within this Commonwealth, who is or shall be of the age of eighteen years, and under the age of forty-five years, excepting persons enlisted into volunteer companies, persons absolutely exempted by law, idiots, lunatics, common drunkards, vagabonds, paupers, and persons convicted of any infamous crime in this or any other state, shall be enrolled in the militia. While the new law did not nullify the militia obligation of most men, it did end the practice of enrolling most free adult white men in town-based companies.[76] Most adult, free, white men continued to have an obligation to serve in the militia when called to do so, but they were enrolled as individuals and not as members of theoretical town-based companies.

Shako with the initials "S R R" on the front plate, probably standing for "South Reading Rifles," which the Washington Rifle Grays were later known as (courtesy Wakefield Historical Society, photograph Nancy Bertrand).

At the time of its demise, the South Reading Infantry Company could justly claim to be one of the oldest military organizations in the nation if not the world, having existed from 1644 until 1840. But most such town-based companies in the enrolled militia were long since moribund, and their existence was mostly myth. The militia company the town maintained in the active militia — the tiny remnant of the Washington Rifle Grays— would not long survive the South Reading Infantry Company. State political and military authorities were grappling with the reform of the militia system, recognizing that the old system had long been broken beyond repair. The final end of the old militia system in Massachusetts in 1840 was part of a national trend. Massive population growth, geographic mobility due to in part to the development of the railroad, and increased political involvement by the lower classes, all worked against the old system. The generally poor showing of the militia in the War of 1812, and the removal of a credible external threat also worked against the concept of a general militia.[77] The adjutant general of Massachusetts during the 1840s, George H. Devereux, complained bitterly about the active militia in his annual reports, noting that "the half-day [muster] in May amounts to little" while "the Fall Review attempts an impossibility."[78] South Reading's Washington Rifle Grays, which by that time was officially designated as Company H of the 4th Regiment, arrived at the 1845 muster with only twenty-three members. By this time, the company was apparently referred to as the South Reading Rifles.[79] Its commander John Wiley, Jr., would long hold leadership positions in military organizations. Captain Wiley descended from one of the first settlers of the town, a sergeant also named John Wiley, and the family had a long tradition of military leadership in the town. But Captain Wiley did not have enough prestige to keep the Grays as a viable organization without other, greater elites throwing their prestige behind the company. Even on paper the Grays numbered a measly thirty-seven members. The adjutant general described it in his report as "depressed."[80] He further opined that it was "in a greatly reduced condition, and, I think cannot long survive."[81]

Dr. Solon O. Richardson (1809–1873), who inherited his father's patent medicine business, and for whom the Richardson Light Guard was named in gratitude for his financial support (courtesy Wakefield Historical Society).

His prophecy proved correct, and in 1846 the Rifle Grays disbanded and South Reading had no company in the active militia.[82] No records have surfaced to explain its failure, but several factors must have contributed to its demise. First of all, the Rifle Grays had been attempting to draw almost wholly from the town of South Reading, while the neighboring town of Reading also attempted to maintain its own company in the active militia, although without much success. Second, the Rifle Grays, while existing for over thirty years, was originally based in another town, and thus at best a transplanted institution rather than a homegrown one. Its officers were usually from the better and older families in town, but little tangible connections between other elites or institutions of the town existed, and thus it was an organization of some residents of the town, but not an institution of the town. And finally, the general anti-militarism that swept New England as a result of popular revulsion against the Mexican American War ended the company, which had long been diminishing in active members. However, the lack of a militia company in South Reading would not last long, and in a few years young men would again talk about organizing a new militia company.

Chapter II

New Militia Company

The construction of the Boston and Maine Railroad through South Reading in 1844 began to alter the town fundamentally.[1] The railroad did what nature did not: directly link South Reading to Boston. The first railroad line in town came directly from Boston, with its first South Reading stop in the village of Greenwood, about two miles south of the town center, then proceeding to the west of the main village, where another station stop was built, and then onward toward Reading, Wilmington, and other towns to the north. Regular service began in 1845. Soon another line branched to the north-east, crossing Main Street just south of the main village, or what was increasingly called the Center. The railroad began the long transformation of South Reading from an agricultural community, into an industrial town, and finally into a bedroom suburb of Boston. At the time the railroad opened in town, South Reading contained about one thousand, six hundred residents, a figure that would double in a decade and a half. Shortly after the rail connections came to South Reading a new type of resident moved to town: the commuter.[2] While South Reading was becoming an industrial town, the roots of its eventual transformation into a bedroom community were just starting to dig into the soil.[3]

If the railroad was beginning to draw the towns in eastern Massachusetts closer to Boston economically, socially, and politically, the commonwealth was becoming more centralized in other spheres. The universal militia obligation envisioned by the framers of the *Militia Act of 1792* had never been a reality, and in much of the nation even the semblance of maintaining the militia had been abandoned after the War of 1812. Massachusetts, like most other states and the federal government, still considered most adult free men to comprise the militia, but also recognized individual companies that formed, trained, and uniformed themselves with little or no state support.[4] Other "militia" companies existed without any official state sanction. Massachusetts law on militia had remained in flux for a decade. Jerry Cooper, in his *Rise of the National Guard*, identified Massachusetts, New York, and Connecticut as the trailblazers in the 1840s in militia reform. Massachusetts and Connecticut created small but permanent bureaucracies within the state government to administer the militia, and gave their state adjutants general the important power to disband militia companies that failed to meet standards. As a result of the reforms, Massachusetts, by the end of the 1850s, had a model volunteer militia that would be the standard to which the early National Guard movement sought to raise the organized militia of all states.[5]

In 1851, under a new state adjutant general, Lieutenant Colonel Ebenezer W. Stone, the Massachusetts Volunteer Militia (MVM) took form, replacing the old active militia. The new law stipulated that volunteer militia companies were to have no more than sixty-four men in all, including officers, non-commissioned officers, and privates. The old law had allowed companies to include anywhere between forty-eight and one-hundred men,[6] so

Railroad junction alongside Crystal Lake, just south of the Center. The line to the left leads to Wakefield's Upper Station, then Reading, Wilmington and beyond. The line on the right cuts across the south end of the Center and leads toward the northeast (courtesy Lucius Beebe Memorial Library of Wakefield).

commanders of existing companies that exceeded the new limit had until the first Wednesday of May to discharge enough members to bring number down to sixty-four, unless the commander would swear he had over that number of men uniformed and active. This provision indicates that some companies carried more members on their rolls than they were capable of actually mustering, and the commonwealth sought to force commanders to purge their rolls of such inactive members. The new law also demanded a minimum number of members, and included provisions for the removal of companies that failed to maintain that minimum. Any company that fell below twenty privates for two years in a row, or failed to submit an annual return for one year, was subject to disbandment.[7] When setting up the new MVM, Colonel Stone disbanded fourteen moribund militia companies that had remained officially in existence under the old militia law,[8] and would disband many more in the future. The new law called for the entire Massachusetts Volunteer Militia not to exceed seven thousand men, although the actual number of militiamen in 1847 amounted to only 4,996.[9] Still, this surpassed the number of men in militia companies in 1845 by 550.[10] In the next decade, many towns would apply for the privilege of maintaining a militia company, but the standards would be much higher than in the past, and many would be rejected. The new MVM placed Massachusetts at the forefront of the movement to replace the general militia obligation with a more professional volunteer militia that would eventually lead to the development of the National Guard.

The genesis of the Richardson Light Guard came from this period of renewed interest in militia in Massachusetts. The idea for a new militia company in the town germinated with three local men, James F. Emerson, George O. Carpenter, and Joseph L.R. Eaton, who began to solicit like-minded men for a new militia company.[11] All three were young men from older families in town, who would themselves become prominent residents in their later years. The Emerson family had a militia tradition, with James's grandfather, Thomas Emerson, fighting the British on the road between Concord and Lexington as a member of the Reading militia. Thomas Emerson, Jr., served as the commander of the old cavalry troop years earlier. George Carpenter, twenty-three years old, was married to the youngest daughter of Thomas Emerson. He was a businessman but had long interest in militia companies. He had joined the Washington Phalanx, an elite militia company in Boston, when he was sixteen and later served as adjutant of the Seventh Regiment, MVM.[12]

Although the original three men remained involved in the new militia company and labored on most of the committees, none initially became officers. On 13 June 1851, forty-eight men, including the original three, as well as John Wiley, Jr., the last commander of the old Rifle Grays, submitted a petition to the governor to have the new company accepted into the MVM.[13] They solicited potential members from the community, and held a public meeting on 1 October 1851 at the Town House, on Church Street between the parish hall of the Congregational Church and the Yale Fire Station, to plan and organize the new company. The reasons underlying the desire for a local militia company were varied, but as the company's official history stated at the fiftieth anniversary of the event:

> This company was not formed out of a love of militarism, which is a desire for war or a thirst for conquest. These citizens believed in peace and thought of the state and nation as devoted to industrial measures, manufacturing interests, and upholding church and school. But the underlying cause of such organization are — a love of adventure, the spirit of comradeship, an instinct of fellowship on one hand, and also a purpose to be ready for any call of duty to defend country, state or town.[14]

Other, more subtle yet powerful forces were also at work. The Mexican American War had recently ended, bringing with it a new era of rising sectional tensions. Much of the North, especially New England, had opposed the war and contributed few regiments to fight it. Massachusetts contributed only one regiment of infantry for federal service during the war.[15] South Reading had contributed but a single soldier to the war. Despite the general opposition to the war in New England, the victory by the United States inspired a wave of nationalistic fever across the country.[16] New Englanders in general might have felt vulnerable to the charge that their lack of participation in the late war showed a lack of patriotism, martial vigor, or courage.[17] So too was the recent lack of a militia company in the town. South Reading had maintained in theory its general militia company from 1644 until 1840, while the more visible Washington Rifle Grays had been disbanded only in 1846. Those years, from 1846 until 1851, were the only period in which the town did not have a militia company until the end of the Vietnam War in 1975. For South Reading not to have a militia company was an anomaly.

Twenty-six men enrolled in the fledgling company at that meeting on 1 October, and they formed two committees, a recruiting committee and a naming committee. Two days later, a third committee charged with securing a suitable armory was added. The armory committee soon selected a room in Academy Hall, located on a small hill just east of the Center. Massachusetts law required selectmen to provide armories to house weapons and any other equipment provided by the commonwealth for the militia,[18] and so three Selectmen for South Reading, Lilley Eaton, Peter Wiley, and Samuel Gould, submitted their report to the governor, saying that they "have examined the room occupied by the newly organized military corps in this town, and our opinion is, it is a safe place for the deposition of the muskets & equipment to be furnished to the corps by the state."[19] Having met the requirements for a militia company, the Commonwealth officially granted "George O. Carpenter and forty seven others" the right to maintain a company in the MVM.[20]

Members paid seventy-five cents for the privilege of belonging to the new company.[21] While not excessive, the amount was not a trifling, in an age when the average Massachusetts farm hand received about $15.34 a month.[22] Dues alone were not enough cover the expenses of the fledgling company, and committee members had to use their own money to carry out its business. However, members hit upon an ingenious method both to raise funds and link the new militia company to the vested interests in the town. The new militia company sought financial assistance and social legitimacy from local elites from its inception. Prominent local men were allowed to become "honorary members," meaning that they could pay membership fees and claim association with the company, but not wear the uniform or train, nor were they expected to serve during times of strife. The Boston First Corps of Cadets, one of the most prestigious militia companies in Massachusetts, had long included honorary members, and probably inspired the practice in the new Wakefield company.[23] The Cadets traced their origins to 1741. The company, which contained several hundred active members, normally served as escort for the governor during official functions. Honorary members of the Cadets outnumbered active members by a factor of three, and were a source of cash for the company, and a way to tie the company to the vested interests and leaders of Boston. The new South Reading company would avoid the fate of sliding into obscurity that had ended the old Washington Rifle Grays. The list of its honorary members included most of the wealthy men of South Reading,[24] men who repeatedly served on boards of directors, committees, and philanthropic endeavors within the town. Thirteen men were originally chosen to be honorary members. Among them were Thomas Emerson, Jr., whose father had been one of the pioneers in the New England shoe industry, James Woodward, who inherited his father's nationally known awl manufacturing business, and Cyrus Wakefield, a businessman whose name would later be adopted by South Reading as its new name. A photo of seven men from 1868, described as "Influential citizens of South Reading" included Emerson and Wakefield, in addition to George Carpenter, a charter member of the RLG, as well as later members Lucius Beebe and Edward Mansfield. Lilly Eaton was, at forty-nine, too old to join the company when it formed, but fifteen members, including two sons, of that influential family served in the RLG during the nineteenth century. Lilly Eaton served as the original trustee of the company, and thus had an important role in it. The new militia company received the largest financial donation from Dr. Solon O. Richardson, then one of the wealthiest and most respected residents of South Reading. Dr. Richardson, forty-two years old when the new militia company formed, never served in the company, but he contributed five hundred dollars for equipping it. To give an idea of the value of Dr. Richardson's donation, rent for an average four room tenement in a Massachusetts manufacturing town in 1860 was about $5.86 a month. In gratitude for this generous gift,[25] and to link the new institution with one of the most respected names in South Reading, the naming committee selected to call the militia company the "Richardson Light Guard" (RLG) in honor of its patron.

As with many other militia companies, the term "light" referred to its self-chosen specialty of light infantry, while the term "guard" was common in volunteer militia companies, and emphasized the defensive ideals of the American militia.[26] The full company voted to accept this name, provided that the naming committee received consent from Dr. Richardson for the use of his family name. Consent was quickly given, and for the next century, the town would christen its local military company, be it Massachusetts Volunteer Militia, Volunteers in federal service, National Guard, or State Guard, as the "Richardson Light Guard."

On Saturday, 11 October 1851, the new militia company held its first regular meeting. A total of sixty-five men placed their names on that first muster roll. Their names show the importance of family ties, as three Eatons were members, along with two Emersons, and five Wileys. About half of the family names of the original members of the RLG had been in the town since before 1700, but the company also drew from newer residents.[27] The militia company provided a means for newer residents to embed themselves into the fabric of the community. Under the direction of Colonel Nathan P. Colburn, of the 7th Light Infantry Regiment of the Massachusetts Volunteer Militia, the members met at Academy Hall and elected the first set of company officers, a captain, and first through fourth lieutenants. Under the Massachusetts Militia Law of 1840, each company of light infantry had a captain, and a first through third lieutenant. In 1846, a fourth lieutenant was added. While a modern U.S. Army infantry company normally has a captain, a first lieutenant, and three second lieutenants, the 1840s Massachusetts militia had one man in each rank. Although a new company, it drew on experienced men for leadership. The first commander of the RLG was John Wiley, Jr., the last commander of the old Rifle Grays. He would serve until 1855, when Emerson succeeded him. Emerson served as commander three different times before he finally resigned from active membership in 1870. The first lieutenant, then a position as well as a rank, fell to Nathaniel S. Dearborn, who had earlier commanded the City Guard of Boston. Carpenter became the company's first clerk, holding the rank of sergeant. He later served as a lieutenant in the late 1850s, and captain in 1856 and again in April 1860. He resigned from the company just before it entered active service for the Civil War. Two years later he became an officer in the Ancient and Honorable Artillery Company, and its commander in 1868.

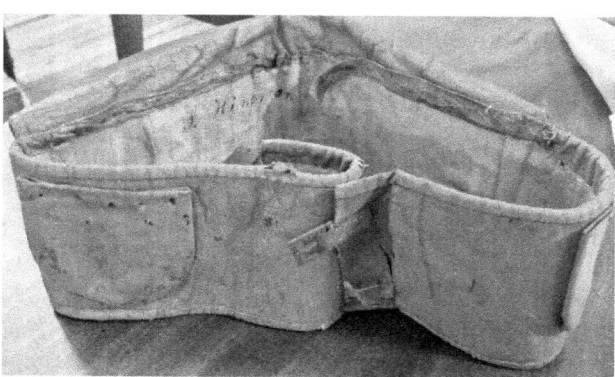

Cartridge belt from about 1840 that belonged to Samuel Kingman. Kingman served in various militia companies for more than twenty years before being elected fourth lieutenant in the newly formed Richardson Light Guard in October 1851, although he resigned two months later (courtesy Wakefield Historical Society, photograph Nancy Bertrand).

The RLG was accepted into the MVM as Company D of the Seventh Regiment, where it briefly joined Reading's rapidly withering Company B.[28] Along with the RLG from South Reading, a total of fourteen new companies joined the MVM in 1851, including companies in the nearby towns of Malden, Stoneham, and Winchester.[29] Having militia companies in neighboring towns essentially deprived the RLG from recruiting substantial numbers of members from those towns.

At first glance, the new RLG did not look all that different from the old Washington Rifle Grays, disbanded less than five years earlier. But the new company was more public, more tied to its home town and to the elite residents of that town. South Reading was beginning a century of profound change, after which the economics of the town, as well as the ethnic composition, would have been unrecognizable to earlier inhabitants. A militia company tied the present to the past, to the men who had originally settled the town, fought the French and Indians in the colonial period, and later the British at Concord, and in the War of 1812. The RLG gave certainty and unity to the town as the future approached.

Seemingly superficial but actually quite necessary to establish the RLG as a going concern, the first order of business was to agree upon a uniform, which members would purchase with their own funds. The first uniform the company chose consisted of a black frock coat, black pants, and black cap "with white cock feather fountain."[30] Bylaws were adopted, musicians chosen, and new applicants for active or honorary membership were voted on. In the fall and winter of 1851, the company began training in drill and the manual of arms. The RLG acquired the services of an instructor who had a reputation in such affairs. After a few practices, the RLG felt itself ready to parade in front of the town for the first time. The event was the first target practice, on Friday afternoon, 20 November 1851. The company marched through the town, stopping at the house of Dr. Richardson for a display of their marching and drill prowess, before continuing to a field on the west side of town, accompanied by Dr. Richardson and several other invited guests. There targets were arranged and practice commenced. On the march back to their armory, the RLG stopped at the circular riding house of Dr. Richardson for refreshments. The marksmanship skills of the members revealed themselves to be decidedly lacking, but the first showing of the company made a good overall impression on the town and the members themselves.[31] The RLG had made its first public appearance and was well received by the town. Dr. Richardson showed the fruits of his civic generosity, and continued to lend his support to the RLG, both by his presence and through his money. The public spectacle of the RLG marching through town and enjoying themselves with local elites while engaged in traditional masculine military activities brought new interest in the company both by members and other townsmen who hoped to become members.[32]

Women were of course not allowed to join the RLG either as active or honorary members, but they did link themselves to the company in other ways. A subscription apparently raised among South Reading women during the first year of the RLG paid for the creation of a company flag. The presentation of the flag was accompanied by a speech by Miss Almira Richardson, sister of Dr. Richardson. While the flag had disappeared by 1901, Miss Richardson's speech

A marionette wearing a recreation of an early uniform of the RLG. The marionette was made by the initial first sergeant, James W. Rutter, who served in the company until 1857. The image appeared in the volume the town published on the history of the Richardson Light Guard in 1901.

survives, and her explanation of each symbol on the flag shows the values and expectations of the town for its new militia company. The design on the flag, painted by local resident and member James Burt, was based on the flag of Massachusetts, with several other symbols added. The new RLG banner had distinct designs on each side, with one depicting an arm holding a sword, as on the seal of the Commonwealth of Massachusetts, with the motto of Massachusetts rendered in English as "By the sword he seeks peace under liberty."[33] Thirteen stars crowned the flag, which also included the sun pushing out clouds, the symbolism explained as showing the expansion of the original union to then unknown numbers. The growth of the nation, as well as the existence of the RLG, represented progress to the residents of the town. The scene was further decorated by cannonball, drum, sword, and flags. The reverse also had the sun, clouds, and stars, but included an eagle in the center, a member of the company, and a portrait of Dr. Richardson.[34]

Button from a Richardson Light Guard uniform (courtesy Wakefield Historical Society, photograph Nancy Bertrand).

After the first year, the RLG settled into the normal routines of the life of a militia company in peacetime. In April 1852, the RLG first paraded though Boston, on the occasion of the visit to that city by the Hungarian revolutionary Louis Kossuth.[35] The RLG as a body received fifty dollars as payment from Massachusetts, the first payment the RLG received from the Commonwealth. Other parades followed. Although Dr. Richardson never belonged to the company, his high-school aged son became a member in June 1852. The younger Solon O. Richardson and his classmate, a younger Thomas Emerson, were appointed as "markers." Markers, a position rarely used in military formations, marched near the head of a company on the march — normally right behind the first platoon. Upon the command for the column to make a ninety degree turn to the right or left, a marker would run to a point at the head of the column on which the marching soldiers would pivot. Being too young to be expected to carry a musket, the two young militiamen were given spears made of dark wood, topped with a golden tip and a small silk flag with the letters "R.L.G." in gilt. Since no other company in the regiment or even brigade to which the RLG belonged had markers, the boys served for the entire regimental or brigade column during larger parades. Thomas Emerson, one of eighteen Emersons who served in the company during the nineteenth century, would eventually serve in the U.S. Navy during the Civil War, while the younger Richardson, one of ten men with that surname to serve in the company during the nineteenth century, would remain an active member until 1860, and for the rest of his life would remain an enthusiastic booster for the company that bore his family name.

Shortly after mid-century, the town of South Reading had grown to about thirty-two hundred residents, of whom just over six hundred were voters.[36] Most had been born in Massachusetts, although a few had been born in Ireland or England. The town, like Massachusetts as a whole, was overwhelmingly white.[37] Only one black family resided in the town, a barber and his wife and four children. They moved to South Reading from Maryland a couple of years earlier and would not remain in the town for long.[38] More numer-

Part of Dr. Richardson's estate, with his round riding house clearly visible. Here the members of the Richardson Light Guard enjoyed the hospitality of Dr. Richardson after their first shoot. The view is looking west. The Methodist Church is on the right, and the unfinished Saint Joseph's Catholic Church is in the middle. The estate would soon be developed into a residential area (courtesy Lucius Beebe Memorial Library of Wakefield).

ous were the Irish, who had begun to settle in the town during the late 1840s, mostly as a result of the famine in their homeland. In 1850, about fifteen Catholic families lived in the town, and the day after celebrating the first recorded Catholic Mass in the town, they bought some land west of Main Street, fronting on the railroad right-of-way. Shortly afterward, the Catholics erected a chapel on the property.[39] As the first Irish Catholic immigrants began to trickle into the town, other newcomers — railroad commuters — powerful men who did not draw their wealth or standing from South Reading, were building elegant homes off Main Street. In the face of these changes to the economics and ethnicity of the town, the new militia company was a celebration of the roots of the town and its old families, and increasingly a way for newer residents to become part of the community.

Along with bringing commuters to South Reading, the railroad provided the town with a way to ship bulk goods economically to Boston and the coast. One of the first extraction industries to set up in South Reading that could not have existed without the railroad was the ice industry. While the town had small scale ice harvesting, storage, and distribution before the railroad, it was for the local market. The railroad made ice a commodity that could be marketed far beyond the borders of the town. With two large freshwater ponds near the center of town and rail lines running very close, the town soon had several companies in the business of harvesting ice in the winter, and either storing it for sale in the summer, or, more profitably, shipping it to Boston for export. Boston had been on the forefront of the international ice trade, and towns like South Reading benefitted from the ability to ship ice via rail. The shores of Smith Pond and Reading Pond were soon marked by boxy, windowless buildings—cork-insulated storehouses where ice would sit until the summer, when it had value. Ice work was seasonal for most of its labor force, a source of extra income rather than steady employment.

To some extent South Reading, and even more so many of the towns to the north and west of it, had always been dependent on agriculture, but the collapse of the New England farm economy in the early to mid nineteenth century as the Ohio Valley opened to agriculture slowed population growth, and in some areas of New England the population declined.[40] In the two years between 1852 and 1854, South Reading lost three eligible men from its Enrolled Militia. Since 1851, the term "Enrolled Militia" referred to those men eligible for militia service. Men who belonged to actual companies, such as the RLG comprised the Volunteer Militia. For the neighboring town of Reading, the change was far more dramatic, at least on paper, for it lost one-hundred and twenty-nine men over the same two year period.[41] However, this change reflected the separation of the old Second Parish of Reading, what became the town of North Reading, in 1853.[42] North Reading, with a far smaller population base and a lower population density than South Reading or even Reading, never maintained its own Volunteer Militia company. The town was much poorer than South Reading, with few industries to supplement whatever farming could be accomplished on its largely miserable soils. While the RLG usually included a few members from North Reading, the center of that town was over five miles away from the center of South Reading, and not directly linked by streetcar to South Reading, and thus North Reading had few of the ties to the RLG that Reading maintained. South Reading, with a larger population, manufacturing base, and direct railroad link to Boston, avoided the fate of its former Second Parish and experienced population growth in most years throughout the latter half of the nineteenth century. Still, South Reading could only support one company in the Volunteer Militia, unlike larger cities and towns such as Boston and Worcester, which often supported two or more. When towns maintained more than one company, the companies were often distinguished by politics or ethnicity. Worcester in the 1840s had the Worcester Light Infantry, which was filled by Democrats, while Whigs in the city formed the Worcester Guards. The rise of companies filled by Irish immigrants, such as the Sarsfield Guards and the revamped Columbian Artillery in Boston, and the Jackson Musketeers of Lowell, so incensed Governor-elect Henry Gardner that he attempted to have all such companies disbanded.[43] Smaller towns such as South Reading could not be so discriminating in accepting members, although the expense of belonging did serve to keep out working class residents until after the Civil War.

The RLG drew its members predominantly from its home town, but after a few years began to include a large contingent of members from the town of Reading, a short three miles away, and linked directly by train. At the 1850 muster of the organized militia, the adjutant general of Massachusetts remarked that "a very orderly company, and very handsomely equipped, was present from Reading." However, this company, like countless others in the early nineteenth century, had a very limited life span, and would be totally gone by the 1855 adjutant general's report.[44] The town of Reading would never again attempt to maintain a separate company in the militia,[45] and instead martially minded young men from Reading would be drawn to serve in the organized militia of South Reading, a larger, wealthier, and older town than Reading. For several generations, Reading would contribute the largest number of members from outside of South Reading to the RLG. This ability of the RLG to draw members from two towns greatly increased its survivability. The introduction of streetcars in the latter half of the nineteenth century made this arrangement even more practical. While the RLG was at heart an institution of the town of South Reading and would remain so, the town of Reading would also take an interest in the company, and this institution remained a tangible link between the two towns, which had formed a single town for over a century and a half.

At its inception, the RLG belonged to its members, and to prominent residents of the town of South Reading. The town of South Reading, the Commonwealth of Massachusetts, and the federal government occupied a diminishing hierarchy of influence. Over the next century, each level of government would take its turn as the most prominent factor in the company. But in the first decade of the company, the RLG really belonged to its members and wealthy benefactors. Changes underway in South Reading — the new railroads that crossed the town — were already making those benefactors more wealthy and more numerous. In the years following the Civil War, economic forces that had been gathering for a generation were to burst forth and bring to prominence a new kind of elite. Dr. Richardson could never have guessed when he received the honor of having the new militia company named after him that such honors were soon to be overshadowed. In the new economics of New England, and the cultural norms that they would spawn, men like Richardson would be far surpassed by new elites, whose status came not from their family name and education but almost wholly from their wealth, wealth of a kind that dwarfed the wealth of an average resident of South Reading, and made even men like Dr. Richardson look humble by comparison.[46]

The railroad made towns like South Reading attractive to Boston's upper middle-class, who could purchase or build a larger home outside of Boston than inside. Towns like South Reading had many of the charms of rural life without the isolation and poverty found in rural communities farther from large cities. The rail line made a daily commute to work in the city feasible and even pleasant, and the area of South Reading between the Center and the railroad station began to fill with stately homes belonging to this new kind of resident. Unlike the former Second Parish, the population of South Reading had long been absorbing outsiders in enough numbers that by 1850 few families could trace their residency in the town back more than a few generations. However, the new upper middle-class railroad commuters differed substantially from the older residents and earlier newcomers. They came to South Reading not to farm the land, work in the shops, or in other ways draw their sustenance from the town. Instead, these newcomers came to buy or build a home, maybe send their children to the local schools, but in many ways they lived lives detached from the community. They were as likely to buy their goods in Boston as from local merchants. Their bedrooms were in South Reading, but their lives were in Boston.

Cyrus Wakefield came to South Reading as one of these new residents, buying a colonial-style house on Main Street in 1851, the same year the Richardson Light Guard formed. Forty years old, he typified the Yankee success story, with a life that could provide the plot for a Horatio Alger book. Mr. Wakefield showed the potential for social and financial advancement impossible in the United States before the nineteenth century and never possible in much of the world. He was born in 1811 on a farm in Roxbury, New Hampshire. He moved to Boston as a young man and served as a clerk in a couple of businesses, including a grocery. After a failed partnership in one grocery, he opened another on his own, this time employing his brothers as clerks. His involvement in the grocery trade took him to

Boston's dockyards. While there he noticed piles of rattan, a vine-like palm plant that grows wild in the Malay Peninsula and coastal Southeast Asia, which ships plying the Southeast Asian trade often used as packing material on their return trips. Cyrus Wakefield learned that he could obtain the rattan cheaply, and resell it to local chair makers who used it to weave seats. When his grocery folded in 1844, he began to work full time in the rattan trade. In 1856 he began to manufacture the split cane in the United States using hand-powered machines he developed. He developed uses for the whole rattan—cane, pith, and shavings—and stopped throwing away what had previously been considered waste. He also developed more uses for the material than just its previous role in the making chair seats. Imitating in part Oriental designs, Cyrus Wakefield developed the rattan style of furniture that became a rage in the United States in the late nineteenth and early twentieth century.

Initially, Cyrus Wakefield established his rattan businesses in Boston. In 1851, he took advantage of the new commuter rail system of eastern Massachusetts to become a pioneer of sorts by becoming one of the first commuters to settle in South Reading. His reasons for picking South Reading are obscure, but his sister had gone to school in the town, while his wife, the daughter of a wealthy retired sea captain, came from neighboring Lynnfield. Some distant ancestors had been among the early residents of the town. But despite these tenuous ties, without the railroad such a move would have been unthinkable. Most likely he settled in South Reading because it was close enough to Boston for easy commute and far enough away for land to be relatively plentiful and inexpensive. Unlike most railroad commuters, once Cyrus Wakefield settled in South Reading, he began to make the town his own, first by buying whatever real property he could, much of it south of the main village. Within a few years of his settling in South Reading, Mr. Wakefield became the dominant resident of the town. Some farmers owned more land within the borders of the town, but no one owned more in the main village area. In 1855 he purchased several acres close to his house known locally as Green's Mill, which included a few small industrial buildings, some small ponds, and an old grist mill. The site had been the homestead of original settler Jonathan Poole, and thus had probably been the site of a grist mill for two centuries.

In this setting Cyrus Wakefield established what would later become the Wakefield Rattan Company. Within a few years of his move to South Reading, he would build new factories for his company in the town at the former Green's Mill, just a short walk from his house. He moved most of his operations to the town, and his factory soon became the largest employer in town, and its smokestack could be seen throughout. He must have much liked the Second Empire architectural style, for the house, downtown commercial buildings, and public buildings he constructed in the town bore the trappings of the type, with mansard roofs and ornate cupolas, and would influence the design of other buildings in the town, such as the new high school in 1871. But Cyrus Wakefield would dominate the town much more so than just through his taste in architecture. Through his funding of fire companies, service on civic and commercial boards, real estate holdings, and general public spirit, he dominated South Reading within a decade of his arrival. Perhaps Cyrus Wakefield saw the vast untapped wealth in idle former farm workers as Midwestern farms demolished the New England agricultural economy. Men whose fathers had farmed the local lands, and recent Irish immigrants, provided a stable work force for the burgeoning rattan industry. Within a few years he became the nation's largest supplier of caning and finished rattan products.[47]

Wakefield's first residence in South Reading, a typical colonial house on Main Street, soon proved too modest for such a prominent citizen. Twelve years after arriving in

Wakefield, he moved into his second and final home, which he had built for himself and his wife. His new estate, at the southern end of the Center on Main Street, was both elegant and public. His estate was not tucked away in some forgotten corner, but exposed on the principle street in town for all to see his wealth and power. The estate sat on a triangular area formed by the divergence of Main Street and the railroad line near the southern end of the estate. Geographically it sat in the center of South Reading. The main house dwarfed the homes of former elites such as Yale, Beebe, and Richardson. Indeed, Richardson's house looked quite modest compared to the neighboring Wakefield estate. Besides the main house, the estate also included a barn of roughly equal size and grandeur, a greenhouse, orchard, gardens, an ornamental pond, and other out buildings. The relative newcomer had already upstaged the former elites of the town, but he was

Cyrus Wakefield, 1811–1873. Businessman, philanthropist, and civic booster, he came to dominate South Reading more than any other man of his time (courtesy Wakefield Historical Society).

far from finished. Earlier prominent men such as Yale drew their wealth primarily from South Reading and eastern Massachusetts. Dr. Richardson had outclassed earlier elites and drew his wealth from all over New England, where his patent medicine was popular. But Cyrus Wakefield drew his wealth from national markets and international trade, while employing hundreds of residents. Even more than Dr. Richardson, Cyrus Wakefield had wealth and status that outclassed earlier elites. As a newcomer to the town, and a rising man of importance in town affairs, Mr. Wakefield quickly became an honorary member of the RLG. Associating himself with the RLG allowed him as a new resident to connect himself to many of the older names of the town. Tying its fortunes to such rising elites of the town allowed the RLG to do what most volunteer militia companies could not do—survive.

In August 1853, the RLG spent three days in Salem at its first muster as part of the MVM, at the annual encampment. Massachusetts has pioneered the annual encampment for its volunteer militia in 1849, and found it to be a great spur to creating and sustaining volunteer companies.[48] Not wanting to look lacking at company drill in the presence of the other companies of the MVM, prior to the camp the RLG drilled on the streets around the armory three evenings a week to sharpen their skills. For this first camp, each member had to pay six dollars—roughly a month's rent—for the privilege of going to the training camp. In Salem, most of the training took place on Winter's Island, in Salem harbor. The RLG participated in both company and battalion drills for most of the first day. On the second day they participated in divisional marches. The men of the RLG could compare themselves to the other companies of the MVM. The camp gave them the feeling of belonging to something large and important, and out of the normal experiences of their fellow townsmen who were not in the RLG. On the last day in camp, the governor came to review his troops.[49]

The RLG still mainly depended on the funds of active and honorary members for the financial support of the company, but the commonwealth slowly became more involved in local militia. At the beginning of 1854, companies in the MVM could not longer accept or discharge members as each company saw fit. The RLG had up to then held a vote on whether to accept new members or discharge active members who asked to leave. Under new regulations, the general commanding the brigade to which each company belonged had to approve all such actions.[50] Aside from small details such as that, the RLG, like all companies in the MVM, generally ran itself. The remainder of the decade was spent drilling and marching, holding elections for officers when one resigned, and making small uniform changes, such as in 1854 when the company voted to acquire bear-skin caps. Each year the company underwent three days of field training with the MVM, at various towns throughout eastern Massachusetts.

As the RLG gained in acceptance and prestige, the town itself was becoming more complex, creating both private and public institutions that reflected and aided the growth of the town. While the first bank in South Reading had opened in 1833, the Mechanical and Agricultural Institution with Thomas Emerson as its president, it was replaced in 1854 by the South Reading Bank, a state bank. Again Thomas Emerson became the president, while the town's unofficial historian and longtime selectman, Lilley Eaton, the original trustee of the RLG, became the cashier. The directors of this most important institution for the growth of the town again shows the power of a few individual men of the town — the elites of the town — and again demonstrates just how ensconced the RLG was into that elite. The first directors of the bank, Thomas Emerson, Lucious Beebe, Samuel Gardner, George Carpenter, Edward Mansfield, and Cyrus Wakefield,[51] were all men connected to the RLG as active members or honorary members, and usually had relatives in the company too. In 1856, South Reading created a committee charged with organizing a town library. While the men chosen for this endeavor reflects on their status, they again show how well the RLG had become entrenched among the more prominent and civic-minded leaders of the town. Of the six men on the committee, two were members of the RLG, including George Carpenter, while three others had relatives in the company. A year later, with the library a going concern in the first floor of the old Town House, a board of trustees was established to oversee the operation of the library. Most of the family names of the six men chosen for that position can be found among the names of the earliest settler, names like Evans, Eaton, Sweetser, and Wiley. Most of them had either active or honorary membership RLG, as well as several relatives who served or would later serve in the company. The Chairman of the Library Committee, Lucius Beebe, had been an active member of the RLG since 1856, one of five Beebes, including his son, who would eventually serve in the company. In May 1859, the RLG voted to accept Lucius Beebe as an honorary member. The same year, former commander John Wiley was elected to represent the twentieth Middlesex District in the state legislature.

That first decade, untested, with only the borrowed heritage of previous militia companies from the town, was the most uncertain in the life of the RLG until after World War II. As the 1850s drew to a close, the Richardson Light Guard had survived almost a decade, becoming a prestigious institution of the town. Most local elites donated their financial support and became honorary members, and most future elites served at least a few years as active members of the company. Mr. Wakefield was firmly entrenched in a rapidly industrializing South Reading. National events, in the growing sectional crisis, cast some uncertainty over the future, but the prosperous town of South Reading took comfort that its martial institution would prevail in any future strife.

Chapter III

Volunteers

At the beginning of the Civil War, the Richardson Light Guard was the only militia company in South Reading. Several members came from Reading and a few from North Reading—towns which had no militia companies of their own—as well as some of the other towns in the immediate area, but the bulk of members were residents of South Reading. The likelihood of war hung over the nation since the election of Abraham Lincoln in 1860, followed that December by the South Carolina's announcement that it was no longer part of the Union. The decision of the lower South to follow South Carolina's lead in the first few months of 1861, combined with President Lincoln's assertion that states did not have the right to secede, made war all but inevitable.

A few days before the 12 April 1861 Confederate attack on Fort Sumter, in Charleston Harbor, South Carolina, Massachusetts governor John Andrew sent to all companies in the Massachusetts Volunteer Militia a notice of alert, asking them to prepare for a call to war. The men of the RLG were eager for the call. Company leaders agreed that the ringing of the bell of the Congregational Church would be the signal for assembly. When news came of the attack on Fort Sumter, men prepared themselves and awaited the call to defend the Union. Slavery had been abolished in Massachusetts in 1783, and since the 1830s South Reading had its anti-slavery societies, mostly centered on the Baptist Church. The neighboring town of Reading claimed the first women's anti-slavery society in the nation. Not all residents appreciated the efforts of the anti-slavery society in bringing in controversial speakers, but the town did note the absence of violence in the town over the issue.[1] The town voted overwhelming for Abraham Lincoln in the election of 1860, with two-thirds voting for him, and that support would grow through the long years ahead.[2] The men of the RLG saw themselves as a force of good in a nation torn asunder by forces of evil. Even forty years after the event, when emotions had cooled somewhat, the chronicler of the RLG would refer to early 1861 as the time of "the darkest and most revolting instances of treason and national corruption."[3] The people of South Reading had no sympathy with the rebellion, and saw their mission to restore the Union as righteous. Having volunteered and trained in peacetime, the men of the RLG now would have their chance to prove to themselves, their fellow townspeople, and the memory of their ancestors, that all the drilling, marching, and training had not been for show or empty glory. The RLG awaited the call from Governor Andrew.

At noon on 19 April 1861, the call came, and the bell in the tower on the Congregation Church rang as the signal to assemble. Not all members were within earshot of the bell, although in an age without the sounds of cars or televisions, when most men worked either outdoors or in small factories near the center of town, a large number of the members would have heard the bell ringing. Men fired guns into the air as an additional signal for

Painting by Joseph Payro of Church Street just before the outbreak of the Civil War. From left to right, the buildings are the First Parish Congregational Church, the parish hall, the Town House, Yale Engine House, and a blacksmith shop. The lake can be seen on the right, the Common in the foreground. By the late nineteenth century, all buildings but the church would be removed, and the church would be torn down and rebuilt in a "Byzantine-Romanesque" style, and the land between the Common and the lake would become in effect an extension of the Common (courtesy Wakefield Historical Society).

the company to assemble. Word spread by wire and travelers to the outlying areas and surrounding towns and by 2:30 that afternoon, the company was assembled at the armory. The RLG had recently moved from the former armory, located on the second floor of a bank building across from the railroad station, to their third armory, on the third floor of a commercial building on Main Street, in the Center. The company marched up Main Street through the Center to the Common. There the soldiers ate a meal that the town had prepared for them, and then marched to the railroad station under Captain John W. Locke, a charter member, for the start of their voyage to war. For the people of the town, as well as the members of the RLG, the affair had the air of a celebration, with crowds, music, patriotic speeches, tearful farewells, and an invocation from the Rev. E. A. Eaton, from the Universalist Church, as the town sent her sons to what was believed would be a short war. The RLG left South Reading around 5:30 that evening, bound first for Boston. Upon arrival in the city, they marched to Faneuil Hall, escorted by several townspeople who made the trip with them, and bedded down in the attic.[4]

    The company that left South Reading that day was among the best that answered President Lincoln's initial call for troops. Unlike later-raised forces during the Civil War, the Richardson Light Guard that entrained in April 1861, had existed for almost a decade, and had strong continuity of members. Most had been in the company for several years and only a few had recently joined. Four times during the Civil War companies bearing the name "Richardson Light Guard" left South Reading for war, but the first had the deepest roots and the most cohesion at the start of its tour of service. Later incarnations might become more battle-hardened through longer and bloodier service, but none would match the first in unity at the start of service.[5] This first three-month tour took the RLG to the First Battle of Bull Run. The second response lasted only a few days due to a disagreement between the governor of Massachusetts and federal authorities, and the RLG never got beyond Boston. The third response lasted nine months and took the RLG to the ramparts of Port Hudson, Louisiana. The fourth response brought the RLG to the defense of Washington, D.C., for one hundred days.[6]

The RLG originally enlisted for three months of federal service in April 1861. Nothing in the U.S. Constitution specifically authorizes state units to be mustered into the army. In the Constitution, militia and army are separate entities. However, during the French and Indians wars colonies established the practice of creating regiments formed from companies composed of individuals drawn from the militia. These regiments were then sent on expeditionary service. State-raised regiments augmented the Continental Army during the War for Independence, and the practice was reinforced by repeated employment during the War of 1812 and the Mexican American War, as well as during some Indian wars. Such regiments were hybrid federal-state forces organized by the states under the militia clauses of the Constitution. When in federal service such units fell under the army clauses, although states continued to have the right to commission officers in these state units. Such units served in federal service for an agreed on length of service, and when that period expired, nothing could keep an individual soldier in federal service if the man did not volunteer again.

Every man in the Richardson Light Guard individually volunteered to enter federal service, and thus the company entered federal service as a cohesive unit. Nothing legally compelled the men to accept federal service. In theory, any member of the RLG could have declined to volunteer and while technically such a man would have remained in the state militia, with no organized unit in the town, his militia obligation would have been purely theoretical. But resisting the pull of a tightly knit organization and town would have been difficult if not impossible for members. South Reading men dominated the company, with all the officers, noncomissioned officers, and musicians coming from the town. Of the sixty-seven privates, thirty-seven came from South Reading, and another nineteen from Reading.[7] Of the eleven privates not from South Reading or Reading, four came from Melrose, three from North Reading, and one each from Andover, Boston, Lynnfield, and Woburn. The length of service, a mere ninety days, would have made most excuses moot. Employers would hold positions and public and private charities would look after family for the next three months. The town had pledged to support the family of any resident who died in service. The experience of this incarnation of the Richardson Light Guard would become part of the mythology of locally raised forces. The unit would leave, fight, and return as a unit to a glorious homecoming in the town. The men who answered Lincoln's first call for soldiers saw themselves as the new minutemen, analogous to the men who marched to Lexington and Concord eighty-six years earlier.[8]

The RLG would of course not fight as an independent company. In peacetime it had been Company E of the 7th Regiment of the Massachusetts Volunteer Militia. However, a quick reorganization designated the wartime RLG as Company B of the Fifth Regiment, Massachusetts Volunteers, for its three months of federal service. The new Fifth Regiment was an amalgamation of companies from the Fifth and Seventh Regiments of the MVM. Joining the RLG in the Fifth Massachusetts were the Salem City Guard and the Salem Mechanic Light Infantry, the Charlestown City Guards, the Haverhill Light Infantry, and the Concord Artillery and the Charleston Artillery, which despite their names, were actually infantry companies. The company from Medford had no unique name, nor did the company from Boston, which was the only company in the regiment created specifically for the war.[9] Regiments, normally commanded by a colonel, were the building-blocks of nineteenth century American armies. In theory an infantry regiment contained usually ten companies, each with a wartime strength of eighty to one-hundred men, giving the regiment a full strength of about one-thousand soldiers. In reality, due to wounds, sickness, deaths, desertions, and detached service, few regiments ever contained their full strength and continually shrank as campaigns wore on. In the nineteenth century U.S. Army, battalions were a subset of a regiment containing two or more companies, existing as a temporary formation. Brigades, usually commanded by a brigadier general, were two or more regiments, whereas a major general usually commanded a division or a corps. Brigades, divisions, and corps existed only as wartime formations, and while they had numeral designations, they were usually identified by the name of their commander.

On Friday, 19 April, the men of the RLG witnessed thousands of other Volunteers pouring into Boston, but they had no time to gawk due to their being fully occupied drawing uniforms. Word of casualties suffered by the Sixth Regiment, which had left Boston only days earlier, as it marched through Baltimore made sharp the seriousness of the task the men faced. The next day was filled with the seemingly chaotic task of drawing muskets and haversacks. Blankets and knapsacks had not yet arrived and would have to be shipped later to catch up with the regiment.[10] Getting all their equipment proved quicker and more efficient than they realized at the time, and the next morning the men were up before dawn and preparing to leave Boston. The trip south was a grand affair for the Fifth Massachusetts, at least until the regiment reached Maryland. While passing by railroad through the Massachusetts cities of Framingham, Worcester, Springfield, and into Connecticut, they were met with cheers, blessings, music, and the ringing of church bells. Especially memorable were the vast quantities of food each town thrust at the soldiers. The *Boston Transcript* published a letter from a member of the RLG recounting his amazement at the emotion of the crowds as the regiment passed through Springfield, Massachusetts.[11] On Sunday evening, the regiment reached New York City, and was fed at the Lafarge house. Following dinner, six companies of the Fifth Regiment, including the RLG, boarded the steamship *Ariel*. The men bedded down as best they could on the decks, and the ship left port before light the next morning. From New York the steamer took the regiment by Fortress Monroe, thence to Annapolis, Maryland. Along the way many of the men fell prey to that common malady of people not used to being on ships: sea sickness. They disembarked in Annapolis on Wednesday morning, during a heavy rain. The regiment found itself unable to use the federal barracks in that city and spent the night getting what sleep they could on the street.[12]

The next day, after waiting until seven in the evening, they marched about a mile to the Annapolis railroad station. From there they boarded railroad cars and moved about twenty miles to the junction of the Baltimore and Ohio Railroad. Some of them had to ride on open flat cars, and the crowded conditions of the cars inspired some of the men to put their weapons into the stake irons of the cars to keep from being pushed off. The men rode about twenty miles, to the junction where the Washington branch joined the main line of the Baltimore and Ohio. At the junction, all soldiers heading toward Washington detrained and marched another twenty-two miles toward Washington. The fear of Confederate saboteurs, who were rumored to be tearing up tracks toward the capital, forced this tactic. In some areas the sandy nature of the roads and the lack of sleep fatigued the men greatly. Five or six miles into the march, rumors of an impending Confederate attack swept through the column. As a yet unbloodied unit, the thought of a scrape with the "Secessionists" roused them to a state of expectant excitement, but no attack came and the march continued.

When they got within ten miles of Washington, the men stopped at a camp where a unit had previously built rude huts using the split rails of a fence. Although not luxurious accommodations, the men of the RLG were appreciative for these huts. After an all-too-brief nap, the men awoke, boarded trains, and arrived in the center of Washington at 1:00 P.M. on the 26th of April. Although by later standards of the Civil War the ordeal of the previous few days, with nothing to eat but ham and bread, would seem relatively light, for new soldiers in the spring of 1861, it seemed grueling.[13] The movements of the company were usually mysterious to the men who made them, unaware at the time of their place in the federal army's plans, or of local conditions. For the men in the ranks, as well as the company and probably regimental officers, the journey was a confusing series of marches, changed plans, sleep lost, and meals missed.

On 27 April 1861, the RLG found itself quartered at the Treasury Building in Washington. The building was new and damp, which caused many members of the RLG to develop coughs during the month they stayed there.[14] In the early, chaotic days of the war, the federal supply service was unable to provide properly for all the soldiers rushed into the capital. As a result, rations were "poor and scarce," resulting in a lot of the men spending all their money buying overpriced food at local markets.[15] On 29 April, four days after arriving at the Treasury Building, the company's baggage arrived, allowing the men to set themselves up properly. That same day, the regiment was formed for an inspection by President Lincoln. For probably all of the men, this was their first chance to see the president. They would see him again two days later, on 1 May. On that day, the regiment marched to Jackson Square for official mustering into federal service. Since the War Department did not recognize the ranks of third and fourth lieutenants, the men in the RLG holding those ranks were demoted to sergeant.[16] As soldiers of the United States, the regiment marched past the White House to be reviewed by the president. At least one member of the RLG thought the president cut a finer figure in person than his photographs portrayed.[17]

A few days after the RLG was mustered in, Captain Locke brought a package to the company containing copies of the *South Reading Gazette*. The men ran toward the pile of newspapers from home, and in doing so almost knocked over the table and with its inkstand on which Captain Locke had placed them. While the men were glad to see that the newspapers showed that the people of South Reading were solidly behind the Union, they were disappointed to find nothing about the RLG specifically in the newspapers, but that situation would change soon enough.[18] An accident three days later demonstrated the lack of professionalism in the soldiers, or perhaps simply a hazard of the trade of war. "B. L.," most likely Private Byron Lord, of Greenwood, shot himself in the foot when he dropped his pistol and it fired. The wound left him a scar but otherwise he was ready to return to full duty in a week.[19] Private Lord was only twenty-one years old when he and his brother joined the RLG for the three months of federal service. Other than that minor wound, the only ailment suffered thus far were colds, from which many of the men continued to suffer, blamed on the damp conditions of the Treasury Building.[20]

The men had been away from South Reading for eleven days and were now federal soldiers. As with many new soldiers, especially those mustered in for a short term of service, they yearned to face the enemy. Most of their time in the capital was filled with drill and target practice, although soldiers were able to obtain passes with relative ease.[21] A week after mustering into federal service, Fourth Sergeant George W. Aborn got up early to cross a bridge over the Potomac River connecting Washington to Virginia. The Confederate flag flew from the Marshall House in Alexandria, which could be seen clearly from Washington.[22] Sergeant Aborn was quite aware that he was on what he considered enemy territory.

He seemed a bit disappointed that all "disunion troops" had been withdrawn and instead federal troops guarded the area to prevent smuggling and the movement of intelligence. The walk, he confessed, made him feel "nice as a pin."[23] On one occasion five members of the RLG went over Long Bridge into Virginia, where Private Henry W. Eustis shot a hawk in flight. After retrieving the trophy, two nearby Confederate soldiers on picket duty commented on the skill of the man who shot the bird. Perhaps not being able to resist a small taunt, Corporal James M. Sweetser indicated who had shot the hawk and then added that he was "the poorest shot in the company."[24]

The townspeople of South Reading continued to send articles of comfort and necessity to the RLG while in federal service. A meeting by the members of the RLG held while in the Treasury Building sent a message of thanks to be published in the *South Reading Gazette* extending gratitude to the town in general, but especially to the "ladies" of the town for the clothing and other comforts sent to them, and in return the members pledged to "never disgrace the ladies, our town, our state, ourselves or the *clothes*, by any misconduct before the traitors...."[25] The men also sent their thanks to their former commander and company founder, George Carpenter, for his labors on behalf of the company, and to Samuel Littlefield and James Woodward for the furniture those men had sent to them. Littlefield had served in the company since its creation, and was only recently discharged for illness.

By 8 May, life had become more tolerable for the men of the RLG. Most important, the food situation had improved. The men had a rare treat of beans, baked overnight in the ground. Although not as moist as the ones to be had at home, they were a welcome treat, rating higher in notice than even the beef they had that same day.[26] But most of the time spent in the area around Washington consisted of drilling, shooting practice, and fatigue duty as pickets and in building new defenses, one dubbed Camp Andrew after the governor of Massachusetts, and the other Camp Massachusetts. Occasionally they had some contact with Confederates, such as an exchange of words with some rebel pickets across the Potomac. On 16 June, Dr. Richardson and his son, along with George Carpenter, Thomas Emerson, and a few other men from South Reading, departed Boston for Washington to visit the RLG. Rumors abounded at home that the RLG had "not been properly cared for by the Regimental officers."[27] Dr. Richardson had been charged by the town to "look after the wants and provide for the comforts of the members."[28] The visitors managed to provide a meal for the men far above the fare normally served to soldiers, and the men were grateful for this show of support.[29] The South Reading men still found the experience of being so far from home in a war exciting. One man acquired a type of heavy shackle used on slaves, and it later became an object of curiosity back in South Reading. Eventually the RLG and the entire Fifth Regiment acquired the mission of providing pickets for Alexandria, Virginia, although Camp Massachusetts continued to be their main base.

On 15 July, more visitors from South Reading visited the RLG. Again George Carpenter was one of them, this time joined by another former member, James Carter, and two other men, William Atwell, whose son would later serve in the company, and John Aborn, who had relatives then serving. The three months for which the RLG and thousands of others had enlisted into federal service were fast coming to an end. If the federal government was to make use of the vast army of short-term soldiers it had been building, it had about two weeks to force the Confederacy to battle. The men of the RLG would not be returning to South Reading without having "seen the elephant" though. Plans were underway at the highest levels of government to force a showdown with the army of the Rebellion gathering in northern Virginia that would quickly end the war. The day after the visit, the Fifth

Massachusetts moved out on the campaign that would take them to what would later be known as the First Battle of Bull Run.

The campaign started with much excitement. Men packed three days' rations, and moved out in light marching order toward Centerville, Virginia. Along the way they linked up with several other Union regiments, and formed into the First Brigade, along with the Eleventh Massachusetts, the First Minnesota, and a battery of light artillery.[30] The Fifth Massachusetts spent the first night about seven miles from George Washington's home, Mount Vernon. The next afternoon, after a difficult march through natural and man-made obstacles, the column reached Sangster's Station, on the Orange and Alexandria Railroad. While camped at the station, the men of the RLG could hear the sounds of battle from nearby Blackburns Ford, where they would later learn, the First Massachusetts was being mauled. Late that afternoon, the column continued its march toward Centerville, with the Fifth Massachusetts in the lead, and the RLG leading the Fifth. Centerville was reached after dark, in a heavy thunderstorm. The men camped in a field under continuing rain, but were aware that they were part of something large, as they gazed out on the campfires of some 30,000 Union soldiers.

The next day the men awoke to one of their own alerting them to a herd of sheep heading their way. The rest of the regiments had also become aware of the sheep, and thus a mad scramble was on to capture a few for breakfast. In the end, the RLG managed to capture three of the animals. On 20 July, still unbloodied, the men of the RLG began to explore their surroundings. A few went into the town of Centerville, about a half mile from their camp site. From there they ventured to the grove near Warrenton turnpike crossroads, where the First Massachusetts had been engaged the day before. One of the men picked up an unexploded six pound shell from the site, and shipped it home where it would long remain on exhibit in the armory. The shell would cause some local notoriety on 3 September 1891, when a barn in which the shell was being stored, owned by former member James Eustis, caught fire, and the shell exploded.[31]

The next day was a Sunday, but the men had no time for divine services. Instead they were roused at 1:30 in the morning and started to leave camp an hour later. However, the mechanics of moving such a large force caused the RLG to stop again and wait until about 4:30 in the morning before beginning movement again. The day dawned clear, and the movement of the large army through Virginia, with flags waving and bayonets shining created a sight long remembered by those who were there. The Union Army seemed invincible, capable of sweeping away anything in its path. The column moved through Centerville to the Warrenton turnpike, heading south over Cub Run, toward Sudley's Ford. They passed two units going the other way, "not New Englanders,"[32] whose time in federal service had expired and thus were returning to Washington to be mustered out. While technically legal, the men of the Fifth Massachusetts, who still had more than a week remaining in service, found little to admire in the returning units. Around six o'clock in the morning a Union battery started firing, but received no reply from the Confederates. The men of the RLG knew that a large Confederate force was near, and they felt the tension as they moved ever westward. About an hour before noon, the men of the RLG found themselves in open country under oppressive heat, and the sound of cannon was soon followed by volleys of muskets. The brigade came in sight of the battle. Blankets and other impediment were cast aside, and the Fifth Massachusetts was ordered to advance double quick. The men noticed the dead and wounded from both sides as they moved forward. The heat and lack of water added to the unpleasantness of the day. One company from the Fifth Massachusetts took a position almost in front of the Confederate batteries, and after an hour of waiting, the Fifth was ordered to charge the enemy rifle pits to their front.

In the ensuing fight, the commander of the Fifth Massachusetts, Colonel Samuel C. Lawrence was wounded and had to be removed from the field. The color bearer was killed, and casualties were spread through the regiment. In all, nine men were killed and another twenty-three captured.[33] One man of the RLG, James Griggs, was wounded in the elbow and captured. Griggs had joined the RLG for the war, and would later serve in the post-war Regular Army. Another man, Joseph Eustis, was wounded in the hand. Eustis, one of three from the family to serve in the RLG, had been in the company since 1854. After the sharp fight, the Fifth Massachusetts received new orders, to support a Zouaves regiment about to charge a Confederate battery.[34] As the Fifth Massachusetts crossed a creek to join the Zouaves, a Union battery came through the middle of the Fifth, causing disorder in two companies. Eventually the Fifth fell in to the left of the Zouaves, but heavy fire on the Zouaves' left caused that regiment to retreat with half their strength dead or wounded. For the RLG, and the entire Fifth, the retreat of the Zouaves signaled a general retreat, as the Union Army began to fall apart. However, Major John T. Boyd, who assumed command after Colonel Lawrence's removal, restored order, and the Fifth began a proper withdrawal.[35] The Fifth retreated to Centerville, but an hour later the order came to begin the twenty-five mile march to Washington. On the 22nd, in a heavy rain, the Fifth Massachusetts reached Camp Massachusetts in Alexandria, where they passed a restless night. The next day, at around noon, the regiment began the final nine-mile leg of their trip to Washington, which they reached around four in the afternoon, wet, hungry, and filthy. Sometime on the march, the RLG realized that three of its men were missing. Some men mentioned that they had seen Sergeant George Aborn, a member since 1858, killed in battle. Griggs and Frank Tibbetts were also listed as missing, although no one seemed to know their fate.

News of the battle arrived in South Reading within hours via telegraph. Letters from members of the RLG began arriving within a few days of the battle, with some recipients claiming that the letters "bore the smell of battle."[36] Rumors abounded of who was wounded, and general consensus held that although three men were thought to be missing, they would soon turn up.[37] The RLG had few days left in their enlistment. They remained in Washington for five days before beginning their journey home on Sunday, 28 July. The men paraded in Washington for the last time, and were addressed by Colonel Lawrence, who was recovering from his wound. At one o'clock in the morning, the regiment boarded a train for the ride home The Fifth Massachusetts reached Boston in the early afternoon of 30 July. While the Third and Fourth Regiments had returned ahead of the Fifth, those regiments had not been in battle, and thus the return of the Fifth was still a massive occasion in Boston. After a short parade to Boston Common, the Fifth was released from federal service in a ninety-minute ceremony beginning at about ten minutes after 4:00 in the afternoon. State headquarters of the MVM offered an official order stating that the "memories of the men of the Fifth who have fallen in the great cause, and whose bodies lay moldering in the soil of Virginia, Massachusetts will ever hold in grateful remembrance."[38] After a collation, the RLG boarded a special train for South Reading.

The town had expected the company to return to South Reading perhaps on Thursday or Friday. News of the Sixth Regiment leaving Washington on Sunday reached South Reading on Monday. Tuesday was thus a flurry of activity in the town as it prepared to welcome home its sons. Each train arriving from Boston brought a rise in excitement followed by a let down, as each reported that the RLG was still in Boston. Finally a special train arrived at six o'clock bearing the RLG.[39] Despite their inauspicious first battle, the RLG returned to South Reading to a heroes' welcome. Before the men could return to their homes, they had to endure the hospitality of the town. The return of the RLG to South Reading was not a private or family affair, but a town event. The return of the RLG would be public, with the soldiers officially and enthusiastically welcomed back from their campaign. At the train station the men listened to a flowery speech from the chairmen of the board of selectmen, and then were escorted to the Common, to the sound of church bells and cannons. While it is tempting to surmise the effect of the cannon on the nerves of the returning soldiers, observers of the occasion noticed, or recorded, no unusual reaction from the men of the RLG. At the Common, under large tents, the men listened to more flowery speeches and some music, and were finally allowed to go home.

Those initial three-month soldiers saw only one major battle, Bull Run, which was a Union defeat. Later battles would dwarf First Bull Run in numbers of men engaged and casualties. The war became something longer and bloodier than anything most Americans thought it could. In total, South Reading would send more than five hundred men into military service, but the bulk went with the RLG or were former members of the RLG. Soon after the original RLG left South Reading for their three months of federal service, a three-year company began recruiting. This new company became Company G of the 19th Massachusetts Regiment, and was later redesignated as Company E of the 16th Regiment. It fought in the eastern theater as part of the Army of the Potomac. The instigator for creating this new company with a three-year commitment was John Wiley, Jr., of South Reading, who had earlier been the last commander of the old Washington Rifle Grays[40] and the first commander of the Richardson Light Guard. He commanded the RLG until 1855, but was again elected to lead the company in 1857. He had served only a year during his second term when he was elected to the rank of major by the regimental officers and rose out of the company. He resigned from the regiment in order to recruit a new company for a new regiment. Captain Wiley soon began drilling his new company, to be offered to the commonwealth as a Volunteer company for federal service. Although this new South Reading company did not bear the moniker 'Richardson Light Guard," the overlap in its leadership with incarnations of the RLG before and after the war blurred the distinction, and the experience of this company would be included in the official history of the RLG published in 1901. Of the approximately one hundred and thirty-six men who served in the company, only fourteen had previously served in the RLG.[41] The low numbers of former RLG men in the new company stemmed in part from most of the able bodied and martially inclined men of the town were at that time serving in the RLG on federal service. However, almost fifty of the members, more than half, came from South Reading. The company left South Reading for Camp Cameron in Cambridge on 25 June 1861 and left for the front on 17 August, serving first at Fortress Monroe, Virginia.[42] As Company E, 16th Regiment, Massachusetts Volunteers, the company remained in federal service for the next three years, and fought as part of the Army of the Potomac.[43] Captain Wiley would not serve the entire time, being discharged on 26 August 1863.[44] Serving as officers in the 16th Regiment were second lieutenants James Oliver and John Eaton of South Reading, both of whom had earlier served in the RLG. The 16th drew all its companies from Middlesex County, although its commander was Colonel Powell T. Wyman of Boston, in Suffolk County, who had been educated at West Point.

The early war years were a time of overt demonstrations of patriotism throughout South Reading. A drill club formed to train male residents in the rudiments of soldiering.

New flags were flying from most buildings. The Fourth of July celebration for 1861 was larger than any before it, with the high point of the celebration a flag rising on a new flagpole that had been installed on the Common. The drill club, the Yale fire company, and Captain Wiley's new company all participated in the celebration. With two companies from South Reading in federal service, the town had well met its quota for troops. South Reading was credited by the commonwealth with having the sixth highest number of residents in federal service — two officers and one-hundred sixty-three privates— out of fifty-two cities and towns, and most of the higher numbers came from much larger cities such as Boston, Worcester, and Springfield.[45]

For the RLG proper, the end of its first three months of federal service had been followed by a couple of months of uncertainty over the future of the company. The RLG was detached from the Fifth Massachusetts, which reverted to its former status as an MVM regiment. The Fifth Massachusetts would later serve a nine-month tour beginning in August 1862, and a hundred-day tour beginning July 1864. However, as its official history notes, little continuity between the first and latter incarnations of the Fifth Massachusetts existed. Only seven officers and men of the entire regiment served in all three tours, and no man who served with the RLG in the first tour served any subsequent tours with the regiment.[46] In the fall of 1861, a committee was formed in South Reading to decide who was a member of the RLG and who was not. The committee ruled that only men who signed the company's constitution could be considered members, and members who had fought as part of the company during the three months on active duty would be issued a new uniform free of charge. The men who had joined the company only to serve in the three months of federal service would be admitted to the rolls as honorary members. The reconstituted company again became Company E of the 7th Regiment, MVM. As part of the 7th Regiment, MVM, the RLG would have little time to bask in the glories of past service. Following the crisis in the Union Army following the defeat of General N.P. Banks at the Battle of Winchester, Pennsylvania, on 25 May 1862, President Lincoln called for more troops from loyal states. On 27 May, the men of the RLG again left South Reading to present themselves in Boston for federal service. However, Governor Andrew desired to muster them into federal service for only three months, while the federal government wanted state-raised regiments to muster into federal service for the duration of the war. After waiting in Boston for a few days, the crisis in Washington had passed and the 7th Regiment, MVM,

Epaulets of Captain John Wiley when he was the commander of G Company, in the 16th Massachusetts Volunteer Regiment during the Civil War. John Wiley was the last commander of the South Reading Rifles and the first commander of the Richardson Light Guard. He had been elected to regimental major in March 1861, but resigned to organize a company in a new regiment formed for three years of federal service, to which he was elected captain (courtesy Wakefield Historical Society, photograph Nancy Bertrand).

was sent home without federal induction. The RLG marched home on the 31st, stopping in Malden, about halfway between Boston and South Reading, for a lunch at the Town Hall. The men took it all in good spirits, referring to the incident as the "Evacuation of Boston."[47]

The summer of 1862 saw the much anticipated return of two of the RLG members who had been taken prisoner during First Bull Run, Aborn and Griggs. The men had returned to Boston around 1 June, but Aborn had been so ill and weak that he remained in Boston with relatives for two weeks before he was able to return to South Reading. The reception for the former prisoners of war, held 14 June 1862, in many ways surpassed the reception given the main body of the RLG upon its return after its three months of federal service. The prisoners had been gone a good deal longer, and had not been surrounded by their fellow townsmen during their ordeal. The men were greeted at the train station by bands, the RLG, town civic and religious leaders, and hoards of ordinary townspeople. Dr. Richardson greeted them first, in his role as chairman of the reception committee. The men were taken by carriage to the town common, where they listened to speeches and prayers of thanks for their safe return.[48] The community had welcomed back its lost sons. After the speeches had finished, the schoolchildren sang, the poets versed, and the bands played. Sergeant Aborn gave his recollections of getting captured and his treatment as a prisoner of the Confederates. The three men from the RLG had been taken to Richmond after the battle and confined in a tobacco shed known as Libby Prison. Aborn and Tibbets were later sent to New Orleans, and still later to a military prison in Salisbury, North Carolina, before being paroled, while Griggs spent his time confined at Tuscaloosa, Alabama. Tibbetts eventually reached the hometown a few weeks after Griggs and Aborn.

Hardly had the festivities ended when on 4 August 1862 President Lincoln called for troops to serve nine months. Again the men of the RLG offered their services and the company became Company E of the 50th Massachusetts. The regiment had about five weeks before being mustered into federal service, and used the time to drill and recruit to war-strength. The second in command of the 50th was Lieutenant Colonel John W. Locke, who had been one of the founding members of the RLG, and had served in the company ever since. Because the current commander of the RLG, Captain Henry D. Degen, was appointed regimental quartermaster, the RLG held new elections for company officers at the armory on 12 September 1862. Samuel F. Littlefield was elected captain, and would be the man to take the RLG to war again. Captain Littlefield was a charter member of the RLG and had served in the company ever since, except for when illness forced him to resign at the start of the Civil War. At thirty-six years of age he was one of the oldest men in the company. All of the officers and all but one of the noncommissioned officers came from South Reading.[49]

The rank and file of this incarnation of the RLG held more folks from outside the town than was common in peacetime. The town had attempted to spur local men to join by offering a bounty of one hundred dollars for residents of the town who enlisted in the RLG for the nine months.[50] This was accepted by vote at the Town Meeting on 25 August 1861. On 8 September the town further voted to pay $100 to any soldier who was a resident of South Reading upon his being honorably discharged, provided he had not earlier received a bounty. Of the seventy-two privates, forty came from the South Reading, and none from Reading. Most of the others came from neighboring towns such as Lynnfield and Melrose, towns where the RLG specifically recruited, but also from relatively distant towns such as Hingham and Topsfield. Although the RLG remained at heart a South Reading military company, the war had made necessary the recruiting of members beyond the town's borders.

Most of the outsiders, especially those from more distant communities, had joined the RLG solely for the war. They lived too distant to make attending peacetime armory drills practical.

The average age of the men who marched off from South Reading in the fall of 1862 was twenty-six years and three months. Forty of the men had wives. Twenty-five of the men listed their occupation as cordwainer — probably referring to occupations in the leather industry but also possibly shoe making — with another twelve listing farmer. However, shoemaker was listed as a separate occupation, which only one member claimed. Of the entire regiment, only thirty men listed their occupation as cord wainer, although a further 187 listed shoemaker, plus thirty-one shoe cutters, twenty-eight boot makers, seven each shoe dressers and shoe stitchers, and three shoe manufactures. The remainder listed assorted trades and skills.[51] The company was very similar demographically to most of the other companies in the regiment, and not very different from the earlier company that had served for three months, except that it had no members from Reading or North Reading. For the only time since 1854, the town of Reading raised its own militia company, this one specifically for the nine-months of federal service. None of the men in this new Reading company came from South Reading. Instead it drew predominantly from Reading, with fifty-two of its members calling that town home, with another twenty-three from North Reading. Most of the rest came from Wilmington. The Reading company would serve alongside the RLG in the 50th Massachusetts as Company D.[52] Unlike the three year company recruited by Captain Wiley from South Reading, the Reading nine-month company would not be included in the official history of the RLG published in 1901.

On the afternoon of the day that the RLG elected its new officers, described as a "beautiful autumnal day,"[53] it departed South Reading for Camp Stanton, in Boxford, about twelve miles to the north. No mention of the festivities accorded earlier send-offs exists. Unlike the hurried mobilization in the spring of 1861, this new mobilization was much more methodical, and the RLG was going to be spending at least several weeks relatively close to its hometown. South Reading had one hundred, eighty-seven men serving a three-year term, while sixty-three — mostly the men of the RLG — served for nine months. Additionally, the town of Reading had one-hundred and fifty-four in for three years, with another fifty-four serving for nine months.[54] Another two men from South Reading served in the Union Navy. North Reading, with a population base less than twenty-percent of South Reading's, also had two in the Navy, but unlike South Reading, North Reading had no militia company of its own and thus residents were more inclined to look to other ways of serving in the war.[55] With the RLG in service, the number of remaining men eligible for military duty in the town shrunk to thirty-two, while Reading, from a smaller population base, still had forty men available.[56]

After spending a week on state service, the RLG, as part of the 50th Massachusetts Regiment, mustered into federal service on 19 September. Camp Stanton was a relatively large cantonment, holding in addition to the 50th Massachusetts, the 41st, 47th, and 8th Massachusetts Regiments, as well as an artillery battery. The infantrymen were issued Springfield smooth bores, which had already become obsolete. The official history of the RLG, written over four decades later, states that the issued weapons indicated that the Fiftieth was slated to perform tasks such as building fortifications and digging trenches rather than participate in front-line combat.[57] They would spend almost ten weeks at Camp Stanton, with their days filled with drill, and their evenings with passes, singing, and camaraderie.[58] The men of the RLG were unaware of their impending role in the war, but federal military planners had allotted the 50th Massachusetts to join a large force of Union soldiers under the command of General Nathaniel P. Banks, with the mission of taking Port Hudson, on the Mississippi River, about twenty miles above Baton Rouge.[59]

On the morning of 19 November, the 50th Massachusetts Regiment assembled on a bleary, rainy day, to board trains to begin their long journey to the front. Passing through South Reading the men looked out their windows to see the locals cheering them. In Boston, the regiment marched to the Worcester depot to board another train, with residents of the city cheering them all the way. From Norwich, Connecticut, the men were loaded onto steamers for the voyage to New York City. Their stay in the city lasted but one night before they marched to a camp on Long Island. Still, many members of the regiment managed to avoid the officers and the guards and were able to see some of the sites of the city in the short time available. The next day they marched out to the Union Race course, at Jamaica, where Camp Banks had been established to gather the regiments to be sent to Louisiana.[60] While at Camp Banks, their days were filled with more training, but many soldiers managed to sneak off to nearby New York City when the charms and temptations of the big city proved overwhelming. They remained on Long Island until 10 December, when companies A, K, and E boarded the *Jersey Blue* for the voyage south. Their time on federal service one-third over, the Fiftieth Massachusetts was finally heading toward the seat of war. The voyage on the *Jersey Blue* became a bit more exciting than most of the men would have liked. The ship was described as a "miserable old hulk," leased to the government by Cornelius Vanderbilt, who had to come on board to inspect his doubtful craft.[61]

The ship was so narrow that some men had to remain in their bunks all day for lack of space. A storm off the coast of South Carolina almost wrecked the ship. The three companies of the 50th Massachusetts, including the RLG, went ashore on a small island near Fort Walker, a Confederate stronghold, to await the arrival of a new ship to complete their voyage. The men of the RLG found the trip more interesting at this point. The site of palmetto trees and plantations, as well as the warmer weather, lent an exotic air to the voyage. Food on the *Jersey Blue* had been miserable, consisting mainly of 'salt horse and hard tack, marked "B.C." (which denoted its antiquity.)'[62] But on the island the men feasted on fresh bread from government bakeries, as well as potatoes bought from local inhabitants. Far more interesting to the Bay Staters was the local plantation culture. The trapping of cotton cultivation and the palm and orange trees stuck in their minds, but most memorable of all was the presence of so many black people. Although a few black people lived in South Reading, they were so rare that little in the way of a separate black culture existed in the town. The Southern plantations, with the large and ornate main house and simple slave cabins left a deep impression on the men. Black people came to the soldiers to sell sweet potatoes and oysters. One evening the soldiers attended a "negro prayer meeting," and even took note of the unknown song they heard that evening, a spiritual called "Oh Lord Remember Me." One hesitates to speculate on the corresponding curiosity of the black people at the sight of the white Union soldiers from New England at their prayer meeting.

Finally, on 31 December 1862, the RLG and two other companies of the regiment boarded the former slave ship *Guerrilla*, formerly known as the *Mary Kimball*, for the remainder of their journey to New Orleans. The first night back on board turned out to be unpleasant, and rough seas and rough soup for the evening meal prevented most of the men from keeping their supper down. A few days later, some missing sugar led to a complete inspection of all personal goods, although the sugar was not recovered. On 8 January, the ship sighted land for the first time since leaving Hilton Head, South Carolina. The land turned out to be the Bahamas Islands, and soon afterward, a Union Navy vessel fired a shot over the bow of the *Guerrilla*, bringing it to for an inspection to make sure it was not a smuggler attempting to break the blockade or a Confederate naval vessel.

On 10 January, the first member of the RLG to die in service expired during the night from illness. The victim was Sergeant George H. Green, who had served in the company since 1855. After a ship-side service, the body, sewn into a canvas bag weighted with shot, was buried at sea. The remainder of the journey, including riding out a violent storm in the Gulf of Mexico, passed and on 16 January, the *Guerrilla* reached New Orleans. The city had been under Union control almost since the start of the war, although reminders of Flag-Officer David G. Farragut's successful campaign to take the city remained. Still, as a measure of security, no one was allowed to leave the ship that first night as it sat dockside. However, despite the posting of armed guards to ensure that the members of the 50th Massachusetts remained on board, members of the RLG were soon walking the streets of the Crescent City, and exchanging postage stamps and the like with the natives for oranges and cakes. The locals were suspicious of silver, but would accept almost any other medium of exchange. The next day the *Guerrilla* was towed up the Mississippi River about five miles, where the men disembarked and camped. The men were fascinated to see that because of the long practice of building levees alongside the river, the surface of the river stood about twenty feet higher than their bivouac site.[63] As a result, the bivouac site was usually damp, and the men feared the spread of sickness. Still, the sights of the area intoxicated the men, although the immaculate plantation homes were all serving as regimental hospitals.[64]

After spending about two weeks on the site, the men of the RLG and the two other companies of the 50th boarded a steam ship for the three-day journey upriver to Baton Rouge. When they arrived in Baton Rouge, General Banks had just arrived to replace General Butler as the Union commander of the Department of the Gulf, which in practice meant that he had responsibility for all occupied lands in the lower Mississippi. Although a former governor of Massachusetts, the men still saw him as just another political general ill-fit to assume command of so large an operation.[65] In Baton Rouge, the 50th Massachusetts fell under the command of General N.A.M. Dudley of the 2nd Brigade, 1st Division, 19th Army Corps. The 50th Massachusetts was brigaded with another Massachusetts regiment, the 13th, plus the 161st and 174th New York, along with the 2nd Louisiana Volunteers. Union strategy called for taking control of the Mississippi River to split Texas, Arkansas, and most of Louisiana from the rest of the Confederacy, and give the Union the ability to use the river to move armies and supplies between the Gulf and the Midwest. Union victories had given federal forces control of the entire river except for the stretch between Port Hudson, about twenty-five miles upriver from Baton Rouge, and Vicksburg, some 110 miles farther north. The 50th Massachusetts would be part of the federal force assigned to take Port Hudson.

On 12 March, following an inspection by General Banks, the RLG began its march toward Port Hudson, along with the rest of the division. On the evening of the 14th, the men of the RLG found themselves about five miles behind the lead elements of the column, traveling on a road next to the river. The same day, the 53rd Massachusetts made a reconnaissance of the area from Baton Rouge toward Port Hudson. After driving out some Confederate pickets, they captured some cattle and brought the animals back to federal lines.[66] That night, some of the cattle were driven into the regimental area of the 50th Massachusetts and the men enjoyed fresh meat for the first time in a long time.[67] Their proximity to Port Hudson became apparent that night, when the sound of cannon from Union forces already engaged in the campaign to take the fort woke the men. A few hours later, still in darkness, the RLG moved out to protect the rear of the besieging army from an expected Confederate attack. The explosion of a steamer on the river brought some excitement, but no Confederate attack came. The following day, the RLG marched back to Bayou Montecina, pushing out any Rebel units in the area. The hard labor and heat of the campaign brought a treat in the form of a whisky ration, but this respite only presaged some hard campaigning ahead.

On 18 March, the 50th Massachusetts began its movement back to Baton Rouge, blocking the roads behind them with felled trees. Official orders told the men that the campaign was almost over as Port Hudson had been effectively neutralized, but the large preparations the men witnessed told them that more fighting was coming. Once in Baton Rouge, the men were given one hour to wash, drop their gear, and draw rations before they embarked for Port Hudson. On the transport, the river boat *Morning Light*, the men fell asleep, exhausted from the constant marching and laboring. When they awoke, they found that the ship had broken through a levy and was stuck fast in a flooded plantation. They would spend the rest of the next day getting the boat free and back into the main channel of the river.[68] Late in the day the *Morning Light* reached the bank opposite near Port Hudson and the men disembarked.

The 50th took possession of Winter's Plantation on the west bank about five miles below Port Hudson. The men used the deserted slave cabins as their billets. The RLG spent much of their time in the Port Hudson area stripping the surrounding countryside of contraband in the form of slaves as well as cattle, sugar, copper and brass, and other expensive commodities for the federal government. Although a Parrot gun from the Confederate gunboat *Gennesee* lobbed shells at them, the RLG lost no men during the operation. Through most of May, the RLG performed similar chores for the besieging army around Port Hudson; they performed picket duty, foraged, and guarded important points. But they saw no real combat. At the end of March, they returned to Baton Rouge.[69] Back in Baton Rouge the regiment continued to drill and perform picket duty, which the men enjoyed as it meant a change of scenery and perhaps a little excitement. On 2 May, the RLG was performing picket duty on the Greenville road about two miles north of Baton Rouge. Lieutenant Warren was in command. At about ten in the morning a rider came in from the north. He turned out to be a Union cavalry man, and was part of a two-regiment brigade of cavalry which had left Grant's Army near Vicksburg seventeen days earlier and had been raiding while seeking contact with the Union army near Port Hudson. Two cavalry troops were sent out, and returned in the middle of the afternoon with the brigade, its prisoners, and hundreds of escaped slaves.[70]

The men were relishing their good fortune of serving in the rear, in the relative comfort and safety of Baton Rouge, when they received unexpected orders to go to Port Hudson.[71] On May 26, the men from the RLG marched 18 miles to the front and began to participate in the fighting around the entrenched Confederates holding Port Hudson. Although the men of the RLG were infantry, they were first put to work manning Parrot guns belonging to the 21st Indiana Heavy Artillery, that unit having sustained casualties high enough to render them combat ineffective. Private William C. Eustis recorded in his diary that the regimental adjutant ordered them to assist the "Jackass Regiment," so-called because the guns were pulled by mules.[72] For many of the men involved, it was their first taste of real battle. From that vantage point, the men of the RLG watched the failure of the first general assault, while constantly under counter-battery fire from the Confederate lines.

One solid shot cut two of the men in half. The men of the RLG worked all morning and into the afternoon without food or respite, but were served a ration of whiskey and bread in the early afternoon. The Indiana soldiers, in gratitude for the assistance, would not eat until the men from the RLG were fed.[73] The weight of lead in the air forced the men to spend the next two weeks using their shovels far more than their guns or rifles as they dug trenches that brought them close enough to the Confederate lines to be able to converse with their opposites.

On the 14th of June, the 50th Massachusetts, including the RLG, entered battle. The losses on the Union side were high, but for naught. The defeat was blamed on the lack of coordination among the various divisions, which were supposed to attack from several directions at the same time but instead attacked piecemeal, allowing the defenders of Port Hudson to shift and defeat each attacking division in turn.[74] Despite these defeats, the lot for the defending Confederates was worse. Their supply of ammunition was limited, the food supply had dwindled to the point that the men subsisted on mules, dogs, and rats, and the constant digging of the Union soldiers had brought the lines ever closer to the defenders. But as May drew to a close, the Union commander, General Banks, faced a serious problem. The enlistments of twenty of his nine-month infantry regiments were coming to an end.[75] This would cut his combat power almost in half. The men of the RLG, and the entire 50th Massachusetts, looked forward to returning home and leaving the monotony, danger, and unceasing heat of the Port Hudson campaign and returning to the pleasantries of New England. The 50th Massachusetts was the only nine-month regiment in the brigade.[76] Tensions between the nine-month Volunteers and the three-year Volunteers rose as each side thought the other slacking in its efforts.[77] General Banks used all means fair and unfair to keep his army together until the fall of Port Hudson. On 30 June, the original nine-month enlistments of the men in the 50th Massachusetts expired.

Such abrupt terminations of periods of service of volunteer regiments were common in the American military before the twentieth century.[78] In a legalistic society, an enlistment contract was a binding contract, and following the expiration of its terms, the members of the 50th Massachusetts had every legal right to quit the campaign and expect transport home immediately. General Dudley summoned all staff and line officers of the 50th to his headquarters and laid his case before them. Recognizing that the climax of the campaign was near, and that their departure would place additional burdens on the Union command and imperil the successful conclusion of the campaign, the men of the 50th Massachusetts voted to extend their term of service an additional fourteen days. Such a gesture was not empty—American men and regiments in similar circumstances have quit battlefields under similar circumstances before. All the men of the 50th extended, including the men of the Company E, the RLG. Here the strong bonds between the men and their bonds with their home communities played a decisive role; none wanted to return to their hometown alone, before the units with which they left town returned. However, other pressures might have inspired this generosity on the part of the soldiers of the 50th Massachusetts. The Fourth Massachusetts had mutinied a few days earlier, and General Banks had disarmed the regiment, stripped its officers of their ranks, taken the unit's colors, and had two companies marched away under guard. The 48th Massachusetts was also close to mutinying, and Banks had some members arrested. Banks told the men of the 50th that he would have them shot if they mutinied.[79] General Dudley addressed the 50th, appealing to their pride in their home state, and promising them that Port Hudson could not hold out long.[80] While probably other pressures were brought to bear to encourage the 50th Massachusetts to extend, neither campaign histories, the regimental history, nor the RLG's history, recorded any threats or improper pressure on the soldiers in the 50th to extend.

On 7 July, while in trenches facing the Confederate defenses, the men of the RLG received word that Vicksburg had fallen to General Ulysses S. Grant on the Fourth. When the Confederate soldiers in the trenches opposite called out to inquire over the reasons for the shouting in the Union lines, they learned of the fall of Vicksburg and that Port Hudson remained the sole Confederate-controlled point on the whole Mississippi River. Later that day the men of the RLG watched the Confederate soldiers approach the Union lines under a flag of truce, followed the next day by the surrender of the defenders. On 9 July, the 50th Massachusetts marched into Port Hudson and helped process the seven thousand Confederate prisoners remaining in the post. The devastation inside of Port Hudson left a deep impression on them men as they witnessed the effects of the Union artillery on the compact area. On 17 July, their baggage catching up with them, the men of the RLG enjoyed their first change of clothing in two months. On the 29th, the 50th Massachusetts left for home on the steamboat *Omaha*. Their journey took them up the river which they helped conquer, passing through Natchez and Vicksburg. From Cairo, Illinois, the men switched to cattle cars for the journey across the North. Ten men of the regiment died by the time the rail phase of the trip began. At stops across the North, in Mattoon, Illinois, Terre Haute, Indianapolis, and Cleveland the local residents feted the returning warriors.

On 11 August 1863 the 50th Massachusetts reached Boston. After eating at the Beach Street barracks, the regiment marched to Boston Common, where the men were given leave until the 24th. For the men of the Richardson Light Guard, the remainder of the day became a minor ordeal. A large group of people from South Reading met the RLG at Boston Common and accompanied them to Boston's North Station for the trip to South Reading. But at the upper station, a huge mass of friends, family, and well-wishers from South Reading and the surrounding towns created pandemonium at the track side. Only with great difficulty could the officers and sergeants form the company for the march to the Town Hall. There the men had to endure another banquet, complete with welcoming speeches by prominent residents of the town. However, good sense took over and the official welcoming ceremonies were brought to an early conclusion from the by-then obvious desire of the returning soldiers to go to their homes.

Of the company that left South Reading almost ten months earlier, eight men died of wounds or disease during the campaign. None were captured and none deserted.[81] Following their return to civilian pursuits in the late summer of 1863, the men of the RLG again began to re-form themselves as the town militia. Many of the residents of surrounding towns who had enlisted for the nine-months of federal service ended their active participation in the company due to the impracticality of attending meetings and drills. However the nucleus of the company remained, and although enthusiasm for drill had slackened, the existence of the company was continued. This relatively sleepy state of affairs lasted less than a year when again Governor Andrew again called for the organized militia of the commonwealth to assemble for federal service.

The call to federal service, coming on 11 July 1864, came as a result of Confederate General Early's invasion into Maryland, which brought his force into a position to threaten Baltimore and Washington, D.C. This renewed Confederate threat brought a general panic to the federal government, which led to a new call to loyal states for soldiers to serve one hundred days. Governor Andrew again called for the organized militia to come to Boston for mustering into federal service. The Richardson Light Guard answered the call once more. The men formed at their armory in South Reading and began the short march to the railroad station. Again the leading citizens of the town showed up to speak a few words of encouragement, and a throng of townspeople filled the streets to see the men off. Captain Littlefield remained as commander, while many of the rank and file were men who had served in the nine-month tour a year earlier. However, for about half of the men who marched out of South Reading in July 1864, this would be their first experience in war. Of the ninety-six men in the company, fifty-four were from South Reading, including all of the officers and most of the noncomissioned officers. The Town Meeting had raised to $125 the bounty it would pay for residents who helped fill the town's quota of men under Lincoln's latest call.[82] The 1864 incarnation of the RLG was a younger company than previous, containing six men who had not reached their eighteenth birthday.[83] Reading gave most of the men from outside of South Reading, with seventeen men. The remaining came from other surrounding towns, such as Melrose, Saugus, Lynn, and a couple of men from Boston and North Reading.

After the short train ride to Boston, the RLG marched across the city to board street cars for the ride to the state mobilization site, Camp Meigs, which had been established in 1862. Camp Meigs was in Readville, which was a village inside the boundaries of Hyde Park.[84] The Richardson Light Guard was assigned to the Eighth Massachusetts Regiment, as Company E. Mustering in to federal service came on 19 July. On the 26th, the regiment was ordered to proceed to the defense of the city of Washington as fast as possible. Their journey to the seat of war took them through New York City, Philadelphia, and on to Baltimore. As residents of the latter city had greeted one Massachusetts regiment in 1861 with gunshots and thrown bricks, the men were apprehensive as they disembarked from their railroad cars for the march across the city to board another train. However, no violence occurred, and instead they saw a sullen yet peaceful city. The trains to take them farther south did not materialize and so the men of the Eighth Massachusetts set up their bivouac. The next day they went into camp at Manakin's Woods, and were assigned to the army's Eighth Corps. With the prospect of immediate campaigning slim, the men were allowed to go to Baltimore's Druid Hill Park for a day. Baltimore was mostly pro-Confederate and news of Eary's raids had emboldened some of the Confederate sympathizers. The day at the park became tense as residents taunted the soldiers with Confederate emblems and the singing of *My Maryland*. Soldiers warned the crowds not to sing that song, so of course the civilians began to sing it with gusto, and a minor riot began. Fortunately the men of the RLG got out of the scrape with only a few cuts and bruises.

The threat to Washington, D.C. turned out to be less than feared in early July, and the Eighth Massachusetts served most of their one hundred days guarding Baltimore from raids by the Confederate Harry Gilmore, who was then operating in the area, and protecting part of the Northern Central Railroad from Confederate raiders. In early November, at the close of their hundred days service, the Eighth Massachusetts again boarded trains for the trip back to Boston, where the men were placed on furlough following brief ceremonies on Boston Common. The Richardson Light Guard marched to North Station and boarded a train heading north. The RLG reached South Reading long after dark, and was thus spared the ordeal of the public welcoming home ceremony. Also, the shortness of their service, lack of battle participation, and general war-wariness of the people of South Reading probably also contributed to the muted return of the RLG in November 1864. Still, the town stood solidly behind President Lincoln, giving him 86 percent of the local vote in that year's election. The men of the company spent only a few days at their homes, before they had to reassemble for the trip back to Camp Meigs in Readville for their official mustering out of federal service on 10 November. Although the war had another half year remaining, the Richardson Light Guard no longer participated as a unit. Many former members continued to serve in the federal armies in long-service regiments, but as a corporate body, the RLG remained at home, under the command of Captain Littlefield, ready to answer the call again in a crisis, although perhaps not as naively about the supposed glories of war as it had been in years past.

After the war ended, the commonwealth tabulated the numbers of soldiers each town contributed to the war, and the numbers of residents from each town who died in the war. South Reading contributed a total of 505 men for the war, although some men were counted twice or more from reenlisting. A total of eighteen men were killed in battle, while another forty-two died of disease during the war.[85] South Reading supplied a total of twenty-eight officers for the war, while neighboring Reading supplied another thirteen.[86] The majority of these officers had served either with the RLG during the war, or if serving as officers elsewhere, such as in the 16th Regiment, had served in the RLG before the war. Additionally, eight South Reading men joined the Regular Army, most in March of 1864. Of the eight, three had earlier served in the RLG, and another did so after leaving the Regular Army.[87] Thus, the RLG had served its purpose in peacetime as well as war. It had prepared residents to fight to preserve what they felt was worth preserving. The RLG, although not linearly the same unit that had marched from what was then Reading to fight the British in April 1775, had shown itself to be a worthy successor to those earlier militia companies from the town. The RLG had become a permanent institution of the town, far less likely to fade in obscurity as had the old South Reading Infantry Company or the Washington Rifle Grays. The RLG in the Civil War had been the vehicle through which South Reading fought the war, and would be the vehicle through which the town would remember the war in the ensuing decades.

Chapter IV

Gilded Age Militia

From the end of the Civil War until the Spanish American War, the Richardson Light Guard, newly designated as Company A of the Sixth Regiment, MVM, maintained an existence similar to that of many militia companies of the period: members drilled, elected officers, paraded, camped, and participated in marksmanship competitions. Equally important to their status in their town were their balls and banquets. They performed no emergency services to town, state, or nation, but maintained the traditions of militia and expressed the martial spirit of their hometown. The post-war RLG drew its members from both veterans and, increasingly, men who wished they were veterans. The Richardson Light Guard gave the men of South Reading a reason to leave their homes for masculine company and to bond with their fellow townsmen.[1] For both residents of the town in general and members of the RLG in particular, the militia company maintained the link between the town's present and its distant and increasingly mythologized past, when hardy pioneers and settlers fought Indians, French, and British. During these years South Reading continued to undergo profound changes as older elites were surpassed by new ones, and economic and demographic forces begun before the Civil War gained momentum. Through these changes, the RLG remained a visible link to the certainty of former times, and a source of continuing civic pride for the town.

Drills during the last two years of the Civil War were spotty and records incomplete. This lack of activity reflected weariness with the war and the absence from the town of many of the martially inclined young men who instead fought on distant battlefields. With the return of peace in the spring of 1865, regular drills commenced almost immediately. New members who had not served during the Civil War were attracted as much or even more than veterans of that struggle. In post–Civil War America, young men who had not stood their time in the ranks could find their masculinity questioned. Those who came of age just as the war ended felt cheated of their opportunity to test their mettle in battle. The RLG gave the men of South Reading a chance to taste again or for the first time the camaraderie and masculinity of belonging to a military organization.

At one of the first post-war drills, in May 1865, the RLG counted one hundred members. Maintaining such a large roster of active members would prove difficult in the decades ahead, but membership always remained higher than fifty. Samuel Littlefield, one of the original members, and who commanded the RLG as Co. E of the 50th Massachusetts in 1862, and as Co. E of the 8th Massachusetts in 1864, was again elected captain. He remained commander until 1872. One of the first actions of the post-war RLG was to vote for each man to purchase a new uniform at a cost of $25. Fortunately, the members voted at the next meeting to rescind this plan due to the hardship of such an expense, for shortly afterward Massachusetts adopted a standard uniform to be issued to militia companies, with

The Richardson Light Guard formed up on Main Street during the Fall Parade, 1866, shortly after the end of the Civil War (courtesy Wakefield Historical Society).

the commonwealth absorbing most of the cost.[2] Each uniform cost $31.53, with the state paying $20, the town contributing $311 in total, and so each man in the RLG had only to pay the remaining $6 for his uniform.[3] Attired in their new standard uniforms, the RLG marched in a grand review of the MVM held in Boston for President Ulysses S. Grant in 1869. The adjutant general noted that some companies of the MVM bought their own, more stylish uniforms for special occasions, but the RLG seems to have forgone this luxury for a while.[4] A photograph of the RLG taken in 1877 in front of the Richardson home shows the men in their state-supplied uniforms. Under the command of Captain Albert Mansfield, the men pose in dark, double-breasted uniforms. A large white plume decorates the top of their hats. The officers carry swords; the rank and file are armed with rifles. Captain Mansfield, one of nine members with that surname to serve during the nineteenth century, was unusual in that unlike most commanders, he came from neighboring Lynnfield, a town from which only a small percentage of members came. But Captain Mansfield had served in the company since the end of the Civil War, and perhaps aided by family ties, had been elected commander in 1876 and served as such for two years.

Despite the new support from the commonwealth, the RLG continued its earlier practice of electing honorary members as a way of courting local financial and social support. In January 1867, the younger Dr. Richardson, Mr. Wakefield, and Colonel James F. Mansfield were elected.[5] Colonel Mansfield had enlisted in 1861 and remained in the Army of the Potomac throughout the Civil War. He had been wounded slightly at Gettysburg, and eventually left federal service as a lieutenant colonel. He later joined the RLG and served as a lieutenant. While the younger Solon O. Richardson apparently had a classical rather than a medical education, and never practiced medicine, he was still styled "Dr. Richardson." As in the past decade, honorary membership benefitted both the company and the honorary members. Those with the honor were able to ally themselves with a prestigious and masculine institution, while the company could show itself intertwined with the most respected names in the town, and get some needed funds. However, money from local elites such as Richardson and Wakefield were soon overshadowed by funds from the town and even the Commonwealth of Massachusetts. The town continued to bear some expenses for the RLG, paying $150 a year for armory rent, which was about the state average.[6]

While the town and commonwealth started to take over more of the financial burdens of maintaining militia companies that residents and members had previously borne, new organizations would also compete for members. The decades after the Civil War saw an increase in volunteer fire companies in South Reading. In 1868, the town's only fire company, equipped with Yale Number 1 and its two hose reels, all of which were hand drawn, responded to three fires. While this number of alarms would soon seem quaint as the town grew in the coming decades, it seemed high enough at the time to justify the expenditure of $947.50 for a hand-drawn wagon to carry ladders. A second volunteer fire company was recruited to man the new ladder wagon. The new fire company became Washington Hook and Ladder Company Number 1 in 1871, with the company's equipment housed in the basement of the Town House, next to the brick station housing Yale Number 1, on Church Street between the Common and Reading Pond. Residents of Montrose, one of the smaller villages in the northeastern section of the former First Parish that remained part of South Reading, did not feel adequately protected by the two fire companies in the main village. Thus in 1871 they purchased an 1852 Hunneman hand tub fire engine from the town of Charlestown for protection. The tub cost $300, and a new wooden building to house the engine cost another $2000. A new volunteer company to man the tub, which was not part of South Reading's fire department, was christened "C. Wakefield No. 2," after the town's most prominent citizen.[7] Volunteer fire companies drew their members from many of the same segments of the population as volunteer militia companies, and offered many of the same attractions: male-bonding, community involvement, a chance for heroism, and of course, the opportunity to wear a uniform. While nothing prevented a man from belonging to the militia as well as a fire company, each additional membership diluted the attention a man could devote to each.

As the Centennial approached, the two most honored Americans in South Reading were George Washington, and Cyrus Wakefield. Heroes were local or national. In spite of this invisibility of the commonwealth when recognizing heroes, the government of Massachusetts would increasingly assume the burdens of maintaining the Massachusetts Volunteer Militia that had been borne by the towns, elites, or by the members themselves. Massachusetts continued the trend toward increasing the professionalism of the MVM in the decades after the Civil War. In 1867, the annual period of field training required of all units in the MVM increased from three days to five.[8] A decade later more major changes came when Massachusetts took another hard look at its militia in 1878, similar to the period of reforms three decades earlier that had given birth to the MVM and the RLG. The reforms in Massachusetts were part of a national trend of reform of state militia that eventually gave rise to the National Guard.[9] In many areas of the country, fear of labor unrest drove some of the revived interest in militia but the MVM had largely remained aloof from such onerous duties. The Massachusetts was ahead of most states in providing infrastructure for its organized militia. New laws reorganized the Massachusetts Volunteer Militia companies into regiments of three battalions, with four companies to each battalion. Each company was allowed a maximum of sixty-two soldiers during peacetime. New positions for officers at the battalion and regimental levels ensured that ambitious officers had opportunities for advancement. Companies deemed inefficiently — usually those that contained few active members or rarely met for training — were mustered out. The RLG passed its state inspection, and was placed at the head of all companies in the state. The commander of the RLG, Captain Charles F. Woodward, received special mention in the report the Regular Army inspector made to the Secretary of War.[10] Captain Woodward would have a long association with the MVM, until the Spanish American War.

Concurrent with this new interest in the MVM by the commonwealth came more standard issued uniforms. In 1878, Massachusetts augmented the uniforms of militiamen with a fatigue uniform for field service.[11] The commonwealth also issued new dress blouses in 1880, and in 1884, Prussian-style black helmets, to the entire MVM.[12] This increased support by the commonwealth was welcomed by the men of the RLG, and brought new uniformity to the MVM. However it was a sign that the RLG, as with similar town militia companies, was increasingly becoming an institution of the Commonwealth of Massachusetts rather than its hometown.

Residents of the town were also finding new ways to express martial enthusiasm. Foremost was Post Number 12 of the Grand Army of the Republic (GAR). The GAR, a veterans' organization whose members had served in the Union Army or Navy during the Civil War, would remain a powerful voice in local and national affairs until age thinned its ranks into oblivion after the Great War. However, not all former members of the RLG had served in the Civil War, and so in 1884, the Richardson Light Guard Veterans Association formed.[13] Any former member of the RLG was eligible to join the Veterans Association, regardless of whether they had served with the RLG in war or peace. Members paid three dollars a year for the privilege of belonging to this group, which was soon dubbed the Fine Members Association.[14] These, and other veterans' organization that would later form in the town, filled a need from former soldiers and militiamen, but boys too young for service in the RLG also had opportunities to participate in military activities when in October 1885, military training began at the high school.[15] The company of Wakefield cadets, along with companies from Reading and Andover, formed a battalion of the Second Massachusetts School Regiment. Members and former members of the RLG helped train the cadets. In turn, many of the cadets would develop a liking for the military and would later join the RLG.

The very look of the town was undergoing constant change. In 1871, the Baptist Church burned down. The growing Baptist community, strengthened in the past by its association with the anti-slavery movement and its work among Swedish immigrants in town, decided to build a new church on a far grander scale. The new church building, dedicated on 11 December 1872, proudly proclaimed that the Baptists had become a major force in the town. The new building was the largest church in town. It occupied a lot fronting Main Street, one block south of the high school, across the street from the Common. The lot had formerly been the site of Hale's Tavern, long a focal point for secular recreation. The new building's asymmetrical front, with its sharp steeple on the northeast corner, outshone the nearby Congregational Church. The Congregational Church had been radically altered by modifications in 1837 and 1859. Initially it been a traditional Puritan meeting house of the type built between 1710 and 1800, with its main door in the middle of one of the long sides and facing the road. The pulpit stood opposite the front door. The bell turret was on a square tower attached to the west side of the building. Later, the building was rotated 90 degrees, and its main entrance placed on the gable end, near Church Street. The pulpit was moved to the opposite gable end. With its new steeple, the building came closer in appearance to the type of Congregational Church built between 1800 and 1825. [16]

Perhaps nothing marked the changes underway in South Reading during the Gilded Age more than the change in the name of the town. Local elites had often attached their family names to public institutions within the town, usually as a reward for generous financial support. Dr. Richardson attached his name to the militia company, while other elites had gotten their names attached to fire companies, streets, or the library. But the leading citizen of the town by the time of the Civil War, Cyrus Wakefield, attached his name to South Reading far more visibly and permanently.[17] The town had already changed its name several times since its founding, from Lynn Village, to Redding, Reading, South Reading, and finally, in 1868, it underwent a final change of name. The people of South Reading were never happy with the implications of the name borne by their town since 1812.[18] The main village in South Reading predated the contemporary town of Reading[19] and residents resented that the former Second and Third Parishes of their town got the moniker "Reading," while the older First Parish became "South Reading" when the town divided in 1812.[20] Even the name of the most prominent natural feature in the town, the large pond lying north of the main village long known as Reading Pond, had been renamed Lake Quannapowitt in 1847. The name came from James and Mary Quannapohit (or *Quonopohit*), two of the five people whose names were on the deed of 1686 ceding much of the land of the old town of Reading from Native to English control. James Quannapohit was recorded as a Christian Indian who, along with Job Kattenanit, served as spies among the Nipmucks for Massachusetts Bay during King Philip's War. They later served as scouts for colonial armies.[21] When, in late 1867, Cyrus Wakefield offered a lot on Main Street and thirty thousand dollars to build South Reading a new Town House as a memorial to the men from the town who died in the Civil War, residents of the town accepted his offer and voted to change the name of the town to "Wakefield" in gratitude.[22] Having one thousand or so of those voters employed by Mr. Wakefield, which made him by far the largest employer in the town, certainly helped ensure the name change.[23] On 20 January 1868, Cyrus Wakefield's offer was formally accepted and on 4 July 1868, the town of South Reading became the town of Wakefield.

The lot for the new Town Hall sat roughly between Cyrus Wakefield's estate and his factories, and directly across Main Street from the Richardson estate. The act changing the name of the town passed the Massachusetts General Court on 25 February 1868. The town held a gala Fourth of July that year, marking the occasion as the change of name, although legally the change went into effect on 30 June. The new Town Hall turned out to be far more elaborate and expensive than originally planned, and Cyrus Wakefield eventually contributed around four times his original pledge when he turned the new building over to the town for one dollar on 22 February 1871. Other elites in the town also contributed to the new Town Hall, although their contributions were overshadowed by that of Mr. Wakefield. Dr. Richardson, for one, contributed one thousand dollars to provide furnishings for the new structure. Lucius Beebe donated the books for the new community library, located inside the new Town Hall. Despite the additional gifts from men like Richardson and Beebe, the exponentially higher sum that Cyrus Wakefield was able to contribute to the project highlighted the relatively diminished status of the older elites. Even the RLG, which so conspicuously bore the name of the town's formerly most prominent citizen, soon

The high school cadets on the Common, 1886. The bandstand, built in 1885, and the fire house, demolished in 1891, are visible in the background, with the lake behind them (courtesy Wakefield Historical Society).

established their armory in some of the rooms of the new Town Hall. Cyrus Wakefield stood at his personal summit of power and influence in the town.

The ceremonies for the formal turnover of the new building to the town occurred on Washington's birthday in 1871. Most businesses closed early in anticipation of the festivities, which began at two o'clock in the afternoon. Although February in New England can bring a variety of weather conditions, most of which are not conducive to outdoor festivities, the day was clear and cool, but not too cold. Main Street thronged with residents, visitors, and schoolchildren, all awaiting the opening of the new Town Hall. Cyrus Wakefield's mansion, just down the road and across the street, welcomed friends and the curious all day. When the doors to the new Town Hall opened at two o'clock, some two thousand people thronged into the building, filling every space with a person.[24] The new building was indeed a grand monument to the town's Civil War dead, but much more so it was a monument to the man who paid for it. Cyrus Wakefield had hired the architect who designed it, and the style must have pleased the man who paid the architect's fee. The building, with three aboveground floors and a basement, matched the 1863 mansion of the Cyrus Wakefield Estate, with its mansard roof of the Second Empire style. Inside the auditorium of the Town Hall, where the selectmen would sit on the raised platform at the front during town meetings, were two portraits flanking the selectmen's stage. To the audience's right, George Washington's portrait harkened to the founding of the nation. To the left, a larger-than-life portrait of Cyrus Wakefield himself stood in equal dignity. The former town of South Reading was Cyrus Wakefield's town. At least for the time being.

In 1870, before the Town Hall had been completed, Cyrus Wakefield built the Wakefield Building, later known as the Taylor Building, directly across the street from the residence of Dr. Richardson and one block north of the new Town Hall. The Wakefield building was

Mr. Wakefield's Town Hall, with the Miller Piano building to the rear, and the Taylor Building to the left. Only the Taylor Building, now called the Wakefield Building, is still be standing, although it was heavily damaged during the arson plague of the early 1970s, and rebuilt without its fourth floor (courtesy Lucius Beebe Memorial Library of Wakefield).

a commercial building filling an entire block in Wakefield Center, with storefronts on the first floor and offices in the upper floor. In style it matched the new Town Hall and Wakefield Mansion. Although Dr. Richardson recorded no comments on the new edifice, the four-story block, on the east side of the street, prevented the early morning sun from landing on his home. Reminders of the town's new leading citizen stood all around Dr. Richardson at the end of his life; the new Town Hall kitty-corner from the Richardson home, the Wakefield Building across the street, the rattan industrial complex just to the east, the Wakefield estate to the south, and the very name of the town over which the Richardson family once held the position as most illustrious citizens. Yet Dr. Richardson seemed to bear no grudge against this latecomer who came to so dominate the town — he had been the instigator and main financial backer of the movement to commission the portraits of Washington and Wakefield that hung in the Town Hall.[25]

On 31 August 1873, Dr. Richardson paid a visit to the home of his more prominent neighbor, Cyrus Wakefield, whose presence literally as well as figuratively overshadowed Richardson's. Dr. Richardson had been in poor health for the past four years, after suffering "an attack of paralysis" while in Boston on business.[26] While at the Wakefield Estate, he was stricken with what was termed apoplexy — most likely a stroke — and succumbed a few hours later. The Richardson Light Guard had been away at camp in Framingham when Dr. Richardson died, but hurried back to their hometown in time to escort the body of Dr. Richardson to the cemetery for interment. Dr. Richardson was buried in Mount Auburn cemetery in Cambridge, joining many other Boston-area elites in one of the most prestigious cemeteries in New England.[27] With its namesake and chief private benefactor gone, one of the most personal bonds between the militia company and town had severed. However, Solon O. Richardson, son of the late doctor, former member and honorary member, would continue to offer the moral and financial support of his family to the company and town.[28] He would go on to serve the town as selectman for two terms, and spend another two terms representing his town in the state legislature. He would remain one of the company's most loyal boosters until his death in 1922, and would live in Wakefield his entire life in the house in which he was born.

Cyrus Wakefield soon followed the elder Dr. Richardson to the grave, dying on 26 October 1873. His final year would be an exhausting struggle against worldwide economic forces. The fire in Boston on 9 November that destroyed around sixty acres of the city hit Cyrus Wakefield, who owned much commercial property in the destroyed area, especially hard. To make economic matters worse an especially deadly equine epidemic, carried by mosquitoes, arrived in late spring of 1872 and by that fall had killed almost a quarter of all horses in the United States. The sudden drop in the horse population severely hampered the ability of American manufacturers to move goods locally. On 19 September 1873, the New York stock exchange crashed, ushering in the Panic of 1873, the worst depression the United States had known until then. As the sole owner of the rattan business, Cyrus Wakefield stood in danger of losing his every material possession. In response, in the middle of October 1873, the Massachusetts legislature formally granted the Wakefield Rattan Company a charter, severely limiting the personal liability of Cyrus Wakefield. Still, as the only holder of the millions of dollars worth of stock, his personal stake in the company remained immense. At the first stockholders meeting, on 14 October 1873, Cyrus Wakefield became president of his company. He would remain in this position for only eleven days. On Sunday morning, 26 October, Cyrus Wakefield suffered a fatal heart attack while reading a newspaper at his home. He was buried in a relatively modest plot in Lakeside Cemetery, on the western shore of Lake Quannapowitt.

By the time Cyrus Wakefield passed away, the economic crisis had wiped out most of his personal fortune. His wife, by using the wealth she had inherited from her father, managed to keep her late husband's financial straits from becoming public information. But Cyrus Wakefield had played his hand at the right time: his gift of the Town Hall, which led to the renaming of the town in honor of him, came when his wealth and status were at their highest. And while his flesh and wealth were transitory, his name would remain permanently attached to the town. The very civic life Mr. Wakefield had led would give the town a memory of a public-spirited man, a local hero, which would last for several decades after the death of the man. The Wakefield Rattan Company did not die with its founder, and would soon rebound from the economic difficulties that had probably been at least part of the reason for Mr. Wakefield's heart attack. Within days of Wakefield's burial, the executers of his estate and the directors of the newly organized company sent an urgent message to Cyrus Wakefield II, nephew and namesake of the late founder, recalling him to the Boston headquarters of the company. The younger Cyrus Wakefield had lived in Singapore since his uncle sent him there in 1865 as his personal representative at the source of rattan to ensure that only quality material was sent to the Wakefield factories. While in Singapore, the younger Wakefield showed a natural inclination toward business, and began buying coffee, pepper, tin, and spices on his own for export to New England. Cyrus Wakefield II apparently enjoyed life in the British colony at the tip of the Malay Peninsula, for he remained there until summoned in 1873, returning only once before to get married.

Upon his arrival at company headquarters in Boston, Cyrus Wakefield II was elected president of the company. Genius and work ethic often skips a generation, but luckily for the Wakefield Rattan Company, and the town that largely depended on it, such was not the case with Cyrus Wakefield II. Under his leadership, the Wakefield Rattan Company emerged from the business crisis stronger than ever, and rose to new levels of national prominence as it began to dominate the growing field of wicker furniture.[29] Indeed wicker furniture, and with it the Wakefield Rattan Company, came to dominate tastes. The light, airy, and seemingly hygienic qualities of wicker furniture contrasted with the heavy, overstuffed styles of a generation earlier. The craftsmen at the Wakefield factories seemed capable of rendering any practical object in wicker, with chairs, tables, mirrors, baby carriages, and even hot-air balloon gondolas pouring from its factories. When a major fire in 1881 destroyed all but one of the wooden buildings housing the Wakefield Rattan Company in the town, they were replaced almost immediately by a large red brick complex of buildings and production continued with little disruption. By 1894, the eleven-acre complex contained five main buildings of four stories each, along with additional buildings for a boiler, storage, dyeing, bleaching, and other purposes for a total of more than thirty buildings in the complex. Wakefield wicker seemed everywhere, at least at all the better country homes and resorts in the nation. The last photograph of President Ulysses S. Grant, taken at his summer home in the Adirondacks of upstate New York, shows him on 19 July 1885 — four days before his death from cancer — reading a newspaper on a porch while seated on a Wakefield Rattan Company reed chair.[30]

The industrial revolution that brought men like the elder Cyrus Wakefield to prominence also began to change profoundly the technology of war. During the Civil War, the various companies drawn from the RLG that entered federal service had been armed with smoothbore muskets. However, the rifle increasingly became the dominant infantry weapon in the decades afterward. Militia companies even more so than the Regular Army began to focus on marksmanship as one of the foundations of military training. Regular as well as militia officers agreed that marksmanship was a function of intelligence, and that the average enlisted man in the militia was smarter than his Regular Army counterpart. This belief reflected ethnic and class prejudices more than anything else. The enlisted ranks of the Regular Army in the Gilded Age contained mostly immigrants and lower class native-born whites, while blacks served in four segregated regiments. Militia companies tended to include a higher portion of native-born Americans, and more from the middle class. Although some reformers in the Regular Army promoted marksmanship training as a stimulus to intellect and sobriety, its main function became the filling of idle time at remote posts. To militia enthusiasts, the strength of the militia came from the thinking and enthusiastic patriot. A citizen-soldier would not submit to the mindless discipline required for field fire. Instead, the strength of the militia rested on marksmanship and individual initiative. To promote marksmanship, the National Rifle Association (NRA) was founded in 1871, copying the British NRA that volunteer militia officers had founded in 1859. After some shaky early years, the rifle movement gained strength, especially in the Northeast.[31] Massachusetts, which had a long tradition of supporting citizen-soldiers, embraced rifle training for its militia as training for future service as Volunteers.

The RLG had long taken an interest in aimed shooting. Within six weeks of the first assembly of the RLG in 1851, the members held a target shoot.[32] Shooting contests were held irregularly in the years before the Civil War, but were generally neglected afterward, until 1875, when leaders of the MVM took a renewed interest in marksmanship, and formed the Massachusetts Volunteer Rifle Association, which began holding team rifle competitions at the state arsenal at Camp Dewey in Framingham. The Association continued to hold the annual competitions until the state took over. In 1878, the RLG organized its first rifle team, with members drawn from the men of the RLG. The men used the .45 caliber

Springfield breech-loading rifles, which Massachusetts had begun issuing to its militia in 1876. The RLG rifles teams undertook systematic marksmanship training, and soon became one of the most proficient teams in the state, first winning the state championship in 1879. The RLG continued to place great emphasis on marksmanship training, and usually scored at or near the top in state competition. In 1881, the RLG had the most members who had earned the state badge designating a qualified marksman. As a spur to increased proficiency with rifles, a group of leading citizens of Wakefield in 1881 introduced medals bearing their family names for which members of the RLG would compete annually in marksmanship competitions; the "Wakefield," "Richardson," and "Beebe" medals.[33] The Cyrus Wakefield and Solon Richardson making the donations were, of course, the nephew and son, respectively, of their late namesakes. However the days when local elites had the greatest impact on the RLG were passing, and town and state governments were increasingly assuming the burden.

In the summer of 1891, the selectmen of Wakefield authorized the building of a new rifle range. The older rifle range, at the rear of the rattan factory buildings, could only handle ranges of two hundred yards, and the growth of the town into the area behind the targets made rifle training too dangerous.[34] State law required towns to either pay a fine or supply a rifle range for the MVM. Most towns paid the fine, but Wakefield had a long tradition of more actively supporting its militia company.[35] The town of Wakefield was proud of the success of its militiamen in statewide rifle competitions and wanted that success to continue. Until the town could build a new rifle range, the RLG had to travel to the town of Woburn to use the Walnut Hill range for rifle practice. Eventually the town agreed to create a new rifle range at a site chosen by the RLG's armorer, James H. Keough, in an area north of Lake Quannapowitt, known as Cox's Woods, where Keough liked to hunt.

The new town-owned rifle range was situated considerably farther away from the main village, in the northernmost corner of Wakefield. Three ranges were laid out, a 200-yard, 500-yard, and a 600-yard range. The standard in most competitions was 600 yards. The town acquired land from a few landowners, and company and town funds were spent to build the facilities. The land behind the ranges, to the north on the border between Reading and Lynnfield, contained forest, meadow, and swamp, reaching all the way to North Reading. Only the entrance and main facilities were within the bounds of the town of Wakefield. The target areas and safety fans behind them straddled the border separating the towns of Reading and Lynnfield, and were later expanded until the northern extreme of the camp extended into North Reading. The range was soon dubbed the Wakefield Rifle Range. With the creation of this new rifle range, the Richardson Light Guard had the privilege of training at one of the finest ranges in the commonwealth.[36] Perhaps inevitably, the state and later federal governments soon took an interest in and control of what had been the town's range. The fate of the town's rifle range mirrored the fate of the RLG, as it too became an institution of the commonwealth and later the nation, and its bonds to Wakefield were dissolved. But that lay in the future. In 1891, the town of Wakefield had again shown that it supported its militia company fully, and depended little on state funds and not at all on federal support.

However, the RLG would have to compete in more areas than marksmanship in order to remain viable. The decades after the Civil War would see a blossoming of fraternal organizations in much of the nation, and Wakefield, with its compact and growing middle-class population, was especially fertile ground.[37] In 1871, the Soughegan Lodge No. 38, Independent Order of the Odd Fellows, was reestablished after an earlier incarnation folded in

1853. The growing Irish population of the town was reflected in the growth of both the Catholic Church and in an Irish fraternal organization. In 1868, the Catholics moved their former chapel a short distance to serve as a function hall, and began construction of what would become Saint Joseph's Catholic Church. In 1873, with the number of church members well over two thousand and growing, the Catholic community in Wakefield was recognized by the church hierarchy as a separate parish.[38] Residents of Irish descent also had the option of joining Division 26 of the Ancient Order of Hiberians after it formed in 1876, absorbing the local Hibernian Aid Society which originally formed in 1868. But in 1888, the largest and most well-known fraternal organization, the Masons, returned to Wakefield.

Golden Rule Lodge, A.F. & A.M., arose in the fall of 1887 by the efforts of local Masons who attended lodges in surrounding towns. Having no building of their own, the new lodge met in the Odd Fellows Hall. Twenty Masons signed the petition to the Grand Lodge of Massachusetts for recognition of the new lodge. Among the founding members of the lodge were Stillman J. Putney, who served as the chair of the first meeting. Putney had served in the RLG between 1858 and 1860. Also involved in the organizing of the lodge was Fine Member Solon O. Richardson. At the first regular meeting of the lodge, held on 9 February 1888, the master was William D. Deadman, who had served in various incarnations of the RLG between 1862 and 1865, eventually rising to the rank of corporal in May 1865. Of the forty-eight charter members, fourteen were Civil War veterans and most also members of the local GAR post.[39] Samuel Littlefield, still referred to as "Captain Sam and probably the most respected veteran and former member of the RLG in town," soon joined the new local lodge.[40]

Other fraternal organizations followed, with the Knights of Columbus forming a counsel in Wakefield in 1894. The Knights of Columbus, a Catholic fraternal organization, could draw from both the Irish and, increasingly, the Italian residents of the town. New fraternities were formed every few years, with Eastern Star arriving in 1897, the Catholic Order of Foresters in 1898, the Knights of Pythias in 1902, the Fraternal Order of Eagles in 1904, and the Improved Order of Redmen in 1909.[41] This growth of fraternal and charitable organizations would continue until after World War II. Such organizations often offered many of the same attractions that membership in the RLG offered — social and business contacts, camaraderie, rituals, a uniform — and thus like the volunteer fire departments, competed for members. A photo of the Knights Templar parading on Main Street in 1893 looks quite similar to photos of the RLG from the era, although the Knights look a bit older and have more ostentatious hats.[42] In addition to fraternal organizations, residents created more volunteer fire companies in Wakefield. Like militiamen, firemen were volunteers, and had full-time occupations to attend. The volunteers had to be summoned from their homes or jobs through a system of bells and whistles, and their equipment had to be pulled by the firemen from the station to the fire. Horses were impractical for volunteer fire companies, as they would have to be fed and housed, with their stalls cleaned daily. Although the growth of the town certainly made more fire companies based throughout the different areas of the town desirable, their creation without central planning or impetus bespeaks of a desire on the part of men in the town to participate in organizations with serious purposes.

In 1876, a resident named Aaron Butler bought a four-year-old Hill hand engine and formed an independent volunteer fire company based on Main Street, christened "Fountain Number 3." This engine remained in service until 1889, when it was replaced by Volunteer Hose 2. The company sold the old Hill engine to the town of Reading. The town absorbed Volunteer Hose Number 2 into the Wakefield Fire Department three years later.

IV. Gilded Age Militia 73

Volunteer firemen and their engines on Church Street in front of the Yale Engine House. The brick fire station replaced an earlier wooden one that burned in 1859. The sign at the top of the station reads, "Fire our servant not our master." The brick fire station was torn down in 1891 (courtesy Wakefield Historical Society).

Other fire organizations forming in this era were the Wakefield Home Fire Protection Association in 1882, with each member equipped with town-bought Johnson pumps, a short length of hose, and a bucket. Half of the fifty pumps were distributed to residents and business owners in the center, with the other half scattered throughout other areas of town. Members received a fee from the town if they were first to get water on a fire. The village of Greenwood, in the southern end of Wakefield, had long gone without a local fire company. In 1886, the volunteer Greenwood Hose Number 3 formed. It would be absorbed into the Wakefield Fire Department in 1900, although the Greenwood Library Association continued to own the unit's equipment.[43] The regular Wakefield Fire Department also modernized during this period. Yale Number 1 was retired in 1882 and replaced with the town's first steamer, a Silsby. The town did not transfer the name "Yale" to the new fire apparatus, and instead named it after the most prominent living man in the town, becoming "Lucius Beebe Number 1." The hose carriage of the former Yale Number 1 was rechristened the J.H. Carter Hose, and a volunteer fire company recruited to man it in 1885. "J. H. Carter" referred to Parks Commissioner James Carter, who had served in the RLG from 1857 until 1860, and had since belonged to the Fine Members. With all but a few of the firemen in town, whether part of the Wakefield Fire Department or independent companies, being volunteers, the desire for an association became apparent. In 1889 the Wakefield Volunteer Firemen's Association formed, giving members a further opportunity for camaraderie and participation. Group photographs of the Association show several dozen mustached men in dark uniforms with caps striking a masculine pose for posterity. But the town of Wakefield would leave more to posterity than a few group photos to show its masculine civic organizations.

Industry continued to grow in the town during the Gilded Age. In 1884, the Henry F. Miller Piano company moved from Boston to establish its manufacturing plant in Wakefield. The company, renamed the Henry H. Miller and Sons Piano Company following the death of its founder, occupied a six-story brick building behind the Town Hall.[44] Other industries grew larger while new ones were established. The Wakefield Rattan Company continued to grow under the leadership of Cyrus Wakefield II. By the time he assumed the presidency of the company, he looked the part of the wealthy American capitalist. While sharing the name and drive of his uncle, he did not physically resemble him. He was a stout man, with ample whiskers on his face combined with a noticeable lack of hair on his head. Although he was as shrewd a businessman as his uncle, he never involved himself in civic affairs of the town to the same degree as his uncle. Instead, as the father of three young children, he spent more of his time enjoying his family.[45] He attended town meetings regularly and was a faithful member of the Episcopal Church, and was ready to donate to local causes, but otherwise remained more aloof from civic affairs. When the widow of the elder Cyrus Wakefield died in 1877, he moved his family into the Wakefield Mansion.[46] For the town, the arrangement was almost feudal, with the descendant of the man for whom the town was named and who was the largest employer in the town, occupying the large estate just south of the Center.

During the fall of 1887, the younger Cyrus Wakefield became aware of a weakening of his body. He continued to commute daily to his offices in Boston, but his declining health made the trip a heavy burden. On the afternoon of 25 January 1888, he took got off the train at Wakefield station and got into his waiting sleigh for the short ride to his home. As the sleigh reached Main Street, Mr. Wakefield slumped against his driver. By the time the sleigh reached the mansion a few minutes later, Mr. Wakefield was dead. Although many in the town wanted the body to lie in state at the Town Hall, the family insisted that the body remain at the mansion, which was opened to the public in scenes reminiscent of the passing of the elder Cyrus Wakefield fourteen years earlier. The RLG passed a suitable resolution mourning the passing of a benefactor the company considered a true friend. Although Cyrus Wakefield left three children, Cyrus, George, and Annie, all were minors when their father died. Their mother inherited his estate, but the domination of the Wakefield family in the company and town that bore their family name had ended. After the funeral, the body was interred in the cemetery near his uncle, although with a slightly larger red granite memorial.

By the late Victorian age, public aesthetics had undergone a dramatic change. The founders of the town saw the large pond just north of the main village in practical terms—as a water source. The north side of Church Street, which ran east to west between the pond and the Common, had been filled with buildings vital to the young town — a parish house belonging to the Congregational Church, the Town House, fire station, and a blacksmith shop — which together cut off the view of the pond from the main village. During the second half of the nineteenth century most of these edifices had been moved or torn down. The former Town House had been sold and moved in 1873.

With increased growth of the town and the shrinking of open spaces near the center, the Common and lake became increasingly viewed as the focal points of the main village, which was increasingly known as Wakefield Center. A bequest to the town by a former resident in 1883 allowed the town to purchase the land between Church Street and Lake Quan-

The bandstand in the park, or Lower Common, built in 1885 (courtesy Lucius Beebe Memorial Library of Wakefield).

napowitt, in essence joining the Common to the southern end of the lake in a large park. The new park land was called "Wakefield Park," or, increasingly, the "Lower Common." The name "Wakefield Park" was also the name of an exclusive subdivision built in the same period on the western border of the town, probably part of the reason that the park at the southern end of the lake was soon referred to as the Lower Common. In 1885, a bandstand was erected on the Lower Common. The bandstand, built of wood and fieldstone with many elements of the Queen Ann style, was supposedly inspired by a pavilion in Brighton, England. The dome was painted red, and the edifice would eventually become the unofficial symbol of the town. By 1891, only the small brick fire station, built in late 1859 to replace the wooden station that burned, remained between the Common and the new park by the lake. The town tore down the brick fire station that year, giving the northern end of Wakefield Center an attractive triangle of parkland with its sharp end pointing south into the center of town and its large end washed by the southern shores of the lake. The point was extended a bit farther south in 1884, when the Parks Commission created what was originally known as the "Sylvan Grotto," intended to be an "oasis on Main Street."[47] The site was a small egg-shaped plot of land at the fork where Common Street veered west from Main Street. The stone and shrubbery design of the site had been the dream of Parks Commissioner and Fine Member James H. Carter, who lived near the site. At the north end of the site stood the grotto, a half dome constructed of stone about ten feet high. An elm tree stood in the center of the site, while other greenery covered much of the remainder. The site, which residents almost immediately began to refer to as "the Rockery," became the object of ridicule and criticism for its disorderly and wild look. With the benefit of hindsight, the Rockery looked like a pedestal waiting for an appropriate monument.

The final major change in the look of the northern end of the Center occurred almost simultaneously. In the spring of 1890, the members of the Congregational Church cele-

Dr. Solon O. Richardson's home fronting Main Street, facing east. It is shown decorated for the town's 250th anniversary celebration in 1894 (courtesy Wakefield Historical Society).

brated their final service in the old church. The church building dated to 1768. The construction of this church building, heavily modified over the past century, had caused the rift that had eventually led to the separation of South Reading from Reading. Two years later, the church dedicated its new building, a solid gray granite building in a Byzantine-Romanesque style, with a square tower on the southeast corner. Fire destroyed much of the structure on 21 February 1909, although much of the stonework remained standing. The congregation rebuilt the structure much as it had been before the fire, and rededicated it in 1912. With the completion of the new Congregational Church, the upper end of the Wakefield Center had much of the look it would have into the twenty-first century.

This new emphasis on the physical appearance of the public areas of the town coincided with the approach of two landmarks in the history of the town. In 1894, the town celebrated its twenty-fifth year since adopting the name "Wakefield" as its own. For the occasion, the Wakefield mansion, at the southern end of Wakefield Center, was bedecked with no less than twenty-eight large American flags, plus yards of red, white, and blue bunting.[48] The Town Hall was similarly decorated. A larger anniversary came the next year, when the town celebrated the two-hundred-and-fiftieth anniversary of its founding. The RLG, as well as all of the fire departments from the town, along with representatives from Reading and North Reading, participated in the parade held in Wakefield to celebrate the anniversary. One float alluded back to the period before white settlement, consisting of a small teepee — a type of shelter probably unknown to the local Natives at the time of English settlement — with four "Indians," mounted on a horse-drawn wagon. Despite the unauthentic representation of the Natives at the time of settlement, the men accompanying the teepee float could have been ethnically Native American. Some residents of Wakefield did consider themselves to be of Native decent, and for much of the twentieth century, the

The armory the town built for the RLG in 1895. To the left of the armory is the Cutler Brothers grocery, the destruction of which by fire in 1911 would later so damage the armory. The Town Hall is visible to the left of the grocery (courtesy Lucius Beebe Memorial Library of Wakefield).

Bayard family ran a "trading post" at the northern end of Lake Quannapowitt, selling Native trinkets.[49] The celebration lasted most of the year, with parties held, banquets eaten, and a commemorative volume published to mark the event.

The town would soon have to spend more money for civic matters, this time for the RLG. In April of 1894, state inspectors condemned the former skating rink that the town had been renting for use as an armory since 1877. The RLG had to remove all state-owned property and begin searching for a new armory. In the meantime, the RLG was again using a room in the Town Hall, although they had no drill hall. Without a suitable hall of their own, no balls or banquets could be held. The company participated in several celebrations for the 250th anniversary of the town, but could host none. The town appointed a committee, which included most of the leading militia officers in the town, to select a suitable piece of land and erect an armory specifically for the use of the RLG, but one that the town could also use for other purposes.[50] The town, which had previously paid the rent for properties used by the RLG, appropriated $12,000 toward the construction of a new armory on Main Street. The town then issued bonds, with tax monies to be used to pay off the bonds. The town planned to have the new armory paid for by 1904.[51] Throughout the rest of the year, the RLG raised additional funds. Finally, on 15 February 1895, the RLG dedicated the new armory, the eighth armory since the founding of the company and the first purpose-built armory occupied by the RLG.[52]

The design of the new armory implies much about the relationship between the RLG and the town of Wakefield. The castellated style of armory was much in vogue throughout the Northeast at that time. The style arose from a belief that the appearance of a building should suggest its purpose, and a Norman castle suggested military resolve. However, the architecture of the castellated armories was meant to do more than advertise their martial occupants; these armories were meant to function as actual bastions of law and order from which a militia of middle-and upper-class men could re-conquer a city from rioting workers. Many of these new urban armories were designed with high, thick walls, narrow gun-

Top: Early twentieth-century postcard of Main Street in Wakefield looking south. The streetcar rails are visible; the 1895 armory would be near the spot where the tracks disappear (courtesy Lucius Beebe Memorial Library of Wakefield). *Bottom:* Early twentieth-century postcard of Main Street in Wakefield looking north. The Richardson Building, built on the site of the Richardson home, is the red brick building on the left. The streetcar had a large impact on militia in the late nineteenth century — the automobile would have increasing impact during the twentieth (courtesy Lucius Beebe Memorial Library of Wakefield).

port windows in the lower floors, and deep protected doors.[53] The occupants of these urban castles feared the working-classes of their cites. By contrast, the new RLG armory looked more like a grammar school. It was a modest two-story wooden building with clapboard sides, with a slight nod to Second Empire architecture in the roof line. The building was designed by members of the RLG, and they had no fear of the workers of their town. During the same period that the new RLG armory was designed and built in Wakefield, fully castellated armories were built in Boston, Lowell, and Worcester, all substantially larger cities than Wakefield. Before the style went out of fashion in the early twentieth century, Massachusetts built at least twenty-five fully castellated armories, plus several others with some elements of the style incorporated into their structure. Increasingly, the members of the RLG were from the working class, as the alleviation of the need to pay dues and buy a uniform made membership in the RLG attractive to men of less means. At the time of construction of the armory, the RLG had not as yet served strike duty. Instead the men of the RLG saw themselves as a military force to defend the nation and build members into responsible citizens. The men of the RLG might not have much sympathy with radical workers in Boston to the south or Lowell to the north, but in their own community no break between militia and workers existed. The dedication of the new armory followed the pattern of most martial celebrations in the town, with a banquet and ball attended by over one hundred couples. The new armory, largely paid for by the town with most of the labor supplied by members of the company, implied a permanent bond between town and militia company.

However, technology was bringing eastern Massachusetts even closer together, in a way that would both allow the RLG to recruit from farther away from its hometown, but also allow residents to join companies based in other towns. This new technology came in the form of the Wakefield & Stoneham Street Rail Company, established in 1889, which began operations between the two towns in the summer of 1892. The line was such an immediate success that soon other street railway companies, usually owned by the same group of investors, began laying tracks and providing service to other towns.[54] In the fall of 1893, direct service with Lynn began. The following year Wakefield had direct connections to Reading, and in 1895, to Lowell. The growing network of street railways in eastern Massachusetts allowed companies of the MVM to use the system to move quickly to distant training areas, or, potentially, concentrate at scenes of strife. One of the main instigators of this growth of street railways was lifelong Wakefield resident Charles Woodward, who as commander of the RLG had brought the company to place first in the MVM and who had been singled out for special praise in reports to the Army. At the time, Woodward was also the president of the municipal gas and electric company in Wakefield, and this company was soon providing the power for his street railways. Woodward would eventually become president of six street railway companies, all the while maintaining other careers in state politics and the MVM.[55] In his role as regimental officer in the MVM, and as elected representative of the town of Wakefield in the Massachusetts General Court, Woodward would help guide the MVM as it modernized during the Gilded Age.

The town continued to grow both in population and industry during most of this period. The links between the town and the Sherry Wine Bitters ended in 1891 when the younger Dr. Richardson sold the company to a firm in Hartford, Connecticut, and largely retired. Despite the omnipresence of the Wakefield Rattan Company, a new company, the A.D. Jenkins and Company, began making chairs and other rattan goods in the town around 1890. Jenkins learned his craft at the Wakefield Rattan Company's great rival, the Heywood Brothers and Company of Gardner, Massachusetts. With only six employees at the start, the upstart A.D. Jenkins was no real threat to the giants in the industry, but the field of wicker furniture was becoming increasingly crowded and the new company was an ominous sign for the Wakefield Rattan Company. Ironically, the expansion of A.D. Jenkins and Company would allow the new business to rent the entire fourth floor of the Taylor Building, the one built by Cyrus Wakefield next to the Town Hall. But the growth of the wicker furniture industry would soon be checked by forces beyond the control of the Wakefield Rattan Company.

Industrial production in the eighteen-nineties suffered a general slump, and the wicker furniture industry was especially hard hit. The difficulties faced by wicker companies like Wakefield Rattan came not so much from changing tastes or high unemployment, but more specifically from an overabundance of cheap wicker products that flooded the market. Much of this cheaper competition came from prison labor. The year 1893 was disastrous for the Wakefield Rattan Company, and 1894 was even worse. By 1896, the spinning room closed for the summer, while other sections of the factory were working less than full-time. Corporate offices in Boston closed as more of the operation was moved to the company's complex in Wakefield in a cost-cutting measure. The company closed a wooden chair company in Illinois it had foolishly bought in 1895, and then began borrowing money to remain solvent. However, the company was unable to halt the decline in its fortunes. In the fall of 1896, it began what would have been unthinkable a decade earlier, entering into discussions with its arch rival, the Heywood Brothers and Company of Gardner, Massachusetts. In February 1897, the two giants in the world of rattan combined to create the Heywood Brothers and Wakefield Company.[56] With a virtual monopoly on rattan goods, aside from shoddy prison-labor produced items, the continued success of the new company was assured, provided that the public continued to have a taste for quality wicker furniture.

With the merger, economic stability returned to the town of Wakefield. However, the spirit of the times would turn the memories of Wakefield residents back to earlier martial feats. Forty members of the RLG Veterans Association met on 25 March 1898 with other men who had served with the RLG in any of the three periods of federal service at the armory to eat and relive past glories, as they periodically did. Current company and regimental officers came to speak to the honored veterans.[57] With the horrors of the Civil War now safely transformed into a test of manhood and patriotism, the lack of a suitable memorial in the town to the fallen became increasingly glaring. The original justification for the Town Hall as a memorial to the town's Civil War dead had largely been forgotten. The United Daughters of the Confederacy (UDC) had been busy during this period planting Confederate memorials all over the South, and Wakefield was not to be outdone in memorializing its own dead.[58] A recently deceased member of one of the oldest families in town, Harriet N. Flint, left in her will $10,000 to be spent, either alone, or with additional funds, for a Civil War memorial "symmetrical in architecture, beautiful in design — a monument worthy of the true men to whom we dedicate it." The town accepted the money in March 1898, and a committee was created to select a design and locate a suitable spot for the monument. However, events in the larger world would delay progress in fulfilling Mrs. Flint's wishes.

Chapter V

A Militia for Empire

On 15 February 1898, at about 9:45 in the evening, an explosion sank the USS *Maine* in Havana harbor, with the resulting deaths of 268 officers and men out of a complement of 374. Tensions between Spain and the United States over Cuba had been building since 1895, when the latest insurrection to end Spanish rule over the island began. As a response, the Spanish army began a program of concentrating the rural population away from the rebels—a common practice in counterinsurgencies, but one that added greatly to the misery of the Cuban people. An insult against President William McKinley in a private letter the Spanish minister in Washington had written in late December 1897 to a friend then in Havana became public when revolutionaries in Havana stole the letter and sent it to U.S. newspapers, which published it on 9 February 1898. The diplomatic situation was awkward, in that the letter was stolen, but Spain issued an official apology on 14 February 1898.

The explosion and sinking of the *Maine* the next day seemed an example of Spanish perfidy, the culmination of years of Spanish atrocities against the people of Cuba, and now against the United States itself. American newspapers claimed explicit Spanish complicity in the tragedy, and further fanned the flames of war with increasingly lurid accounts of Spanish behavior. Although the actual cause of the *Maine*'s demise was more likely the result of the accidental ignition of coal dust, the U.S. Navy's board of inquiry concluded that an external explosion caused the ship's magazine to explode. Newspaper drawings showed a hypothetical mine planted by Spanish agents against the side of the ship. The Spanish conclusion that an accidental explosion of coal dust caused the wreck was ignored in the United States. The next few weeks were filled with increased tensions between the United States and Spain. At heart was a basic disagreement over the future status of Cuba. While Spain was willing to grant autonomy to the island, independence was out of the question. The United States would accept nothing short of full independence. In the early hours of 19 April 1898, Congress passed a joint resolution recognizing Cuban independence and authorizing the president to use the military to secure that independence. On 21 April, before the resolution had officially been presented to the Spanish government, Spain recalled its ambassador to the U.S. The blockade of Havana by the U.S. Navy began the same day. On 23 April Spain declared war on the United States, which returned the favor on the 25th. For the first time since the ending of the Mexican American War in 1848, the United States was at war with a foreign power.

War with Spain brought a huge outpouring of patriotism and martial enthusiasm to Wakefield. The members of the RLG would test their mettle and find if they were worthy of the name carried by so many men from South Reading into battle during the Civil War. For militia units, entering federal service for war required a legal shuffle that changed their status from state militia into state-raised Volunteers for federal service. Col. Charles F.

Woodward, the recently appointed commander of the Sixth Regiment, MVM, of which the RLG was Company A, offered Governor Roger Wolcott the services of the regiment for the war before any other regiment in the commonwealth. The governor accepted Colonel Woodward's offer, and as a result, the Sixth Regiment, Massachusetts Volunteer Infantry (Sixth Massachusetts) came into existence. This new Sixth, as a practical matter, was basically the Sixth Regiment, MVM, under a new legal standing. The old Sixth was placed on hiatus until the new Sixth left federal service. Simply being chosen for federal service was a great honor for the Sixth Regiment, as the federal government had requested Massachusetts furnish only three infantry regiments and three batteries of heavy artillery.[1] Members of units of the MVM not selected could be bitter over the federal quotas, and potential recruits turned away from non-selected units occasionally broke down in tears.[2]

The Sixth Massachusetts Regiment was unusual in the American military in its composition in that one of its companies, L from Boston, was comprised of African-American militiamen. The company had been organized in 1877 in Boston.[3] The old Massachusetts 2nd Infantry Battalion was a black unit formed in 1866, with one company in Boston and another in New Bedford. During cutbacks in 1876, both black companies were disbanded. Charles Francis, a messenger for the governor and former commander of the Boston company, petitioned to form a new black company in Boston, which formed on 1 June 1877, and became part of the 6th Regiment on 12 March 1878. All members of the company—officers and men—were black. The Regular Army of the period contained four regiments of black soldiers, with mostly white officers. Some states had separate companies or battalions for black militiamen. Only Massachusetts, Ohio, and Indiana included any black units in the initial response to the call for Volunteers. In the second call for Volunteers, black units were accepted from Alabama, Illinois, Kansas, North Carolina, and Virginia largely as a result of black agitation for a greater role in the war. A few blacks served in otherwise white Volunteer Regiments raised in Maine, Iowa, and South Dakota.[4] But having a black company in an otherwise white regiment was a unique solution to the issue of where to place the single black company in the MVM. One of the officers of L Company, Second Lieutenant George W. Braxton, was a longtime resident of Wakefield. Lieutenant Braxton, a native of Portsmouth, Virginia, was thirty-five years old in the spring of 1898 and had served in Company L since 1887, being elected to second lieutenant in 1894.[5]

The town of Wakefield apparently held Lieutenant Braxton in high esteem for his involvement in the militia and other civic organizations. Before the Sixth Massachusetts left the commonwealth, the local camp of the Sons of Union Veterans, in Reading, of which he was a past captain, presented Braxton with a sword. The sword was described to be appropriately engraved and a "fitting testimonial of esteem on behalf of Reading members of the order toward their Wakefield friend."[6] The Sixth had contained the black company for many years and its presence caused no recorded problems with the other companies in the regiment. Indeed, the high number of professional cooks in the company gave it a reputation for having the best food at summer encampments. However, in the segregated U.S. Army of 1898, the presence of the black company caused unease. Most black units were combined with other black units to create all-black separate battalions. No other regiment contained black and white units and the Army was uneasy about setting a precedent.

The town of Wakefield noted with satisfaction that the recently appointed regimental commander, Colonel Woodward, had initially entered the MVM through the RLG, and formerly commanded it.[7] Colonel Woodward had been born in South Reading in 1852 and lived in the town most of his life. His father had been one of the original honorary members of the company in 1851.⁸ Colonel Woodward personified the man who combined a lucrative public civil career, political career, and a parallel career in the MVM. After college, Woodward and his father established a company manufacturing equipment for the shoe industry. After eight years working with his father, he left the company he had cofounded and went into business on his own. He served as president of the Wakefield Gas Company and was instrumental in bringing electricity to Wakefield. But his real business interest was in street railroad companies, of which he was president of seven throughout Middlesex and southern Essex counties, including the Wakefield & Stoneham Street Railway. Some of his other street railways were the Reading & Lowell, and the Woburn & Reading, giving Colonel Woodward's companies a virtual monopoly on the inland routes north of Boston. Six of his railroad employees also served under him in the MVM, five of them in the RLG.⁹ A year later, in 1899, most of the street railroads in eastern Massachusetts were bought by the Boston & Northern Street Railway, which in 1911 became part of the Bay State Street Railway Company. The company at the time had the greatest amount of mileage of any such company in the world.¹⁰ Colonel Woodward had long experience in managing complex institutions and was a well-known figure in Wakefield. Charles Woodward Jr., son of the colonel, would run the street railways in his father's absence.¹¹ Along with his efforts in bringing modern conveniences to Wakefield, Colonel Woodward also pursued a political career, beginning when he served as collector of taxes for Wakefield from 1883 to 1897. He also served as assessor for the town during that period. From 1887 to 1889 he served as the town's representative in the Massachusetts House of Representatives, and chaired the committees on military affairs, armories, and perhaps inevitable, street railroads. He served in the state senate from 1897 to 1898, also serving on the military affairs committee.¹²

Colonel Woodward originally joined the RLG in 1869, and after serving in a variety of positions from company quarter master to treasurer, was elected second lieutenant in 1876. Two years later the members of the RLG elected him as their commander, a position he held until 1882, when the regimental officers elected him to the rank of major. Thereafter he served at the regimental level, always in the Sixth Regiment, MVM. Another regimental officer, Maj. George H. Taylor, was also a past commander of the RLG. Major Taylor had joined the RLG in 1880, becoming its commander in 1888, a position he held until his election to major in 1890. One of Colonel Woodward's new appointments to the regiment was Charles E. Hussey, the superintendent of schools for the town of Wakefield, as the regimental sergeant major. Sergeant Major Hussey had been associated with the RLG for several years, and although not himself a member, had served as toastmaster at many RLG banquets.¹³ Another Wakefield resident appointed as a noncommissioned officer on the regimental staff was Sergeant Arthur L. Wiley. Many members of the Wiley family were veterans of various incarnations of the RLG. Sergeant Wiley himself had lately been a member of the Wakefield High School Corps of Cadets, and was active in the Sons of Union Veterans.¹⁴ Thus residents of Wakefield predominated at the regimental level, and the RLG had been the vehicle through which most of these men had originally entered the military.

The Richardson Light Guard in 1898 was commanded by Capt. Edward J. Gihon, who personified the generation that grew up in the wake of the Civil War. Born in Wakefield on 5 February 1865, he received his education in the town, and worked for a time in the rattan factory after he graduated.¹⁵ His father and an uncle had served in the Union Army.¹⁶ His father's service during the Civil War must have left a deep impression on him; he became active in the Sons of Union Veterans, rising to captain in that organization. The elder Edward Gihon had served as a corporal in the 28th Regiment, mustering in on 13

The back of this 1895 photograph identifies it as "Rifle Team, Co. A, 6th Regt M.V.M., Richardson Light Guard." During the Gilded Age, marksmanship competitions played an important role in the life of militia companies. The decorations on the chest of most of the men attest to their success in rifle competitions. The man in the center of the front row is Captain Edward J. Gihon, who commanded the Richardson Light Guard from 1893 to 1899, distinguishing himself during the Spanish American War. Sergeant F. Gray, left front, would also distinguish himself in the war and would command the company afterward (courtesy Wakefield Historical Society).

December 1861, and reenlisting on 1 January 1864. He had been wounded on 6 May 1864 during the Wilderness campaign, and was listed as a deserter on 15 June 1864. As his son, Edward, was born less than nine months later, his wife's pregnancy might have been a factor in, or the result of, his failure to return to his unit.[17] Captain Gihon originally joined the RLG in 1882, when he was seventeen, and became its commander in 1893. His Roman Catholicism apparently was not a handicap, as religious affiliation had long ceased to be a major issue in the town. He was active in the Elks and the Knights of Columbus. Captain Gihon enjoyed great popularity with the men of the RLG during peacetime; in war he would earn their respect and the high esteem of the town.

At the start of the war, speculation ran high as to just what role the RLG would play. General civilian opinion held that northern regiments would not be used in any tropical invasions, but instead would serve along the east coast, guarding the precious and vulnerable industrial areas from Spanish attacks.[18] Despite this imagined passive mission for the RLG, opinion shapers still saw much potential for the United States as a whole in the war. "In addition to the taking of Cuba, this country will of course capture Porto Rico," opined one columnist in Wakefield.[19] The town and members of the company understood that they would most likely see active service during the war, although the exact nature of that service remained unclear, even to Governor Wolcott. As a precaution, the company and town agreed that the town's fire alarm whistle could be used to summon the members of the RLG in the event of an emergency. A blow of twelve blows followed by twelve more was the agreed upon signal.[20] In the following week, any blowing of the fire horn or the ringing of church bells for any reason set off a minor buzz of excitement.[21]

Despite the distinct likelihood of involvement in war, the traditions of militia continued for the RLG. On Friday evening, 22 April, several RLG officers attended a ball given by G Company in the nearby town of Woburn.[22] The RLG had its regularly scheduled state inspection on the following Monday evening. At 7:30 P.M. on the same evening, Colonel Woodward and his regimental staff assembled at the regimental headquarters in their fatigue uniforms to undergo their state inspection.[23] Colonel Woodward had only recently taken command of the Sixth Regiment and had been scrambling to ensure all positions in his regimental headquarters were filled with qualified men in anticipation of entering federal service.[24] The Salem Cadets, which was a marching band in the Massachusetts Volunteer Militia, became the regimental band to round out the regiment.

Despite the lack of official notification about what role the Sixth Massachusetts would play in the present war, the governor's office must have had some indication that it stood a fair chance of seeing active service and possibly combat. General Curtis Guild, Jr., who came from a prominent Republican family and served on Governor Wolcott's staff as his adjutant general, resigned his position to become the adjutant of the Sixth Massachusetts.[25] For an officer of General Curtis's status to resign from his position as the governor's primary advisor on military affairs, which carried with it the rank of brigadier general, to accept the position of regimental adjutant, which carried with it the rank of first lieutenant, was unprecedented. Guild had thus far lived in an age when most prominent politicians had service records from the Civil War, and all realized that for an aspiring politician to sit out a war would have been a hindrance to a later political career, while honorable service in combat was a boon to getting elected. Guild, with political aspirations of his own, must have sensed that joining the Sixth Massachusetts—in any capacity—was his best chance of achieving some martial glory in the war. However rumors said that Massachusetts Senator Henry Cabot Lodge wanted Guild for other positions in the Army, and thus his position as adjutant of the Sixth Massachusetts was only temporary, until he entered federal service with the regiment.[26] The rumors proved true and eventually Guild transferred out of the Sixth Massachusetts to serve as the inspector general in the Seventh Corps, with the rank of lieutenant colonel. In that capacity he saw the results of what a lack of experienced staff officers could inflict on a vastly enlarged wartime Army.[27] Guild would indeed have a successful political career after the war, and would serve as the governor of Massachusetts from 1906 until 1909.

While preparations were underway for possible federal service, planning for normal militia training had to continue in the event the regiment remained unmobilized.[28] Under a new state law, each regiment of the MVM would serve seven days at summer training encampments rather than the five days that had been traditional in the MVM in past years. Rather than beginning on Monday, summer training would begin on Saturday, meaning that militiamen would spend a Sunday at camp.[29] Additional planning would have to include divine service during summer encampments if the MVM was not to incur opposition from elements of society concerned about the souls of the men, a requirement not necessary under the former five-day training schedule. However the Sixth Massachusetts would not undergo the normal summer training that year. In all, four infantry regiments—including the Sixth—one artillery regiment and the naval brigade from the Massachusetts Volunteer Militia reorganized as Volunteer units and entered federal service for the war. The MVM issued a statement saying it planned to hold the positions for members and units that were away in federal service, allowing them to reintegrate into the MVM on their return.[30] The members of the MVM were technically on temporary detachment from the MVM while they were in federal service. They were granted furloughs from their duties with the MVM until thirty days following their discharge from federal service. If their term of enlistment in the militia expired first, then of course they would have no further obligations to the MVM.[31]

For the town of Wakefield, the RLG was again the main vehicle through which residents went to war. Of sixty-three residents who served in the Army during the war, all but nine entered through the RLG or the Sixth Regiment headquarters. Another five residents served in the navy and two in the Marine Corps during the war. Reading had twenty-nine residents in the army during the war, twenty of whom served in the Sixth Regiment, and one each in the navy and the Marine Corps.[32] Service with a state-raised Volunteer regiment was preferable for most men who desired to participate in the war to joining the regulars, with their stricter discipline, longer enlistments, and, in general, lower prestige.[33] Local businesses, not wanting to appear deficient in their support for their employees who served, issued promises to hold the jobs of workers away in federal service, with some going as far as to promise providing half pay for employees away at war.[34] Fraternal organizations too stepped in to support their members in the RLG. The Knights of Columbus promised to pay all benefits and obligations of its members in federal service.[35] One of the local newspapers, the *Wakefield Citizen and Banner*, consciously compared the present war with the situation "as in '61," at the start of the Civil War, noting that once again all the men of the RLG had volunteered for federal service.[36] The romanticization of the mobilization of the RLG in the spring of 1861 for the Civil War pervaded much of the coverage of the RLG throughout the Spanish American War, but this nostalgia was strongest at the beginning of the war. The same paper noted with satisfaction that many former members of the RLG who now lived in other towns were joining their local militia companies for the war.[37] The RLG had indeed done its mission during peacetime of instilling in its members a sense of duty and patriotism.

In late April the Sixth Massachusetts was notified that on 6 May it would be brought onto state active service in preparation for entering federal service. Farewell receptions were held in both Reading and Wakefield, with the Wakefield reception a very large affair held at the Town Hall on the evening of 5 May. Early the next day the men of the RLG assembled at their armory and marched the short distance to the train station to take them to the state campground. Members of the RLG who cared about their health and were up-to-date on the latest medical theories eagerly sought flannel abdominal bands to ward off the chills expected to bring a healthy man down even in tropical climates, despite his wool uniform.[38] The schools were closed for the day, and many residents came to the station to see the soldiers off. A generation had passed since the RLG had entered federal service, and members and residents seemed to have thought the mobilization camp would be run similar to the annual state training camps. Some residents wanted to accompany the RLG to the camp, but officials warned that anyone not on the muster rolls would not be allowed in camp. Additionally, alcohol was banned from the camp. The men were told to expect stricter discipline during the few days they would spend at the state camp than they were normally under during annual training camps, and probably under even stricter discipline when they actually entered federal service.[39] The train carrying the RLG departed Wakefield shortly after nine in the morning. The first leg of their adventure took them to Boston, where, accompanied by the Wakefield High School cadets, they marched in a short parade for the governor's review, which took them down Beacon Street, passing the State House on their right and the Common and Shaw Memorial on their left. After the short review in Boston, they again entrained for the state camp at Framingham.

On 7 May, the officers were officially accepted into federal service. The three officers were all from Wakefield, although seven NCOs in the company were residents of Reading. The Army accepted the seventy enlisted men in the RLG two days later. Because this was short of the projected war-strength of the company, Captain Gihon returned to Wakefield, and brought back another ten recruits.[40] The formal mustering of the RLG into federal service came on 12 May. Some members would not be able to go because of physical ailments, including Sergeant Roger Howard. He had originally enlisted in the RLG in 1879, when Captain Woodward commanded the company. Howard himself was elected captain and commanded the RLG from 1882 until he resigned the position in 1884 and returned to the ranks. He had almost twenty years of service behind him and was bitterly disappointed when the poor health he had been suffering from for the past two years, since he came down with what was described as "the grip" about two years earlier, rendered him unable to go with the RLG.[41] However, his discharge meant that another soldier could be promoted to the rank of sergeant to fill the position. The federal officer who mustered the men of the RLG into federal service told them they were the finest volunteer company he had seen.[42] The following days saw regiments leaving the Bay State for the war. On the 18th a couple of men came from Wakefield to give to the men of the RLG clothing and shoes donated by the Citizens' War Relief Committee, formed on 2 May 1898, before the RLG had left Wakefield. Through the Committee, prominent citizens who did not serve with the RLG during the war could show their largess and patriotism. Fifteen men were chosen to raise funds for the Wakefield soldiers and their families while the men were away at war. Solon O. Richardson, son of the late namesake of the company, served as chairman, while most of the other men were former members of the RLG, or had relatives in the company. In addition, the GAR post and fire departments also raised funds for the support of the town's soldiers and their families. The men of the RLG left their hometown knowing that their families would be looked after, at least in the short term.

When the RLG left Wakefield, the town did what it had often done since colonial days when the local militia company left on an expedition; the town formed a replacement militia company for local service. Part of this desire for a militia company came from a lingering uncertainty over the vulnerability of the nearby coast. Also, thirteen members of the RLG had not gone with the company when it left for federal service, a few because they decided in the harsh light of day that business or family duties were more pressing, and most because they had failed the medical screening process.[43] Men who did not go with their reorganized companies to federal service were relieved of service with the MVM until their parent companies returned, but such men were welcome to join the new provisional companies.[44] Most of the members of the RLG rejected for federal service were still fit enough for the lower standards required for strictly state service. On 18 May 1898, the adjutant general's office authorized these provisional companies to exist for the duration of the war plus thirty days after the signing of a peace treaty with Spain.[45]

The law creating the provisional militia was substantially the product of Colonel Woodward, who drafted it while serving as chairman of the Massachusetts Senate committee on military affairs.[46] Only towns such as Wakefield, which had an armory vacated by the mobilization of the normal company for federal war service, could have an authorized company as part of the commonwealth's provisional militia.[47] State law required the selectmen of the town in which resided the majority of the petitioners for a company in the provisional militia give their approval.[48] Each provisional company was allowed fifty-eight enlisted men. The neighboring town of Stoneham became the first in the commonwealth to hold a drill for its new provisional militia company on the evening of Monday, 13 June, when the company elected the minister of the Congregational Church as commander. Perhaps spurred by this development in a neighboring town, First Sergeant Gray notified all men who signed the petition for a provisional company in Wakefield to meet at the armory on Tuesday evening, the 21st.[49] The new Wakefield company was designated to be the 23d Provisional Company, MVM. Ten of the men rejected for federal service with the RLG came to the first meeting to organize the provisional company.[50] In keeping with its tradition of having a company for black militiamen based in Boston, a new "colored" company was designated as the Tenth Provisional Company, MVM. Although the *Citizen and Banner* mentioned the names and occupation of the officers of the black company, none were residents of Wakefield.

For members of the RLG who did not enter federal service, their separation from their fellow RLG members must have been a terrible blow to their pride and self-identification. To be able to face their former fellow militiamen after the war ended, they needed to show their continued willingness to shoulder at least some of the burden during the war. With the men in the RLG not included, the town of Wakefield contained an estimated 1,600 men between the ages of 18 and 45 years of age in what was called the unorganized militia.[51] Theoretically the provisional company would have more than enough potential applicants. However, 25 percent of the recruits failed the physical examination all members were required to take, and Wakefield needed to recruit more able-bodied men to maintain its provisional company.[52] Colonel Gordon Dexter of the governor's staff inspected the Wakefield provisional company on Tuesday evening, 21 June. Forty-seven of the sixty-one officers and men enrolled in the company were present, with three unaccounted for. Another three members, however, managed to join the Sixth Massachusetts and left for Camp Alger. But since the minimum number of members needed for each provisional company was forty-one, Wakefield's company still met the minimum standard.[53]

Despite the short-term nature of the twenty-five companies in the Provisional Militia, some pundits believed it had a long-term role to play. After the Civil War, rebuilding the MVM had been difficult as many former members had seen enough military service by the end of the war, and this same scenario might recur at the end of the Spanish American War. The *Wakefield Citizen and Banner* noted that "if Civil War is any criterion, many of the former militiamen, when they return to their homes and firesides may have had all they want of the trappings of war and not care to resume their militia status. In this event the M.P.M.'s would become M.V.M.'s and all would be happy."[54] However this idea was not wholly supported within the MVM, whose hierarchy feared that a movement to make the Provisional Militia permanent might become a threat to the positions held for the members of the MVM then in federal service. The MVM made no plans for the annual encampment in Framingham to include the Provisional Militia, although some members thought paying to raise the provisional militia but not train it was shortsighted.[55]

The provisional company, what would be called a "home guard" company, held a muster on Monday, 11 July, with 41 men. Of them, six came from Reading, continuing the tradition of that town providing men for the RLG.[56] The men in the company voted to adopt the constitution of the RLG, with a few minor modifications, as their own. William E. Gray was elected captain.[57] Captain Gray was thirty-nine years old in the spring of 1898. Although born in Boston, he was raised and educated in Reading, where he still lived. In his civilian occupation, Captain Gray worked for Colonel Woodward as a conductor on the Wakefield & Stoneham Street Railway. He was a past grand master of the Odd Fellows Lodge and past master of Good Samaritan Lodge, A.F.& A.M., of Reading.[58] Captain Gray had served in the RLG since 1877, and had been the company first sergeant when the war started. To his dismay, he failed his federal mustering physical examination. The home guard company gave him an outlet for his frustration. The new company's two lieutenants had also been noncommissioned officers in the prewar RLG. After being elected by the men of the company, Gray awaited official commissioning by the commonwealth, when he could appoint the noncommissioned officers for the new company.

Its first regular drill came a week later, on 18 July, in the armory on an exceptionally warm summer's evening. The men trained without weapons or uniforms. The commonwealth promised to provide uniforms and arms when inspectors believed the company had progressed enough at drill to be able to present themselves as part of the militia of Massachusetts, ready to take on real missions. The new company contained the type of men that home guard companies traditionally contained — those too old, too young, or too burdened, for expeditionary service yet physically fit enough for local emergency service.[59] The mixture of old and young, in their first drill together, reminded some of "old times" of the RLG from years ago.[60] The home guard company gave men such as Captain Gray a martial activity during the war despite his rejection from federal service, and it gave the town of Wakefield a sense that whatever the future held for the RLG and the war with Spain, the town would not be left to the mercy of unruly mobs or, more plausibly to the people of Wakefield, unable to assist in repelling Spanish raiding parties from the Massachusetts coast. However the true martial expression of the town in the current war remained the RLG in federal service, and rumors of its imminent deployment to the American South raised doubts that the New England volunteer regiments would be used primarily along the East Coast. The editor of the local newspaper fully supported the conquest of the Philippines, and opined that the Sixth Massachusetts might take part in that theater of the war.[61]

On 20 May, the RLG boarded a train from Framingham for their trip south. The trip was punctuated by cheering crowds at each station they passed. The Sixth Massachusetts had sustained some of the first casualties in the Civil War when it was shot at by citizens of Baltimore on its way to Washington.[62] The headline story of the 27 May *Citizen and Banner* contrasted the heroes' welcome the Sixth Massachusetts received marching through Baltimore in 1898 with the decidedly unfriendly reception the regiment received in the same city on 19 April 1861, when the regiment was attacked.[63] During the new war, the mayor officially welcomed the regiment, and the city fed the men before the soldiers again climbed on board the train for the rest of their journey. Colonel Edward Jones, who had commanded the Sixth Massachusetts in the Civil War, wrote a letter thanking the mayor of Baltimore for the kindness the city had shown the Sixth Massachusetts in this latest war.[64]

Lieutenant Charles E. Walton, who had first enlisted in the RLG in 1880, resigned on 25 June due to illness. However, he would not be granted his discharge until the RLG reached Charleston, South Carolina.[65] Because Captain Gihon was then serving as acting battalion commander for a month, Lieutenant Gray, newly promoted to first lieutenant, had to act as company commander. Thus both the RLG and the home guard company had a Gray as its commander. Much to the disappointment of the noncommissioned officers of the RLG, the Army assigned J.A. Bailey, Jr., from Billerica, about thirteen miles northwest from Wakefield, as replacement for Lieutenant Walton. The *Citizen and Banner* referred to him an "outsider," and reported the resentment the town felt that the first sergeant was not commissioned as second lieutenant and the other noncommissioned officers advanced in rank. The paper said that residents would have understood if a West Pointer had been attached to fill the position, but the assignment of an outsider rankled them.[66]

The RLG, as with the rest of the Sixth Massachusetts, arrived at Camp Alger, near Falls Church, Virginia, on the 24th. They would remain at Camp Alger until 5 July. The Wakefield men watched as the arrival of the Ninth Massachusetts Regiment brought more Bay Staters into the camp. They spent the middle two weeks of June on provost duty, under the command of Lieutenant Gray. While leaving camp for provost duty, the RLG received the first of several boxes marked "Crumbs of Comfort" that a resident of Wakefield, Rufus Kendrick, sent to them.[67] Major Taylor was absent from camp because he had taken a detachment from the Sixth Massachusetts back to the towns north of Boston, from where the Sixth drew its members, to recruit additional men to bring each company up to its new allotted war strength of 109 officers and men. Major Taylor and Sergeant Harvey G. Brockbank arrived in Wakefield in early June on the recruiting mission. In all, the Sixth Massachusetts established three recruiting stations, one in Wakefield, with the others in Boston and Concord. The mission was to recruit thirty-two men for each company from the towns from where the companies originated, for 384 men in all. Wakefield men had a priority for joining the RLG but men from Reading and Melrose also were trying to join.[68]

Major Taylor told the people of Wakefield, who wanted to know all the details of federal service for the RLG, that the nights in Virginia were damp with heavy dew, and that the mules awoke the whole camp each morning when they started braying at 4 o'clock in the morning.[69] The mountains of administrative details required for each recruit proved difficult and unpleasant for Major Taylor, not least of which was ensuring each man got his proper travel pay and allowances. The men were not sent off to Virginia empty-handed—each man left Wakefield with six sandwiches, six donuts, some cheese, and $.21 to buy coffee, for his first day's ration. After that he was entitled to $1.50 a day to provide his sustenance until he reached Camp Alger. The arrival of new recruits in civilian clothing became a source of amusement to the now long-serving members of the RLG. Massachusetts eventually sent four hundred new uniforms for the recruits sent to the Sixth.[70]

While at Camp Alger the men of the RLG trained and became accustomed to the routines of army life. Rumors swept through the camp on a daily basis; they would soon be sent back to Boston and disbanded, or they would be sent to Washington to act as body guards for President McKinley, or Spain had surrendered and the war was over. For the men of the RLG, these rumors brought no cheer. They yearned to see combat in this war and did not relish the thought of sitting it out in a camp in Virginia. The reports of an American victory in Cuba made many of them despair of ever seeing combat, of having their mettle tested. In the meantime, training continued. During peacetime, most training for the RLG, as with almost all companies of organized militia, as well as the Regular Army, was held at company level or lower, such as platoon or squad. During summer annual training, leaders attempted battalion or occasionally regimental maneuvers. However, on active federal service, the Sixth Massachusetts found enough room at Camp Alger for maneuvers up to the brigade level.[71] Captain Gihon earned praise while serving as the acting battalion commander during regimental drills.[72] Still, the uncertainty of their eventual role in the current war gnawed at the men. They might go to Cuba or the Philippines, or even take part in the conquest of Puerto Rico. On the other hand, they might serve occupation duty in any of those places. Perhaps they would serve as a reaction force in the increasingly unlikely event of Spanish raids or invasions on the East Coast. The least desirable role, as far as the men of the RLG were concerned, was to spend the war training in the United States for a mission that never came. As their forbears had dreaded during the opening months of the Civil War, the men in the latest incarnation of the RLG fretted over the idea of returning to their town having never gotten the opportunity to prove themselves in battle.

A few visitors from Wakefield took the train trip to Virginia to visit the men. Even better were the arrival of packages of treats from Women's Relief Corp of Wakefield. What the men of the RLG really wanted though was for Massachusetts to pay them their wages for their days on state active service before their induction into federal service. The tardiness of the commonwealth caused concern owing to the soldiers' lack of spending money.[73] While in camp the men had regularly been getting packages sent from organizations and individuals in Wakefield, they lacked money to buy fresh fruits and vegetables.[74] On 5 June the paymaster for the Sixth Massachusetts finally arrived and paid each man $16 from the commonwealth for the period between 6 and 12 May, when the Sixth Massachusetts was on state service, and the men expected a similar payment from the federal government for the period since 12 May.[75]

When Sixth Massachusetts left Wakefield, Solon O. Richardson, on behalf of the Citizens' War Committee, requested that the company officers inform the committee of anything the men wanted. The committee worked hard at fulfilling that promise, and soon every man in the RLG got a new bath towel and sponge, which Captain Gihon had requested the committee send to them.[76] A more serious donation came when Clan McPhail, O.S.C., a fraternal organization for Americans of Scottish ancestry, arranged with Captain Gihon to give every man in the RLG a metal identification badge with a number to be used to identify the man if he were found hurt or dead.[77] Other fraternal organizations continued to show their support. The Knights of Columbus sent in early July a new revolver to Sergeant John H. McMahon.[78] Sundays passed with no training, only guard duty and an inspection of quarters. Church services were held outdoors. One Sunday Captain Gihon and an enlisted man visited Arlington National Cemetery.[79] One visiting resident reported back to the town that long marches and hardtack had left Sergeant Harvey G. Brockbank tanned and feeling better than he had in years. Sergeant Brockbank told the visitor that "I do not propose to tell a man that it is a snap to be in the army, for it isn't. It means work and unless a man is willing to chew hardtack he had better stay at home."[80] Still, army life was not all hardtack. The men of the RLG had no complaints with their food while at Camp Alger, which was prepared by their own company cooks.[81]

While still in the United States, the men of the RLG received regular mail from home, which boosted moral. One man got a box from home while at Camp Alger with four cakes and "real homemade doughnuts." Sergeant Oliver from Reading got a box on Saturday, 4 June, and shared his treats with the others. Aside from private efforts, more community-based efforts to assist the town's men in federal service added to the flow of goods to the camp. The H.M. Warren Women's Relief Corp voted on 5 June to send a box to the RLG. The group sought donations of light underwear and food from townspeople, with donations accepted at the GAR Hall that Saturday and Monday.[82] In three weeks they sent one hundred and forty-five articles to the men in the RLG.[83] At a meeting of the Volunteer Hose Company the evening of Wednesday, 15 June, the company created a committee of three to provide relief to the RLG.[84]

As well as efforts by townspeople to support their men in the RLG, the *Citizen and Banner* reported on other residents and former residents who were serving in the war, noting that Azzel Ames, Jr., who formally lived in Wakefield, was an officer in the newly formed First Regiment, U.S. Volunteer Engineers.[85] Another former resident became a hero on 11 May 1898, in the Bay of Cardenas, when the torpedo boat U.S.S. *Winslow* came under Spanish fire and sustained extensive damage. The ship's captain received serious injuries and was unable to remain in command. Another officer was dead, along with some of the crew. The former Wakefield resident, identified as Lieutenant Newcomb of the Revenue Cutter Service, saved the wounded captain, and managed to bring the crippled boat out of the range of Spanish fire and into safety.[86]

But the RLG and the Sixth Massachusetts as a whole were never far from the thoughts of the community, as the town's newspaper published a steady stream of articles, letters, and photos of the experiences of the town's company in federal service. The front page of the *Citizen and Banner* for Friday, 24 June, featured a photo showing the rows of white tents Sixth Massachusetts Regiment at Camp Alger. The photo had originally run in the *Boston Sunday Journal*.[87] Although the black population of Wakefield was minute, the *Citizen and Banner* also took an interest in Company L, the "colored" company in the Sixth Massachusetts, probably from the pride that one of its officers was a local resident. On 17 June the paper noted that, at Camp Alger, "it seems generally conceded that no company is better drilled and disciplined than this one." In keeping with the company's prewar reputation, their food was also considered to be the best in camp.[88] Company L of the Sixth Massachusetts would be the only unit of black Volunteers to fight in the Spanish American War, although others would perform occupation duties.[89]

On 24 May, the Sixth Massachusetts was combined with the Eighth Ohio and the Sixth Illinois to create the Second Brigade of the First Division, Second Army Corps, of the U.S. Army. On 7 June, the brigade met its new commander, Brigadier General George A. Garretson, of Ohio. During June, the Sixth Massachusetts participated in a training exercise known as a sham battle. The sham battle pitted the Sixth Massachusetts against the Eighth Ohio, which was part of the same brigade. Company L of the Sixth Massachusetts captured the regimental colors of the Ohio regiment. Loss of a unit's colors was considered a great embarrassment, but the capture of an opposing unit's colors brought great pride to the capturing unit. However, the entire exercise did not play out in good humor. Some members of the Sixth Massachusetts received injuries in scuffles with the soldiers from other regiments. Some men from a Missouri regiment in the Third Brigade, which was not part of the exercise, got into a brawl with the Massachusetts men, creating a lot of bad blood between the Massachusetts men and the Missouri men. Colonel Woodward became incensed at what he believed to be the improper conduct of several of the Missouri officers involved in the incident and preferred charges against them.[90] Relations between Colonel Woodward and General Garretson had gotten off to a bad start and would only get worse.

By the end of June, Major Taylor had completed his recruiting duties in Wakefield. While in town he attended a funeral for a man from Stoneham in Company H who died of typhoid fever at the Fort Myers hospital and was shipped home for burial.[91] Major Taylor had sent several squads of civilian-attired recruits to Camp Alger over the past few weeks.[92] Nine of the new men hailed from Wakefield, of whom three went into Stoneham's Company H, while the other six joined the RLG. Still, men from all over eastern Massachusetts joined the regiment and some became part of the RLG.[93] The front page of the *Citizen and Banner* for Friday, 1 July, again featured a photo of the RLG in formation at Camp Alger.[94] The same issue published a letter from Captain Gihon asking the town not to send any more comfort boxes because he expected the regiment to move by the middle of the next week, and mail might not reach the men of the RLG for some time.[95] Shortly before leaving Camp Alger, the regimental sergeant major resigned, and was replaced by a Lowell man.[96] The Sixth Massachusetts left Camp Alger on the evening of 6 July with most members believing their final destination was Cuba to assist General William R. Shafter.[97]

In recognition of the war and that so many of the townsmen were away in the Army, the town held a relatively quiet Fourth of July celebration.[98] The headline for the *Citizen and Banner* for 8 July carried the headline that "[t]he Sixth On the Way to Cuba." The accompanying story said that the Sixth Massachusetts had left Camp Alger the previous Tuesday and was on its way to Santiago, Cuba, via Charleston, South Carolina. While the regiment was in Charleston, Lieutenant Walton's resignation was approved and he returned home.[99] However, the remaining men of the RLG, and the entire Sixth Massachusetts Regiment would soon distinguish itself as one of the best of the Volunteer regiments. While most state-raised units spent the entire Spanish American War sweating in the American South, the Sixth Massachusetts was chosen to participate in the invasion and conquest of Puerto Rico. General Nelson Miles, hero of the Civil War and Indians Wars on the high plains, wanted to get a foothold on Puerto Rico as soon as possible after the surrender of the Spanish forces at Santiago, Cuba, and the destruction of the Spanish Atlantic fleet, for fear that Spain might sue for peace before the United States had a legitimate claim to Puerto Rico by right of conquest. Some in government wanted Puerto Rico, because, having a smaller population than Cuba, it would make a better location for American naval bases on the eastern end of the Caribbean. Such a naval base would be even more important should the United States commit to building a cross-isthmus canal, which the opening naval campaign of the war had shown the necessity of having. Miles would take a force of some 3,500 Regulars and choice Volunteers to Cuba to participate in that campaign if still in progress, after which he would move on to Puerto Rico.

The trip from Camp Alger to the port of embarkation at Charleston, South Carolina, was memorable to the men of the RLG mostly for the filthy condition of the railroad cars. Still, the sight of large numbers of blacks as well as whites greeting them and waving flags as their train passed also stood out in the recollections of the soldiers. The men spent the night of 6 July quartered in a cramped and hot warehouse that crawled with fleas and where no water for drinking was available. Many of the men opted to spend the night outside on the cobblestone court yard in the rain instead of inside the warehouse. On the morning of 8 July, the men of the RLG boarded the *Commodore Perry* for the short trip past the bar which prevented larger ships from coming into Charleston Harbor. The ship came alongside the larger, three-funneled steamship *Yale*, which would take the entire Sixth Massachusetts plus a few other smaller units to, it was assumed, Cuba. The men of the RLG waited two hours in the rain on the deck of the smaller ship before they boarded the *Yale*. The ship held about the 1,300 or so men of the Sixth Massachusetts, and a company from the Sixth Illinois, as well as the crew. The *Yale* left at midnight, as soon as General Miles came on board. The trip south proved to be unpleasant, with food episodic and most of the men forced to remain on the open deck.

The *Citizen and Banner* continued to publish letters sent by members of the RLG to their family and friends to allow the townspeople to read about the adventures of the company. Mail took only about ten days from when it was sent until it arrived in Wakefield and was published in the newspaper. A letter from Fred Wiggin of Wakefield, who was serving in H Company, to his parents, written around 12 July as the *Yale* approached Cuba, mentioned that Lieutenant Gray told him that the RLG would go into the line in Cuba if Santiago not yet been taken. If it had been taken, then Lieutenant Gray thought the Sixth would probably be sent to "Porto Rico." Colonel Woodward wrote to his wife that only three cases of sickness occurred on the ship, one case of measles and two of typhoid. Mrs. Taylor received a letter from her husband dated Tuesday, 12 July, saying, "We have just landed on Cuban soil" but Mrs. Taylor and the newspaper editor were unsure if Major Taylor meant that all the regiment had landed or just himself and a few other staff officers.[100]

Apparently only a few officers actually landed in Cuba, accompanying General Miles. The men of the Sixth Massachusetts had each been issued three days of rations and 110 rounds of ammunition, and so they fully expected to disembark and go right into the line, but instead Generals Miles returned to the ship and what training could be done on board continued. News arrived on the 14th that the Spanish in Santiago had surrendered and the fighting in Cuba was effectively over. Again rumors swept through the Sixth Massachusetts that the regiment might be shipped home now that Cuba had fallen. While proud that American arms had prevailed in Cuba, the men felt greatly disappointed by not being afforded the opportunity to participate in combat. However, their disappointment at not getting into the fight in Cuba was soon alleviated when on 21 July, the fleet was ordered to Puerto Rico. After a week of sailing around Cuban waters with no apparent destination, and with appalling food that became a national scandal during the war, the men of the Sixth Massachusetts learned of their new destination. But if the trip to Cuba had been unpleasant, the shorter trip to Puerto Rico was far worse.

Along with being quartered on the ship's open deck, the men had to endure almost constant friction with the ship's crew. Most of the meals contained meat that had rotted while still in its cans. The only drinking water came from the ship's distiller, which dripped out hot water from which the sick and able stood in line to drink from the same cup. Disease swept through the men almost as fast as rumors. Exposed to the rain and sun, many of the men became ill. The sick men were housed in one of the cabins, but without medicine, ice, or even basic sanitation, they had little chance of recovery while at sea. Illness began to take its toll. In the early morning hours on 23 July, Corporal Charles F. Parker became the first member of the RLG to die in service during the Spanish American War. He had served in the RLG since 1889. Captain Gihon insisted on a short and simple funeral from fear that a long and emotional funeral would demoralize the men. Instead, with the regiment mustered around, the men of the RLG stood at the side of the ship, a few Bible verses were read, a song sung, a prayer said, and the body was buried at sea off the coast of Haiti. The playing of "Taps" completed the sad scene and the men of the RLG dispersed to be alone with their thoughts.

Originally, General Miles planned to disembark his forces on Cape San Juan, on the eastern end of Puerto Rico, and proceed toward San Juan, the capital. For reasons not entirely clear, he changed his destination en route and disembarked on the southern shore of the island, near the town of Guanica. Most likely, General Miles believed that opposition from Spanish forces would be less likely in the south, and that his forces would find a friendly local populous as well as sufficient forage for his army. Then again, the entire Cape

San Juan plan might have been a ruse all along, its well-known details intended to lure the Spanish to concentrate there while the American invasion came on the other side of the island. On 24 June, the *Yale* reached its position off the south coast. A landing party of Marines found little opposition from the Spanish forces on shore, and so by mid-afternoon, the men of the RLG transferred to a smaller craft and went ashore. The Sixth Massachusetts claimed to be the first American regiment to get fully on Puerto Rico. By noon, General Miles had around 3,500 troops at Guanica, and the invasion was well begun. Most of the local residents had fled to the hills, so the soldiers established camp and began to shake off the effects of their miserable sea voyage unmolested. During the night, some skittish members of the Sixth Massachusetts engaged and killed a few luckless cows that happened to be grazing in the nearby woods, but officers were soon able to restore order.

General Miles's plan called for four separate American columns to cross the island to the northern coast and converge on San Juan while defeating small Spanish detachments in detail and showing the Puerto Ricans the limits of Spanish power. His forces were eventually to drive the Spanish forces into San Juan, and force the capitulation of the capital of the Spanish colony. The Sixth Massachusetts was part of General Garretson's column that was to move east from Guanica to take the city of Ponce, the largest city on the island. From there the column would move north across the island to the coast and link up with the western column to begin the final drive on San Juan.

On 26 July, first blood was drawn from the RLG. The evening before, a company from the Sixth Illinois on outpost duty a few miles out on the road to Yauco had been fired on. Companies L and M of the Sixth Massachusetts were dispatched to augment the Illinois soldiers. Throughout the night the men at the outpost heard shooting, and believed that a general Spanish attack would begin about first light. At about 1 A.M., the RLG, along with four other companies from the Sixth Massachusetts, were sent out to reinforce the position. Around 5 o'clock in the morning, after marching to the rear of the position in the dark, the men of the RLG laid their bedrolls in a pile, and moved forward, with Companies L and G behind them. After marching only a few hundred yards, the RLG came under fire from a Spanish position on a hill to their left, about two hundred yards away. The column had walked into a small ambuscade manned by the Puerto Rican Eighth Volunteers. The enemy occupied banana groves and the hill on the left of the road. Instinctively most of the men of the RLG dropped to the ground as soon as the shooting started. Captain Gihon was laying prone, with his head up, speaking with a man in his company, trying to understand the situation, when he dropped his head and groaned. When asked what the problem was, he laughed, and said that he had been shot "in the seat of my pants." Most newspaper accounts, reflecting the sensitivity of the times, refer to his wound as being in the hip.[101]

Despite his wound, Captain Gihon stayed with his command and deployed his men. Lieutenant Gray led three squads of volunteers to clear a hill, while the remainder of the company provided cover from a corn field. Some Spanish cavalry appeared in the distance, and the men of the RLG fixed bayonets in anticipation of receiving a charge, but the attack never came. The entire battle lasted about ninety minutes.[102] Casualties among the Americans were light. In all, the regiment suffered four wounded during the battle. Along with Captain Gihon, a private from Company K was wounded in the neck, and a private from L was wounded in the right arm. Additionally, Captain J. H. Prior of L was wounded in the hand.[103] Throughout the remainder of the day, the men of the RLG entrenched on the hill expecting a Spanish counterattack. The next day, worn out from tension and heat, as well as a lack of food and sleep, the men of the Sixth were informed by civilians from the nearby town of Yauco that the Spanish had all left. The Americans found the bodies of three men they had killed, and believed another thirteen had been wounded. Not all the companies of the Sixth conducted themselves so well during their first skirmish; some men had initially panicked and ran to the rear. The Army immediately took notice of the discrepancies in the performance of various companies of the Sixth Massachusetts.

In general, the Regular Army officers were favorably impressed with the men of the Sixth Regiment, and Captain Gihon received official commendation for his "gallantry and coolness under fire."[104] The same could not be said for the regimental officers, several of whom were in fact absent from the column, having spent the night on the troop transport ship and had not caught up with their regiment when it came under fire. General Miles began an investigation. Over the next couple of days, as the column consolidated and rested back in Guanica, rumors swept through the RLG that several regimental officers had been relieved. Despite these rumors and the absence of Captain Gihon, who was undergoing treatment for his wound, the men of the RLG, and the Sixth Massachusetts, still had a war to fight. After resting and bathing, the regiment again left Guanica on 30 July taking the same road to Yauco, on the way to Ponce. The RLG sent out the flanking elements and scouts for the column. The men were surprised by the friendly greetings they received from the civilians as they moved through Yauco. The column reached Ponce at noon on 1 August, totally worn out by the march. Blisters had torn the men's feet, and marching became an ordeal.

On 3 August, after a day of rest about a mile beyond Ponce, the old black power Springfield rifles most Volunteer regiments carried were exchanged for the newer Krag-Jorgensen rifles, which was the standard weapon of the Regular Army. Even more welcome was the return of Captain Gihon, just released from the Army hospital in Guanica. However, the rumors of some of the regimental officers being relieved continued to hang over the RLG. Finally, on 5 August, Colonel Woodward, Major Taylor, and others officially resigned for the good of the service. Their resignation was in fact a face-saving method of relieving them from command. Colonel Woodward claimed, upon his return to Wakefield, that the brigade commander, General Garretson of Ohio, had put pressure on him to replace Company L, the only black Massachusetts company, and Colonel Woodward had repeatedly refused this. As a result, General Garretson long sought a way to remove Colonel Woodward. Another officer who resigned, Lieutenant Colonel George H. Chaffin, stated that General Garretson personally disliked the regimental officers, and after the drawing of lots placed the Sixth above the Eighth Ohio, the general constantly sought to ensure the Sixth Massachusetts suffered the most hardship. Because of the hardships, some men of the Sixth Mass were straggling during the march, and the general used this as an excuse to take what Colonel Woodward thought was an unprecedented step and ordered him to appear before a board of Regular Army officers and to be tested as to his fitness for the rank he wore. Because this was highly unusual and Colonel Woodward sensed he was being set up, he resigned his commission instead. Lieutenant Colonel Edmund Rice, a Regular officer from General Miles's staff, was soon commissioned the colonel of the Sixth Massachusetts. As a result of the shake up, the RLG, Company A, lost its company commander, Captain Gihon, who became the acting major of the regiment, and Lieutenant Gray, who became the regimental adjutant. With all the other company officers on detached duty or ill, First Sergeant Charles Bridge, who had been in the RLG since 1892, assumed company command.

The column continued on its mission after a few days in Ponce. On the 11th, the column reached the small town of Adjuntas. The march had been difficult, between the blisters, rain, and lack of shelter at night, and the two-day rest in Adjuntas was quite welcome. When the rain cleared on the 13th, the column resumed its march northward. The column contained only eight companies of the Sixth Massachusetts, as the other four and the Illinois company remained at Adjuntas to garrison the town. The column covered eighteen miles to the town of Utuado in seven hours over a trail so rough the Spanish did not even know of its existence. The men assumed that they were being hurried to take part in a battle with Spanish forces in the town of Arecibo, on the north coast. Upon reaching the town, rumors again swept through the Sixth that the war had ended. That evening, official news arrived that an armistice had been arranged and that the fighting was over.

With the possibility of hard campaigning behind them, the men were given passes to explore the city of Utuado. For the Wakefield men, the small city of a few thousand residents offered a glimpse of what must have seemed an exotic and strange culture. The inhabitants were friendly, the city was clean, and the small cathedral and colorful houses added to the pleasant atmosphere of the stay. More rumors spread through the ranks; they were going to stay in Puerto Rico to garrison the island, they were going home any day now, perhaps they would be sent to the Philippines. But for the next several days, Utuado would be their home. The other companies that had been left behind in Adjuntas soon joined the column, and some of the sick men caught up too. However the incessant rains took its toll on the men. Diarrhea became common, but other ailments also swept through the ranks. Some 10 percent of the RLG went on the sick list. Colonel Rice sought permission to house the men in solid barracks, as opposed to their tents, but that would not be forthcoming. Instead the men spent a day doing what they should have done originally when they first made camp; they dug channels throughout the company streets to allow the water to drain. Still the sick list continued to grow. By 22 August, thirty-two men of the RLG were ill. The Puerto Rican campaign, which so far had been conducted far more professionally than the Cuban campaign, was in danger of succumbing to the same deadly aftermath.

On the 22nd, the RLG finally moved into solid quarters, a schoolhouse, while all but one other company began the trek back to Guanica. The school house was only marginally better than the tents, as it was crawling with rats and cockroaches. Mail from home finally caught up with the RLG, but pay had not reached them since 6 July, and most of the men had no money to buy local foods. Some of the men were too sick to eat Army hardtack, but not sick enough to be admitted to the hospitals. Still, the Army issued the men new clothing and shoes on 2 September, which the men sorely needed by that point. Captain Gihon returned to the RLG, to inform the men that rather than going home, the Sixth Massachusetts would instead have the honor of going to San Juan. However, the regiment remained at Utuado throughout September. Sickness continued to ravage the ranks, with 203 men from the Sixth Massachusetts on the sick list. The wife of Colonel Rice arrived, and worked hard to improve the field hospital, but in the absence of proper medicine, little could be done. On 20 September, the one hundred sickest men from the regiment, including ten from the RLG, were carried to the coast for the voyage home on the *Bay State*. The removal of men allowed some of the remaining soldiers to be promoted, but the company continued to function without commissioned officers. By the end of the month, the regimental sick list was so high — just over half of the men were on it, that military training and drills came to an end so that the well could nurse the sick and perform guard duty.

Finally, in the first week of October, the Sixth Regiment received official word that it would not be going to San Juan but instead would soon begin the journey home. Pay had reached them a few days earlier and this enlivened the men greatly. The hometown continued to support its company, both by following its exploits and difficulties in the town's newspaper, and by raising money to send to the men in the RLG to augment their sometimes delayed federal pay. While the men of the RLG had been paid while in federal service, that money had gotten to the men sporadically, and means for getting money home to their families were not always available. On the twelfth, newly promoted Major Gihon received $250 from the Citizens' War Relief Committee of Wakefield, which allowed the men to buy extra food on the trip home, and to relieve some of who returned home to find their financial situation desperate.[105]

On 13 October, Company A, the RLG, along with Company C, marched the twenty-three miles to Arecibo over bad roads in one day. The heat and afternoon rain made the march worse, and six men from the RLG fell out on the way. One man recorded in his diary that the RLG "began its first march on the island with over one hundred men, ended its last with barley twenty."[106] From Arecibo, on the north coast, the trip took them by slow train to San Juan, from where they boarded the *Mississippi*, a dirty cattle ship outfitted with hammocks for the men. After a few days in San Juan, the ship headed for Boston, which it reached on 27 October after five days at sea. The air got colder as the ship headed north, but the chill and seasickness did not dampen the excitement as the ship came into Boston Harbor. Newspapers were brought to the ship, and the men of the RLG learned that the Wakefield Citizens' War Committee planned a large celebration when the town's sons returned. The wait on the ship must have seemed maddeningly slow. The governor came on board after lunch to greet the men. By 3 o'clock in the afternoon, the *Mississippi* finally was tied to the South Boston dock, and an hour later the regiment marched to a tumultuous welcome through the normally orderly streets of Boston to city hall to pass in review for the mayor of Boston, and onward to the State House so they could perform a similar service to the governor. The celebration in Boston was considered to be the largest celebration since the Civil War.

After the brief parade through the city, the men of the RLG were greeted by a group from Wakefield, composed of members of the Grand Army of the Republic, with some of the Sons of Veterans and other townspeople also in Boston to greet the local heroes. The symbolism was not lost on observers; the old veterans of the town militia from the Civil War welcoming back into the community the latest incarnation of the town militia from newer distant battles. The column of militiamen, veterans, and well-wishers then took the train back to Wakefield, which they reached around dusk.

While the members of the RLG might have desired to shed their uniforms and reunite with family in their homes, those private desires would have to await the completion of official welcomes. As the train pulled into Wakefield station, the waiting crowd began a celebration that surpassed any in memory. Two brass bands played, fireworks and dynamite were set off, and every bell and horn was sounded to add to a deafening cacophony. Loved ones were able to greet the men as they departed the train, but soon the company formed and marched to the armory. Marching ahead of the RLG was just about every uniformed organization in the town; the GAR, the Sons of Veterans, the high school cadets, the various fire companies, the mail carriers, Fine Members Association, the Hibernians, and the War Relief Committee. Although far outnumbered by their escorts, the men of the RLG were the heroes of the day and the ones the crowds had come to cheer.[107] After the men ate from a table prepared by the Citizens' Committee, they were finally allowed to disperse to their homes. Crowds gathered around the homes of some of the officers and men, prompting Major Gihon to come out on his porch and give an impromptu speech praising the steadfastness of men of the RLG under fire. Congress furloughed all Volunteers who had seen foreign service until 2 Jan 1899 so the men were able to enjoy several weeks at home with Army pay. On 9 November, a few weeks after their homecoming, the men of the RLG, as well as other veterans of the Spanish American War from Wakefield, were treated to a more formal banquet in the Town Hall. Many of the regimental officers were present, including Colonel Rice and his wife. Also present was the new brigade commander, Brigadier General T.R. Matthews, and a colonel from the governor's staff. As a link to the past, Captain Samuel F. Littlefield, who had commanded the RLG for part of the Civil War, was also a guest of honor. A week later, the town of Reading gave a similar banquet to the men of the RLG.

Technically, the RLG remained on active federal service through the fall of 1898 and into the winter of 1899. Although this was a busy time for the officers and administrative soldiers, who needed to prepare rolls, make reports, and square accounts, for the rank-and-file, it was a paid vacation. The men reported to the armory in Wakefield for morning and evening roll call. They could eat at the armory with meals provided by a caterer, but were free to do pretty much what they pleased during other times. Rumors continued to spread through the RLG that the Sixth Massachusetts would soon be sent to fight in the Philippines, that the Sixth was soon to be ordered to occupation duty in Cuba — but these proved unfounded.

On 2 January 1899, the RLG met at the armory and founded the "Richardson Light Guard Association of the Porto Rico campaign," with Major Gihon as president. Also at this meeting were two members of the RLG who had remained hospitalized in Puerto Rico when the main body of the RLG left the island. Finally, on 21 January, the RLG took a train to Boston, and at a simple ceremony at the South Armory, received their final pay and were mustered out of federal service. The Sixth Massachusetts Infantry Regiment ceased to exist, although the RLG would continue to exist in the Massachusetts Volunteer Militia. During the war, the Sixth Massachusetts lost twenty-five men to disease, and four men wounded by enemy fire. Captain Gihon and Captain Prior from L Company were the only officers in the regiment to be wounded, the others being enlisted. Of all officers from Massachusetts in the war, five died of disease, and one was Killed in Action — an officer serving with the Second Regiment in Cuba.[108]

Colonel Woodward's reputation in the town and state had effectively ended. Although he had resigned from the Sixth Massachusetts, he was still a colonel in the MVM. The state law that allowed Colonel Woodward to retain his commission in the MVM, while serving as a colonel in the Massachusetts Volunteers in federal service, was created by Colonel Woodward, while he was a member of the Massachusetts Senate and chairman of the committee on military affairs. Upon his return to Boston following his resignation from federal service, he reported for duty as a colonel of the Sixth Regiment, MVM, much to the surprise of state military leaders. This awkward situation, along with Colonel Woodward's military career, ended in late 1898 when he sent his resignation from the MVM to the governor, who immediately accepted it. The colonel had turned over what little state property he had to the quartermaster general two months earlier so resignation was expected.[109] The RLG, the town, and the leadership of the MVM had, it seems, come to accept the official reason for Colonel Woodward's troubles in the war, that he was incompetent as a combat commander, and discounted his pleas of Army prejudice against Volunteer officers and racial discrimination. The adjutant general's annual report simply listed his resignation, and Colonel Woodward essentially disappeared from further mention in the military records of the commonwealth.[110]

The provisional companies of the militia, having spent the short war without being called to active service, were quietly disbanded on 15 April 1899, which exceeded the "peace treaty plus thirty days" under which they were originally mustered by several months.[111] For the men of the RLG, the Spanish American War had shown themselves and their hometown that, despite the decades of balls, banquets, and parades, they were true soldiers and worthy successors to their predecessors. Unlike most militia companies that had entered federal service, the RLG actually served in a campaign. Although the record of some regimental officers who came from the RLG was a bit embarrassing, the company and the town celebrated the exploits of their company, not the regiment to which it belonged. For Wakefield, if not for most of the nation, the Spanish American War had indeed been a "splendid little war."

Chapter VI

National Institution

When the RLG left federal service in January 1899, it reverted to its status in the MVM. Major Gihon was offered a commission as a captain in the newly formed 26th Regiment, U.S. Volunteers,[1] for service in the Philippines, but he declined the honor, citing his wife's health.[2] Instead, he reverted to his state rank and position as captain and commander of the RLG. Massachusetts law regarding militia, which had been largely crafted by Colonel Woodward in his capacity as chairman of the committee on military affairs in the Massachusetts House of Representatives, held all positions in the MVM in stasis while members were in federal service and thus all the men of the RLG reverted to their prewar state ranks and positions when they returned from federal service. Lieutenant Walton had to resign again, this time from the MVM. Captain Gray, commander of the Company K at the end of the war, returned to his rank and position as second lieutenant in the RLG. The resignations of Woodward and Taylor, and the fame of Captain Gihon at the end of the war, ensured that more changes would soon come to the RLG and the Sixth Regiment, MVM. Almost immediately after the return of the Sixth Regiment, Captain Gihon was elected to the rank of major, to fill the vacancy created by the resignation of Major Taylor. He would hold that rank for a scant eight months before Governor Winthrop Crane, probably with an eye toward Gihon's new status as a war hero, selected him to join the governor's staff as assistant inspector-general, with the rank of lieutenant-colonel.

Although Colonel Gihon would serve three terms as selectman for the town of Wakefield in the years immediately following the war, he does not seem to have much more interest in using his military reputation to make a political career. Instead, his postwar interests tended to veterans' groups, including memberships in the Fine Members, and the RLG Veterans Association of the Porto Rico Campaign. He played a leading role in the creation of the United Spanish War Veterans (USWV) out of such local veterans' groups. He founded the local USWV camp in Wakefield in 1905, and a woman's auxiliary in 1906.[3] In 1910 he was elected the national commander of the USWV at their annual meeting in Seattle. Probably most significant for him, his conduct in the Spanish American War had earned him an honorary membership in the Wakefield chapter of the GAR. Colonel Gihon later became one of the founding members, and the first Exalted Ruler, of the Wakefield Elks Club, in 1912.[4] Lieutenant Gray was also promoted almost immediately upon his return to state status to fill the rank of first lieutenant recently vacated, and from there was elected captain to replace Gihon.[5] Elmer Morrison, who had served since 1886 was elected to first lieutenant, and John McMahon, member since 1893, became the second lieutenant. Thus reorganized, the RLG looked toward an approaching milestone in the life of the company.

On 11 October 1901, the men of the Richardson Light Guard gathered for their fiftieth-anniversary celebration. In anticipation of the anniversary, the selectmen of the town appointed a committee to oversee all aspects of the celebration. Plans to mark the milestone had been underway since the previous June, with Solon. O. Richardson, former member and son of the late doctor, presiding over the preparation committee. The committee decided that the town should pay to publish a history of the RLG as part of the celebration. To assemble the book in the allotted time, a bare two months, the committee turned to a local historian, William E. Eaton. Mr. Eaton, a descendant of Lilley Eaton, the original trustee of the RLG, had long served as the town's historian. William Eaton drew upon the prodigious writings of the elder Eaton to complete the book. The finished volume was published locally by the Item Press, and admirably fulfilled its purpose. In its 216 pages, the story of the RLG in war and peace from origins in 1851 until 1901 unfolds. The volume contained many photographs, and a roster showing the highlights of the service of all men known to have served in the RLG. The volume stood as a testament not only to the RLG, but to the pride that the town of Wakefield took in its militia company in 1901. Volumes about a single militia company are exceedingly rare, and the production of the volume demonstrated again that the RLG was a something other than the average militia company.

On the day of the anniversary celebration, much of the town of Wakefield turned out to watch the company parade on a perfect fall day. Many former members of the company also turned-out for the occasion, making it seem more like a homecoming celebration than a military observance. The RLG, under the command of Captain Gray, marched through Wakefield to their armory, with a platoon of the town's policemen leading the way, followed by a military band from Salem, local dignitaries riding in carriages, veterans, and finally the RLG. The men marched through the town center, which had been heavily decorated with patriotic ornaments. The armory itself was dazzling in electric light, with a large portrait of the late Dr. Richardson himself over the doorway,[6] the same portrait that had hung over the Richardson home during the 250th Anniversary celebration of the founding of the town in 1894.

Reviewing the RLG for the occasion were Lieutenant Governor John L. Bates, the adjutant general of Massachusetts, a captain from the U.S. Army, and many officers from other units of the Massachusetts Volunteer Militia and the Grand Army of the Republic. Also on the reviewing stand was Solon O. Richardson, son of the namesake of the unit being honored. Speeches were made, awards presented, and a banquet consumed. Afterwards the assembled militiamen and friends adjourned to the ballroom of the Town Hall — the same Town Hall that earned Mr. Wakefield the honor of the town now bearing in his name — for an evening of further festivities.[7]

The fiftieth anniversary of the RLG was indeed a special occasion for those present, and it was a high point in the life of the company. Although no one would have guessed it on that auspicious day, by the end of the next fifty years, the deep connections between the town and the local militia company would be little more than a memory, with its centennial a small and barely noticed affair. The RLG would not be destroyed in battle, either by failure of courage or destruction by a superior enemy. In fact the RLG would do all that was asked of it and more, but the bonds that held it together and tied it so fiercely to the town of Wakefield were about to be slowly but steadily weakened by automobiles, centralization, professionalization, and a couple of very large wars. For thirty-two years after the last unit bearing the name "Richardson Light Guard" returned from the Civil War, Wakefield's local militia led an existence common to many militia companies in the northeast. They trained, socialized, and stood ready to respond to civil unrest or war. In the two decades after the Spanish American War, the Richardson Light Guard, as most militia units, would evolve into something very different from its nineteenth century namesake.

These changes lay in the future, and Wakefield and the RLG at the turn of the century was still focused on the past and present. Eight months later, the RLG and the town of Wakefield would be at the center of another martial celebration, this time for the dedication of the Soldiers and Sailors Monument. Although Mr. Wakefield's 1871 Town Hall was ostensibly built to honor the town's Civil War dead, Mrs. Harriet Flint believed the town still had no proper memorial to the town's heroes. The bequest from her will had been accepted by the town in March 1898, and the Spanish American War had delayed fulfilling her wishes but a short time. During the war, the committee appointed by the town to carry out her wishes continued its work. The town added almost four thousand dollars to Mrs. Flint's original gift of ten thousand dollars to make the memorial a reality. The completed memorial consisted of a marble column topped by a south-facing soldier holding a flag. At its base a figure stood sentinel on each corner. Three figures were soldiers, representing infantry, cavalry, and artillery, while the fourth, a sailor, represented the navy. An inscription on the side of the column proclaimed that the memorial was dedicated "To the Men of South Reading Who Gave Their Lives to Keep the States United, 1861–1865." One site given serious consideration for the new monument was the "Richardson lot," which was then for sale. The third man to bear the name Solon O. Richardson had moved to Toledo, Ohio, in 1888,[8] and the elder Solon O. Richardson was selling the part of the family property fronting on Main Street. The Richardson house was moved back and turned 90 degrees, as part of an overall development of the estate. Wakefield residents believed the Richardson family would give the town a better deal than it would a private buyer. The lot contained some of the last open space on Main Street in the center of town, and many thought it would make an excellent location for the memorial and a new library.[9] Eventually, the town did not buy the Richardson property, and instead placed the new memorial on the Common. Townspeople had long opposed any attempts to build structures on the Common, and indeed had waged a long campaign to remove several structures such as the old Town House, fire station, and a blacksmith shop on Church Street which separated the Common from the lake. The decision to place the monument on the Common shows the high regard held for the men whose sacrifice the monument was meant to represent.

The dedication of the memorial was a gala event, and again the RLG would fulfill its role of organizing and hosting the event that celebrated the town's martial past. The RLG assembled at 8:30 the morning of 17 June 1902, under its newly elected commander, Captain John H. McMahon, for a parade through the town, where they were joined by the GAR and other civic organizations.[10] Bleachers were set up on the Common facing north across the lake, but the bleachers could not begin to hold the crowd that arrived for the event, and the Common thronged with visitors who turned out in their finest clothing for the dedication. Colonel Gihon served as Chief Marshall for the ceremony. Also present were the governor of Massachusetts, Winthrop Crane, and Congressman Samuel McCall, who made the oration. School children sang, and Wakefield celebrated its part in the Union victory in the Civil War.[11]

The refusal of the town to acquire the Richardson property on Main Street for the monument led to other uses for the land. In 1900, the Richardson house was one of the few remaining residential properties on Main Street. The town's population was approaching ten-thousand and would top eleven thousand at the end of the decade.[12] The growth of the town had made the location too lucrative for continued use as a private residence. The

Ceremony marking the dedication of the Soldiers' and Sailors' Monument (Civil War memorial) on the Common on June 17, 1902. The figure holding the flag on the top of the monument is facing south (courtesy Wakefield Historical Society).

house was moved further back from Main Street along a short street that became Richardson Avenue, which bisected the former Richardson estate, and eventually was converted into a rooming house.[13] After the house was moved, Solon Richardson developed the frontage of the family property on Main Street, erecting the Richardson Building, with several storefronts on the street level with office space on the second story. While the Richardson Building occupied the northern corner formed by Richardson Avenue and Main Street, other commercial properties filled the southern corner. Main Street in Wakefield Center was quickly becoming an overwhelmingly commercial zone. But for the RLG, fundamental changes in the role of militia were brewing at the federal level.

Between the end of the Spanish American War in 1898 and the entry of the United States into the Great War in 1917, the RLG became something quite different from a legal standpoint from its nineteenth-century version. The creation of the National Guard Association (NGA) in the decades after the Civil War in part to lobby for increased federal support for state militia and a continued role for militia companies during wartime ushered in an era of fundamental change in the relationship of most organized militia to the federal government. The roots of militia ran deep in Massachusetts, and officers of the MVM had been among the founders of the National Guard Association. By 1903, the NGA included representatives from the organized militia of almost every state in the Union. In most states, the very term "militia" had long since fallen into disuse and discredit, bringing to mind images of drunken training days, men ill-equipped and untrained, or at best one of snobs in garish uniforms toasting each other at lavish banquets. The MVM had largely avoided

Postcard view of the Soldiers' and Sailors' Monument (courtesy Lucius Beebe Memorial Library of Wakefield).

both of these images and the term "militia" remained an honorable one in the Bay State. Not so in most states, where the term had become so embarrassing that many units called themselves "national guard" without official authorization to do so. The *Militia Act of 1903* authorized, but did not require, states to adopt the term "national guard" for their organized militia. That the first militia company in the United States to call itself the "national

Close up of the Soldiers' and Sailors' Monument. The individual figures at the base represent the infantry, artillery, cavalry, and navy. The words on the base say that the memorial is dedicated "to the men of South Reading who gave their lives to keep the states united, 1861–1865" (courtesy Michael Boucher).

guard" came from the rival state of New York did not encourage Massachusetts to adopt the term for its own militia. In 1824, some companies from the New York militia adopted the title as an honor to the Marquis de Lafayette, hero of the American War of Independence and later commander of the Guarde National during the French Revolution, who was then visiting the United States. By 1916, when federal law dubbed all land forces in the organized militia as "National Guard," Massachusetts was one of the few remaining holdouts. Even when Massachusetts finally adopted the term "National Guard" for its land forces in 1916, it still kept the term "Massachusetts Volunteer Militia" to refer to the entire organized militia of Massachusetts, which contained both the National Guard and the naval militia.

The MVM was one of the state militias most active in the National Guard Association, but supported a federal war-fighting role for the National Guard more than federal financial support or federal interference. The MVM received adequate financial support from the towns and the commonwealth, and the leadership preserved a tradition of military competence so that the MVM did not feel the need federal support or intrusion for existence or levels of readiness that would make it usable in war. Few states rivaled Massachusetts in financial support given to the militia, allowing the MVM to maintain an ambivalent relationship with the rising National Guard movement during the final decades of the nineteenth century. In 1903, for example, the commonwealth spent $327,825 in support of its militia, whereas the federal government contributed only $31,916. In addition, most companies in the MVM also received support from their home towns. Only Connecticut surpassed Massachusetts in the ratio of state funding to federal funding of the state militia.[14] A decade later, Massachusetts spent an average of $104.19 annually on each member of the MVM, the highest amount in the northeast.[15]

The organized militia in most states in the early twentieth century was better organized, equipped, and trained than most previous incarnations of the militia, but the National Guard was still an awkward vehicle through which to expand the army in wartime. The National Guard bore a heavy imprint of state identity among its units. Composition of units was not standardized across the nation or with the Regular Army. Each state's National Guard had its own uniforms, training programs, and organization. Most National Guard units took whatever branch the original members wanted, and were thus heavy in infantry, cavalry, and artillery, and light in support functions. To integrate the National Guard into the federal military service would be extremely difficult. The *Militia Act of 1903* began the transformation of the various state National Guards into a proper reserve force of the army. The act was more commonly known as the "Dick Act" after the bill's sponsor, Congressman Charles F. Dick of Ohio. Congressman Dick served as president of the National Guard Association, and was the Commanding General of the Ohio National Guard. As president of the NGA, Dick had been a strong force in favor of a war-fighting role for the National Guard.[16] The former federal militia law, the *Militia Act of 1792*, had never been followed, and the rise of the National Guard made urgent the need for adequate federal legislation regarding the militia. The *Militia Act of 1903* recognized in federal law for the first time the distinction between the organized and unorganized militia — which Massachusetts had by then recognized in law for over half a century. The *Militia Act of 1903* defined the organized militia as those military organizations, primarily the National Guard, that had been organized and recognized by the states. In addition to the National Guard, a few states also maintained a naval militia as part of their organized militia. A very few states, including Massachusetts, also maintained a small marine element. The unorganized militia consisted of the remainder of the adult male population liable for militia service should the state someday choose to organize it. The new federal law basically implemented on the federal level what Massachusetts had done in 1851 when it created the MVM.

Although most companies in the MVM trained one evening a week, militia units in some states seldom trained at all. The Act specified that units hold at least two mandatory evening drills per month.[17] The *Militia Act of 1903* also required militia units to spend at least five days training in the field annually, as Massachusetts had required since the 1867.[18] This requirement had no effect on the MVM, which had extended its annual field tour to seven days in the late nineteenth century. Each state's militia was supposed to follow the Regular Army in organization, equipment, and discipline.[19] The planners hoped the changes would mold the National Guard into an effective reserve force that the federal army could and would use in war. State governments generally supported any federal move that brought more equipment and money to the National Guard because the states would then have at their disposal a better organized, equipped, and trained militia during peacetime. Massachusetts, while adequately supporting its state militia, did not want to be left out of federal support for it, nor have the MVM prevented from fighting alongside the army during the next war.

The *Militia Act of 1903* began the transformation of the various state militias into the modern National Guard. However, the National Guard remained legally militia. With the

National Guard organized under the militia clauses of the Constitution, it could not be used outside of the United States as National Guard. In a reply to an inquiry from the Secretary of War, Attorney General George W. Wickersham ruled that the National Guard was still militia, and as militia could only be used for the constitutionally sanctioned purposes: suppression of insurrection, repelling of invasion, or execution of laws. The militia could only leave United States territory in pursuit of an invading army.[20] Militia units would still have to reorganize as Volunteers, as they had in the Spanish American War, in order to serve outside the United States as part of the army. Also, governors retained the right to authorize or to deny their state's units to enter federal service. Even with the governor's consent, individual National Guardsmen needed to volunteer for federal service; they could not be drafted against their will.[21] The awkward system for mobilizing the National Guard that had been used during the Spanish American War had been preserved for the time being. With these limitations, the National Guard, or any state militia, was ill-suited to augment the Regular Army in war. Fearing that the National Guard might be pushed aside by a new wholly federally controlled reserve force, the National Guard Association lobbied hard for a solution to the constitutional barrier that prevented the National Guard from serving alongside the Regular Army anywhere in the world.[22]

The struggle of the NGA to find a solution to the constitutional obstacles of using the National Guard in war had little visible impact on the Richardson Light Guard, because Massachusetts was usually ahead of the federal government in legislation regarding the militia. In early 1914, the commonwealth took a large step away from relying on enthusiastic middle-class volunteers with support from local elites to maintain the MVM when it began to pay members of the MVM for up to ten drills per year. Members had been paid for summer encampments for decades, and of course were paid when brought on active duty for state or federal service, but drill had always been uncompensated. But beginning in 1914, militiamen received ninety cents for each "Rendezvous Drill" if forty men from the company attended.[23] However, paying Guardsmen for attending drill bit deeply into the budget of the MVM, with about a third of state funds going for pay.[24] The adjutant general of Massachusetts strongly urged the federal government to assume the burden of paying Guardsmen for drill, arguing that it would bring new levels of professionalism to the militia.[25] For most of the MVM, the new federal involvement in militia nationwide simply standardized much of what they were already doing in the Bay State. The Massachusetts Volunteer Militia had consistently ranked as one of the best state militias by army inspectors, and the increased federal presence in state military forces brought few changes to the force. The commonwealth had been gaining more centralized control over individual companies in the MVM for a generation, and continued to do so. In retrospect, nothing symbolized the transformation of the RLG away from its status as institution of the town of Wakefield than the unplanned movement to its last armory.

The 1895 armory, built for the RLG and owned by the town, burned on 6 July 1911. Around seven o'clock that evening, a violent thunderstorm struck the town. Lightning hit the Cutler Brothers grocery and grain elevator on Main Street, igniting a blaze that threatened even the Town Hall, on the block to the north. Cyrus Wakefield's edifice stood, but the armory of the RLG, much closer on the south side of the Cutler Brothers building, began smoldering from radiated heat. Members of the company, aided by some veterans, arrived and were able to remove most of the state and company property from the building, excluding a few items in the attic, and stored them in a garage on the other side of Main Street. Despite the torrents of rain, the armory began burning. By 11:00 P.M., the

VI. National Institution 109

The ruins of the Cutler Brothers grocery and grain elevator following the fire of June 6, 1911. The 1895 armory is still standing, but the damage is apparent (courtesy Lucius Beebe Memorial Library of Wakefield).

Wakefield Fire Department, aided by fire companies from surrounding towns, extinguished the flames. The Cutler Brothers building was a smoldering heap of ashes and charred wood. The armory still stood, but fire had so damaged its drill hall roof and upper floors that the building was condemned.

A photograph of the RLG taken on an apparently cool day some time after the fire shows the company formed in two ranks, on the east side of Main Street, arrayed between the Town Hall and their former armory.[26] The company is at rest, and the pose of the men is informal. The armory is still standing, and shows little visible signs of damage from the fire, but the mansard roof is missing, and the former Cutler Brothers store is absent from the background. The sergeant on the far right of the company, closest to the camera, has a small banner embossed with the letter "A" attached to his bayonet. In front of the first platoon, and glaring at the unknown photographer, stands Lieutenant Fred H. Rogers. The privates are all smartly attired in canvas legging, white pants, dark tunics, and service caps. The officers and NCOs have dark pants with a light stripe down the seam. The privates and NCOs are armed with Springfield rifles, while the officers carry swords. The lieutenants stand in front of the platoons, while an officer, most likely Captain McMahon, approaches the second platoon leader. The occasion for the formation is unknown, but it shows a serious and disciplined-looking company apparently ready for any challenges the future held. However, the challenge of providing a new armory would no longer be borne by the RLG or the town of Wakefield. Instead, the commonwealth provided the new armory.

Throughout the months following the fire, the RLG remained busy with military events, but the MVM desired to construct a new armory to house the RLG as soon as possible. Strangely, despite the damage to the old armory, and its description as being con-

The RLG, as Company A of the Sixth Regiment, formed up on Main Street in front of the Wakefield Town Hall. The date is uncertain but the Cutler Brothers Grocery that burned in 1911 is absent. The 1895 armory, missing its distinctive third floor, is clearly visible behind the troops on the right. The officer looking at the camera is probably Lieutenant Fred H. Rogers, while the officer walking toward the formation is Captain John H. McMahon (courtesy Wakefield Historical Society).

demned after the fire, it would be repaired and put to commercial uses. Although its distinctive third floor was gone, and its second floor needed extensive rebuilding, the former armory with its drill hall would remain standing and in use through the end of the twentieth century, indeed much longer than the remaining life of the RLG.[27] But the commonwealth was focused on providing a new, state-owned armory for the RLG, not repairing the old one. The days of town-owned or rented armories were almost over in the Bay State. The fire came during a decade and a half when Massachusetts was placing almost all of its militia companies in state-owned armories. In all, Massachusetts spent $3.2 million between 1900 and 1915 on armories, after which only three companies in the entire MVM were not in state-owned armories.[28] The state legislature authorized $55,000 to build and equip a new armory for the RLG.[29] The town received one thousand dollars from the insurance on the old armory, which Major McMahon wanted to give to the commonwealth to use in purchasing land for a new armory.[30] One proposed site for a new armory was situated adjacent to the Heywood Brothers and Wakefield Company factory complex, almost on the site of the former town rifle range. However, after a hurried search, the commonwealth settled instead on a site for the new armory almost across Main Street from the old armory, and in 1911 bought a plot of land in the northeast corner of the old Cyrus Wakefield estate. The 29,805-square-foot-lot fronted on Main Street, and cost $5,660.00. Constructing the new armory cost another $47,056.33, with furnishings an additional $3,194.96.[31] The dedication came on 15 January 1913.

The new armory stood on higher ground than the old one, and was set back from the street. It was a more imposing structure than the former armory but still looked more like a school than a fort or castle. It was a red-brick building with four white columns in the front and a large attached drill hall in the rear. The drill hall, at 109 feet long, surpassed the old drill hall by twenty feet, and was seventeen feet wider.

Plaque inside the 1912 State Armory (courtesy Michael Boucher).

The new armory was the only one in the Massachusetts with a colonial design on the front, but neither the RLG nor the town of Wakefield had much to do with the planning of the new armory. Whereas the former armory had been built by the town for the RLG, the new armory was a state armory, paid for and owned by the commonwealth. Significantly, the front boasted the initials "M.V.M.," which of course stood for "Massachusetts Volunteer Militia," and made no mention of the Richardson Light Guard, nor included the term "National Guard." An armory built and owned by the commonwealth for the RLG could be closed by the commonwealth, and the RLG, or its descendant, could be moved or disbanded by the commonwealth, or later, federal government.

But for the time being, the RLG was still state militia and had obligations to the state in peacetime. In February of 1912, the RLG responded to labor problems in the immigrant city of Lawrence, about ten miles north of Wakefield. A strike in several of the textile mills involved the RLG for two weeks. Troops from the 8th Infantry, MVM, had been on duty in the city since 8 January, but by the 29th, the situation was too unstable for one regiment to handle. On that date, the state adjutant general's office called twelve additional companies, including the RLG, to active service. Captain McMahon received the order by telephone at 3:00 P.M., and began assembling his men. By 8:31 P.M., three officers and forty-eight men of the RLG had entrained for the short ride to Lawrence. The men reached Lawrence during a heavy snowstorm. For the next two weeks, the men of the RLG, along with twenty-one other infantry companies and a cavalry troop patrolled the city until passions subsided. Temperatures fell below zero so the state issued winter boots, mittens, and hats for the militiamen. The men of the RLG were sent out in detachments to guard parts of the city, such as the library, or the power station. One night a detachment of the RLG patrolled the area around the city jail, where a labor leader was being held. When not performing guard duty, most of the service for the men of the RLG consisted of drilling and holding classes at the state armory in Lawrence, while one platoon stood ready as a quick reaction force to be trucked to any location should violence erupt.

When the commonwealth released the men of the RLG from their service after fifteen days, they took a specially chartered streetcar back to Wakefield, their pockets filled with state pay. The Lawrence strike had been the longest period of continual state active duty the RLG had performed to that date. What is even more surprising is that the Lawrence Strike of 1912 was the first strike duty of any kind that the RLG performed. Generally, the widespread strikes in the eastern half of the United States in 1877, beginning in the railroad industry, are normally seen as the catalyst of the resurrection of the state militia, and the real beginnings of the National Guard movement. States needed reliable militia companies when labor disputes threatened a breakdown of order. Despite the widespread industrialization of eastern Massachusetts throughout the nineteenth and early twentieth century, the RLG existed from 1851 until 1912 without ever having to perform this onerous duty. Ironically, the RLG would perform its first strike duty as the age of militia involvement in labor disputes was drawing to a close.

The Lawrence strike was not a harbinger of things to come, and the MVM would move in the opposite direction. Strike duty in general was considered an inappropriate use of militia by state military leaders, and the MVM sought to avoid such a role. The MVM existed primarily to respond to war, with natural disaster a distant second. Service during strikes could be controversial and tended to alienate workers from the militia. The RLG and the MVM as a whole saw themselves as a warfighting force and desired to distance themselves from police duties. In his report following the Lawrence Strike, the adjutant general of Massachusetts wrote:

> I cannot urge too strongly the fact that the military forces of the commonwealth should not be called out to aid the civil authorities until every possible effort has been made by civil authorities to control the situation. The organized militia is not trained to do police work and should not be called upon to do the work of civil officers.[32]

This advice was followed two years later when, in the spring of 1914, labor troubles came to Wakefield. A strike that began in the middle of April by about seventy-five mat workers at the Heywood Brothers and Wakefield Company soon spread until about 400 workers were on strike.[33] Many of the workers were first or second generation Italian Americans, and ethnic violence became part of the unrest when the factory owners tried to bring in ethnically Greek replacement workers from Lynn. A group of strikers stopped the streetcar containing the replacement workers, many of whom were women, and threatened them. The Wakefield police chief arrived in time to prevent further violence, although he confiscated a few guns from the strikers and arrested some.[34] The strikers themselves were split over whether to join the American Federation of Labor or the International Workers of the World. As the strike grew more violent, the Wakefield Police Department found itself overwhelmed, but rather than call upon the RLG, it instead called for assistance from the Metropolitan Police.[35] The height of violence occurred on 8 July, when a bomb exploded in the home of a factory superintendent during the early morning.[36] The town selectmen became involved in seeking an end to the strike, and eventually the State Board of Arbitration intervened. The strike ended with the defeat of labor in mid–August.[37] The very real threat that the company might close the factory in Wakefield and shift all operations to Gardner put labor in a weak bargaining position.[38]

Surprisingly, the RLG seems to have been oblivious to the unrest in its hometown. Its history prepared in 1926 never mentions the strike, but instead mentions only the normal activities of militia in peacetime during the period. The strike itself, although a large issue in the town, does not seem to have been much more than a curiosity or a nuisance to many of the residents. The rattan works were no longer the dominant force in the town, and the largely immigrant face of its work force made the strike a more manageable event for the town. On 5 July, as the strike was entering its most tense period, the RLG left town for its annual week of field training, in Lakeville, Massachusetts, about 40 miles to the south.[39] The RLG succeeded in avoiding involvement in the labor unrest. Instead, the MVM as a whole was more concerned over a potential war with Mexico, although the men of the RLG still found time to initiate new soldiers and play sports during their eight days in camp.[40] Tensions between Mexico and the United States had been rising since 1913, when General Victoriano Huerta seized power. The situation became critical when US forces began occupying the port of Veracruz in April 1914 to prevent a German ship from unloading its cargo of weapons for Heurta's forces. The town newspaper, which gave a column on the front page each day to the strike, opined that war with Mexico was inevitable, and that Massachusetts regiments would soon be off to fight.[41]

As the labor unrest in Wakefield in 1914 had shown, despite their employment in Lawrence in 1912, the men of the RLG still saw themselves as soldiers ready to fight wars, worthy successors to the earlier members of the RLG who fought in the Spanish American War, or in the Civil War. The introduction of drill pay by the commonwealth in 1914 meant that, for members of the MVM of humble means, militia membership was becoming more a financial matter, and long-term service did not pose as much of an economic burden as in the past. Fighting the battles of the nation, alongside the Regular Army, provided the self-image of the RLG. As events were to show, the Regular Army was stretched thin by the increasing burdens placed upon it. The conquests and annexations from the Spanish American War era left the army with additional missions far removed from the continental United States. In addition to manning coastal defenses and posts in the West remaining from the frontier era, almost half of the manpower of the army was stationed in Hawai'i, Alaska, the Canal Zone, Puerto Rico, and the Philippines. While the war that erupted in Europe in 1914 increasingly threatened to involve the United States, the army became seriously committed responding to a small raid on the country's southern border. In early March 1916, Francisco "Pancho" Villa and about five hundred men raided the town of Columbus, New Mexico. In response, President Woodrow Wilson sent an expeditionary force under Brig. Gen. John J. Pershing into northern Mexico to destroy Villa's army. Smaller raids by bandit gangs from Mexico two months later on Texas border towns led President Wilson to call for National Guard units from every state to protect the border. Although on active federal service, the National Guard on the border was still serving as militia and could not cross the border into Mexico.

During June 1916, Massachusetts mobilized four regiments of the MVM for service on the Mexican border in response to the call from the president. The RLG belonged to the Sixth Regiment, which was not mobilized. However, the mobilized regiments included many men who had only recently joined during a recruiting drive to bring the units to full strength. Captain Edward J. Connelly commanded the RLG at the time, having been elected commander following Captain McMahon's election to major in 1914. To train the recruits, Captain Connelly and two of his officers, eleven NCOs, and the cooks of the RLG were ordered to state active service. Captain Connelly and his men moved to the state camp in Framingham, Massachusetts, to train all the recruits while the four mobilized regiments were preparing for their move to the border. When the four mobilized regiments left Massachusetts for the border, they took with them all their tents and other camp equipment, requiring the Sixth Regiment to use its own field equipment to establish the camp anew for the recruits. The detachment from the RLG trained some ninety recruits for two weeks. Their basic training completed, the recruits left Massachusetts to join their units on the border. Lieutenant Fred Rogers of the RLG, along with a few other officers from the Sixth Regiment, was inducted into federal service and escorted the troops to the Mexican border. After arriving at the border with his charges, he remained in federal service through September 1916, and was the only member of the RLG to serve on the border.

As the bulk National Guard served on the border, Congress passed the *National Defense Act of 1916*, which fundamentally changed the relationship between the Regular Army and the National Guard. Through it, Congress tried to correct many of the shortcomings of the militia left unsolved by the *Militia Act of 1903* regarding the National Guard in its long-sought role as reserve of the army. Although the RLG did not participate directly in the Mexican Border campaign, the shortcomings in the land forces of the United States exposed by the campaign and looming American involvement in the Great War would have far reaching consequences for the RLG. The *National Defense Act of 1916* altered the constitutional status of the National Guard when called into federal service. The new law defined the National Guard when in federal service as forming part of the Army of the United States.[42] Under the new law, the president could augment the Regular Army with National Guard for the duration of a national emergency, referring to its use for more traditional militia functions as in repelling invasion or suppressing insurrection. However, the *National Defense Act of 1916* also overcame the constitutional problems of using the National Guard to accompany the army beyond the national borders. The constitutional objections against sending the National Guard outside of the country were solved by having National Guardsmen take a dual oath, which obligated them to answer a call from either the state or federal government. No longer would members of the organized militia have the option of declining to volunteer for federal service. As members of the National Guard, they were required to enter federal service when the president called.[43] When inducted with their units into the Army of the United States, National Guardsmen were, in effect, discharged from the organized militia of their state and brought into the federal army as part of the National Guard of the United States. When inducted into the army during war or other emergency, National Guard units ceased to be organized militia and were thus unhampered by the constitutional limits on the use of militia. Once inducted, National Guardsmen fell under the sections of the Constitution that gave the federal government the power to "raise and support armies" rather than the militia clauses.[44]

Originally, the law intended that National Guard units inducted into the army would retain their unit designations and the basic structure of their regiments.[45] Unlike the experience of the Spanish American War, where the Sixth Massachusetts Regiment, MVM, had to in effect reform as a the Sixth Massachusetts Volunteer Regiment, under the new law, such legal shifting of unit identities would be unnecessary. To allow the War Department to integrate the National Guard into its war plans, the army, under the authority of the president, would henceforth decide which types of units states were to maintain; prior to the *National Defense Act of 1916*, National Guard companies usually adopted whatever branch they fancied. After the *National Defense Act of 1916*, the army, not the members, town, or state, would decide whether a National Guard unit would be infantry, cavalry, artillery, quartermaster corps, or whatever the army needed.[46] In return for vastly increased federal influence over the National Guard, and in recognition of the increasingly serious nature of National Guard service, Congress authorized federal drill pay, and greatly increased federal funding for the National Guard.

Becoming part of the National Guard was not inevitable for the Richardson Light Guard. Although the *Militia Act of 1903* stipulated that all organized militia in the nation be part of either the National Guard or naval militia, state laws were needed to force existing land-based militia units into the National Guard. In some states, particularly in southern New England, a few largely ceremonial units continued to exist outside of the National Guard. Connecticut maintained two companies of the Governor's Foot Guard and one troop of the Governor's Horse Guard, while Rhode Island maintained several Chartered and unchartered militia companies outside of the National Guard. In Massachusetts the Ancient and Honorable Artillery Company, one of the oldest military units in the world, remained separate from the National Guard.

However the Richardson Light Guard had never placed ceremony and social status above martial training, and so the RLG found itself drawn increasingly into the more centralized control of the National Guard. The RLG continued to be a prestigious organization in the MVM, and even maintained elaborate and unique ceremonial uniforms until World War II. But unlike the Ancient and Honorable Artillery Company, the men who filled the ranks of the Richardson Light Guard increasingly came from the lower-middle class, and even working class, of Wakefield. The increased assumption by the commonwealth of the burden of belonging to the MVM made membership more attractive to men of humble means. By the beginning of the twentieth century, members no longer had to pay dues, purchase a uniform, or pay to attend training camps. Benefits, such as drill pay, were welcomed by the men of the company. The idea of forsaking its status as a real organized militia company of the Commonwealth of Massachusetts and evolving into a private drill company, in which members would again pay dues, never had much likelihood of occurring to the Richardson Light Guard.[47]

For the members of the RLG, the *National Defense Act of 1916* had a more immediate impact on the company than did earlier federal legislation concerning the militia. The *National Defense Act of 1916* brought the Massachusetts Volunteer Militia into the National Guard structure. Members had to take the new dual federal and state oath that would allow the Richardson Light Guard to be taken as a body into federal service and sent beyond United States' territorial limits, without the legal necessity of reforming as a Volunteer company.[48] Captain Connelly had the men assemble for a special meeting on 30 June 1916, at which he explained the ramifications of the new oath they were to take. Under the new oath, members of the RLG no longer had the option of volunteering or not for federal service in the event that the federal government activated National Guard units. After taking the oath, all further federal service would be mandatory when called. All of the men present at the meeting took the new oath, although three men who did not attend later declined to assume this new obligation.[49]

The RLG that took the new oath was still an overwhelmingly Wakefield company, with Reading providing most of the other members, but each decade it drew an increasing percentage of members from other towns. Between 1895 and 1903, a little more than 20 percent of the members came from other towns.[50] From 1903 until June 1916 — the pivotal years for the establishment of the modern National Guard — a total of three hundred and sixty-nine men served in the RLG. Of that number, two hundred and forty-three came from Wakefield, and another thirty-five came from Reading. Ninety-one men, or 25 percent, came from other towns.[51] The percentage of men from other towns had peaked for a while, but the trend would continue after the Great War. At the annual tour of camp duty for the summer of 1916, held on Martha's Vineyard, the men spent two weeks training rather than the eight days of previous summers. During the first week the men practiced battalion drills, with sports on Sunday. For the second week, the men trained in tactical exercises over the northern half of the island. The men could not have known it, as they enjoyed the novelty of the boat ride to the island, bathing in the ocean, and other pleasant pursuits of militia training in peacetime, but by the next summer the United State would be at war with Germany and the RLG would find its training much more arduous and purposeful. But such ominous clouds on the horizon did not dampen the spirits of the RLG in this swan-song of the traditions of the nineteenth-century MVM. The company spent a September weekend in the town of Fitchburg helping to celebrate the hundredth anniversary of the Fitchburg Fusiliers, a militia company even older than the RLG. On 12 October, the company, along with several of the Fine Members, held their annual Fall Field Day and Shoot at what was now called the Bay State Range, at the northern extreme of the town. The next week, on 18 October, the RLG held its sixty-fifth anniversary banquet at the State Armory on Main Street. The evening's program featured a speech by the adjutant general of Massachusetts on the benefits that young men and the nation would reap if it adopted compulsory military training. Also speaking were the commander of the regiment, and Mr. Charles Walton, representing the Fine Members. Mr. Walton was the lieutenant who had to resign on account of his health during the Spanish American War. Through the Fine Members Association, he had remained involved in the RLG since his resignation.

Following the speeches, a ball was held in the drill hall. For the event, the hall had been decorated with American flags and bunting, which hung from the metal frame that spanned the interior ceiling and held the roof up. The ball began at 9:30 in the evening, with a grand march led by Captain Connelly and his wife, and Mr. Walton and his wife. Following the march, the couples attending the ball assembled for a formal group photograph. The photograph shows some three hundred people, formally attired in uniforms, tuxedos, or gowns, posing stiffly for the photographer, who took the photograph from the balcony overlooking the drill floor.[52]

Throughout the fall of 1916, the company remained busy. Drills were increased to almost two evenings a week, with the extra drill earning the new federal drill pay. Privates earned one dollar per drill — a substantial increase over what the commonwealth had been paying them — which was paid every three months. To receive the pay, at least 60 percent of the company had to attend drill, and at least ninety minutes had to be spent actually on the drill floor training. The RLG met all the standards for the pay. A public drill was held in December to exhibit the talents of the company to the people of Wakefield. But the largest event on the calendar was a trip to Washington, D.C., to march in the inaugural parade for President Wilson, who had recently been reelected. For almost a year the company raised money through contests, exhibitions, and other fund raisers, with the members themselves providing the balance. The RLG spent five days in March 1917, traveling to and from Washington by train, sightseeing, and marching for the president along with other units from Massachusetts. While in Washington, one of the lieutenants, Edgar B. Hawkes, left to join the 16th U.S. Infantry on the Mexican border.[53] By the time he rejoined the company that June, the United States would be at war with Germany and the days of banquets, shooting contests, and public drills would be on hiatus and the RLG would be on active service preparing to enter the army with the rest of the National Guard to fight in the world war.

Chapter VII

In the Army Now

The inclusion of the Richardson Light Guard into the National Guard had little immediate impact on the company, but the world wars would emphasize this new standing of the RLG as an institution of the nation more than the commonwealth, and increasingly weaken the bonds that tied the company to the town of Wakefield. Whereas previous incarnations of the RLG formed in the town, left the town for war or strife, and then returned to the town, the demands of modern war shattered the links between town and company. Companies bearing the name "Richardson Light Guard," or its heritage would leave Wakefield in 1917 and again in 1941, but the RLG would be no longer be detraining as a company in Wakefield for a grand homecoming when federal service ended. During the Great War and World War II, the RLG on active federal service would get reorganized, split, and filled with draftees until the Wakefield, or even National Guard, character of the company became tenuous at best. The world wars jumbled men and units until lineages became almost pure mythology. The RLG, as well as countless other National Guard units, would not recover from the experience.

In mid–March 1917, Colonel E. LeRoy Sweetser, the commander of the Sixth Regiment, told Captain Edward J. Connelly, commander of the RLG, to recruit to the anticipated wartime strength of one hundred men. The RLG found recruiting to wartime strength difficult before the actual declaration of war. A call on 22 March for thirty new members brought in only ten potential recruits.[1] Letters to patriotic groups in surrounding towns, posters, and advertisements in the local newspaper were all employed to entice men to sign up for service with the RLG. A parade through the center of town on 27 March, complete with bugle calls to attract attention, brought the company up to eighty-six men and two officers; still below the anticipated wartime strength.[2]

The day before the parade, the town meeting created a War Relief Committee, which included many former members of the RLG and other notable residents of the town, such as the younger Dr. Richardson. The Committee pledged to raise $5,000 to aid families of RLG members in the event the company was called into federal service, and to buy a truck for the use of the company. The truck, a Stuart Motor Truck, was presented to the company on 2 April after a local mechanic prepared it for operation. The purchase of the truck demonstrated the relative wealth of the town of Wakefield, and allowed the town to continue to provide tangible support for the RLG. At the time, most armies of the world, including the vaunted German army which was then occupying most of Belgium, large sections of France, and almost all of Poland and the Ukraine, was tied to horse transport. The truck allowed the RLG to make almost daily runs to the State Arsenal in Framingham for equipment.

In the late afternoon of 30 March, Captain Connelly got the telephone call he had expected for several weeks; the RLG was being called to active service. However the Commonwealth of Massachusetts and not the federal government was bringing the RLG to active duty. Using telephones and motorcycles, most of the men were alerted and they began reporting to the armory on Main Street. Most of the company was ready by 7:00 P.M., although telegraphs had to be used to reach five soldiers who were scattered around the commonwealth. The company marched up and down Main Street that evening, in part in an attempt to inspire recruiting among the more adventurous of the crowds that had gathered to watch. This attempt brought in seven new members.

Although on active duty, the RLG remained in Wakefield for the next week drilling, training, and attempting to get their equipment in order. The United States had not yet declared war. Men who lived in town were allowed to go to their homes at night to sleep; the others slept in the armory. A local restaurant received a contract to feed the men, allowing the cooks of the RLG to concentrate on packing their equipment rather than attending to the daily grind of procuring food, cooking it, and ensuring that all the equipment was cleaned by the end of the day. Many of the soldiers did not avail themselves of the restaurant though, and instead took some or all of their meals at their homes. On 1 April the RLG again paraded through the town center in an effort to recruit. In this endeavor they were joined by Company I from neighboring Stoneham, as well as a fife and drum band. That evening, the RLG attended a patriotic rally in the town of Melrose, which bordered Wakefield to the south, at that town's Memorial Hall.[3] Melrose had only a battalion headquarters company, and thus the RLG could potentially attract new recruits from that town. At the rally, the RLG was joined by Company C of the Sixth Regiment, from Lowell, as well as Lieutenant Fred Rogers, who had become an officer of Company M in Everett since his return from the Mexican Border.[4]

In the days before a centralized training system for new National Guardsmen, each company was responsible for providing basic instruction for recruits. Fortunately, with the Bay State Rifle Range — the former Wakefield Rifle Range — so close, the RLG was able to use its ranges for training new members in marksmanship. The company was still an overwhelmingly Wakefield company, with one hundred and fifty-one of the men who served between June 1916 and August 1917, coming from Wakefield, and another nineteen from Reading. Only thirty-nine — 18 percent — came from other towns, mostly neighboring Melrose, Stoneham, and Lynnfield.[5] Most members of the RLG therefore spent their first week of active service sleeping in their own beds, eating at home or in a restaurant, attending rallies, and training in the armory and at the nearby rifle range. This rather pleasant form of soldiering would not last long.

On 5 April 1917, President Woodrow Wilson brought most of the National Guard onto federal active duty to patrol transportation, industrial, and communication centers in each state. The next day, on 6 April, Congress declared war against Germany. As with previous wars, the town of Wakefield and the men of the RLG reacted enthusiastically to the declaration of war. Mexican Border service involved the RLG only incidentally, while the pursuit of "Pancho" Villa's army by the U.S. Army in northern Mexico had turned into an indecisive and politically difficult occupation. By contrast, the entry into the Great War came as a clear and moral mission. The Great War would give the men in the RLG an opportunity to test their mettle, to see if they too were worthy successors of their fathers and more distant ancestors who fought in the Spanish American War, the Civil War, and indeed all the way back to the founding of the town. In those early days of direct United States involvement in the war, the town of Wakefield again saw the RLG as the institution through

VII. In the Army Now   119

The RLG leaving Wakefield for Fort Revere on April, 7, 1917, and the start of their World War I service. Four trucks were used to transport the men, with three borrowed from local businesses (courtesy Wakefield Historical Society).

which the town would fight a war. The scope of the war and the increased complexity of the army and society would soon alter that perception.

With the declaration of war, the army's Coast Defense authorities in Boston became increasingly concerned about the vulnerability of Boston Harbor to a German seaborne raid. Fort Revere, on the peninsula containing the town of Hull, defended the southern approaches to the harbor. Although the coast defense artillery unit stationed there, a Regular Army unit, could be expected to provide competent service in the event of attack on Boston Harbor, the vulnerability of the fort to sabotage or attack by a landing party caused increasing concern. Eventually the army decided that a National Guard infantry company would augment the fort's defenses sufficiently. Colonel Sweetser, the commander of the Sixth Infantry Regiment, was asked to provide his best company to help defend the fort. He chose the RLG.[6]

On the morning of 7 April, the men of the RLG assembled at the armory to board trucks for the trip south through Boston to the South Shore and Fort Revere. Although the company had its own town-donated truck, more than one were needed to transport the company and its equipment to Fort Revere. Fortunately, the RLG was able to draw more support from its home community. The Fine Members Association, which merged with the Public Safety Committee for the duration of the war, created a Transportation Committee, which convinced several manufacturers from the town to donate the services of their trucks for this patriotic purpose. The Miller Piano Company, which had its factory just behind the Town Hall, lent one of its trucks for the occasion, and the sight of the town's soldiers departing the town in Miller's truck, with its sideboards still proclaiming "Henry F. Miller Boston Pianos," could only bring positive results for the piano company.[7] Advertising more linked to the immediate task was also present, with one of the borrowed trucks sporting a recruiting poster encouraging young men to enlist in the RLG "NOW."

The send-off of the RLG was more emotional than festive, although the company was only going about forty-five miles away. Despite the lack of prior public notice of the departure of the RLG from Wakefield, the area in front of the armory and the old Wakefield mansion next door filled with relatives and well-wishers. The residents—civilians and soldiers alike—bundled against the early morning spring chill. Some waved small American flags, but all expressed in their faces the seriousness of the task ahead. The uncertainty of the future and the likelihood that the company would not be returning for good before going overseas to the bloodbath in Europe must have hung heavy in the air. Still, that the RLG was chosen, before all other companies in the MVM, to augment Fort Revere was a matter of some pride. The convoy reached the fort about 1:30 P.M., after a cold and wet ride for many of the men in the open beds of the borrowed trucks. The captain commanding the fort assigned a storehouse and a gymnasium to the RLG to establish their quarters. The men cleaned out the buildings and began to set up their living quarters while the cooks got busy establishing a kitchen to prepare the evening meal.

Their first days at the fort were cold and snowy, but the men ran a connection to the post's steam plant for heat and strung up wires to provide electricity for light, so all was not gloom and shivering. But the men of the RLG were not at the fort to make themselves comfortable, but to aid in its defense. The men began standing tours of guard duty at the fort, and in detachments at points as far as sixteen miles away securing the water works, the town of Hingham's pumping station, and the cable station, which contained the fire control system for the fort. For the soldiers not on guard duty, a program of exercise, drill, training, classes, and inspections ensued. The men dug trenches around the fort for use in case of attack, and participated in unscheduled tests of their ability to respond to a raid. Despite the teary departure from Wakefield, and the cold of the first days at Fort Revere, the early weeks of the war turned out to be relatively easy for the RLG. Relatives and friends of the men visited often, and many brought gifts. On 11 April, the Company A Lady Associates formed from wives, mothers, sisters, and girlfriends of members of the RLG. Captain Connelly's wife, Kathryn, presided over this support organization, which regularly sent a shipment of fruit, candy, and other treats for the men. Other military-connected organizations in Wakefield, such as the USWV, as well as fraternal organizations, also sent packages to the company.

While the time at Fort Revere in many ways insulated the RLG from the other activities going on in the nation as the small peacetime U.S. Army reorganized itself for a large war, implications of the future intruded into the company. This war was not going to be like past wars, in which the participation of the RLG came in relatively short bursts. This war threatened to be long, and would require an enormous American commitment. The federal government, rather than states, would organize this war effort. Two men were ordered discharged from the RLG so they could resume their civilian occupations, which were deemed vital to the war effort. One worked at an ammunition plant, the other for a firm that made valves for submarines. Three other men were discharged because they were needed to care for their dependents, while one was transferred to another company. Discharges for dependents and other reasons continued to cause men to trickle out of the RLG. In May two sergeants left to attend Officers Training School at Plattsburg, New York. These men eventually became officers in the National Army, and would not return to the RLG.[8] Another man left to enter West Point as a cadet. However, another change brought a familiar face back into contact with the RLG. The Regular Army captain who commanded the fort was promoted and reassigned, and his place was taken by another Regular Army officer, Captain Walker, who was once a lieutenant in Company H of the Sixth, from neighboring Stoneham. On 19 May the army ordered each company to recruit to the new wartime strength of 150 men. Captain Connelly detached Lieutenant Rogers and a corporal and private to return to the armory in Wakefield to recruit new men. The RLG would remain a presence in Wakefield in other ways. On 30 May, Captain Connelly returned for the day with six squads of soldiers to escort the GAR at the annual Memorial Day festivities. However, Captain Connelly needed army permission to do so, and the members of the RLG had to be back at Fort Revere that same day. By 3 June, Lieutenant Rogers was able to send 54 recruits to Camp Darling, in South Framingham, for training. Two sergeants and four corporals were assigned to Lieutenant Rogers to train the new soldiers.[9]

Male residents of Wakefield who were of military age and had previously resisted the call to join the RLG had reason to reconsider joining in May. In response to a request from President Wilson, Congress passed the *Selective Service Act* on 18 May 1917, after a careful study of the failures of conscription during the Civil War. The Act authorized the president to call a half a million men to military service immediately, and another half a million in the future. Initially, voluntary enlistments in the Regular Army and National Guard continued, with the National Guard offering the advantage that recruits would in theory be serving with men they already knew from their community. On 15 December 1917, the army stopped accepting enlistments from men between the ages of twenty-one to thirty, which was the age range of men liable to the draft. On 8 August 1918, the army stopped all voluntary enlistments, followed by the Navy and Marine Corps the next month. In all, about two-thirds of those whose served in the U.S. military during the Great War were drafted as individuals through Selective Service.[10]

The mission of protecting Fort Revere lasted until 6 June 1917, at which time the RLG rejoined the Sixth Massachusetts at Camp Darling, where the new recruits were added to the company and a program of regimental training begun. After six weeks, the Sixth Regiment marched to the newly constructed Camp Devens, in Ayer, Massachusetts. The new camp was named for Brevet Major General Charles Devens, Jr., who served as a Volunteer from Massachusetts during the Civil War and was severely wounded three times. Camp Devens was about thirty miles from Wakefield, but only slightly less than that from Framingham. The twenty-five-mile march took three days under hot and dusty conditions, and the RLG arrived at the new camp as the only company in the regiment with all men present. Unlike Camp Darling, Camp Devens was a federal camp, built for the present war. Clearing the land and constructing the camp had begun only on 18 July.[11]

Although on active federal service, the RLG had so far still been serving as organized militia. The National Guard as a whole, including the RLG, mustered into the United States Army as per the *National Defense Act of 1916* on 5 August 1917. Under the National Defense Act of 1916, National Guard officers held federal commissions in addition to their state commissions, and enlisted men took a dual oath to both their state and the federal constitutions. These changes were necessary to overcome the constitutional barriers to using militia beyond U.S. borders. On 23 August, Lieutenant Edgar Hawkes, who had been in the company since 1902, became the first man from the RLG to go overseas when he was detached from the company and sent with an advance party to France. The men of the RLG, now officially soldiers in the army, were allowed to return to Wakefield on 25 August on a twenty-four-hour furlough before returning to Camp Devens for overseas deployment.

Before the train pulled into the station at the lower end of Wakefield Center, the fire alarm blew the old military call. Most of the town turned out to greet the soldiers, with the Special Committee of Twenty-Five, a subset of the Committee on Public Safety, forming up to greet the soldiers. The president of the committee, Charles E. Walton, one of eight Waltons to serve in the RLG during the nineteenth century, was the same Charles Walton who had served in the RLG from 1880 through the Spanish American War, when he had to resign due to illness.[12] The members of the committee were dressed formally in coats for the occasion, with most wearing panama hats. With five policemen in the lead, the committee, followed by the RLG, paraded up Main Street into the heart of the Wakefield Center, where the parade turned around and marched back down Main Street to the armory. Main Street was again bedecked in American flags for the occasion. At least one of the spectators was also wearing an army uniform, possibly a resident of the town who had been drafted and was now home on leave, and perhaps wondering if he too would not be better off had he joined the RLG when he had the opportunity.[13] At the armory the men of the RLG were released, although they returned that evening for an informal reception for the soldiers and their "lady friends."

The next morning, the company assembled again at the armory. The short march of the RLG to Wakefield station began at ten in the morning, but crowds had been gathering in the Center much earlier. The occasion, although of far more import than the send off to Fort Revere at the beginning of April, had more of the traditional airs associated with earlier times when the RLG left its hometown. Main Street was clogged with residents dressed in their Sunday clothing, while buildings public and private were draped with flags and other patriotic decorations. Significantly, the farewell address was given by the priest from the Catholic Church in town, reflecting how entrenched in this former Puritan town the Catholics had become, and also the impact the heavy Irish immigration into the town was having on the membership of the RLG. Under the command of Captain Connelly the company marched from the armory, up Main Street to the Common, and from there took a sharp left to the main railroad terminal in town, Wakefield's Upper Station. Escorting them were the GAR, the Boy Scouts, the USWV, and other groups in various uniforms. At the railroad station, family and friends of the soldiers wept openly as the men of the RLG boarded the train. The trip from Camp Devens to Wakefield and back again prior to the RLG embarking for Europe was not paid for by the army or by members of the RLG. Instead, the Fine Members Association bore the cost, and even cancelled their annual banquet in order to hold the reception for the RLG. It was a fitting tribute for the RLG from the Fine Members, for it showed the spirit of sacrifice for the men during wartime, and the advantages that a military company with deep community roots like the RLG enjoyed.

Despite those deep and solid roots in the town of Wakefield, the RLG now belonged wholly to the army, and the army would decide how to best use the RLG, and the Sixth Massachusetts Regiment to which it belonged, and indeed all National Guard units. The army was building new divisions for the war, and major transformations were underway in all army units. The National Guard was not balanced, with infantry units predominating, and a lack of other arms or supporting units. Many regiments were destined to be broken up and its companies retrained to form new units for divisions under the new tables of organizations. Many Guard leaders throughout the nation tried to use political pressure to preserve some regiments, but for most, once the decision was made, reorganization was inevitable.[14] To the dismay of the men of the RLG and the town of Wakefield, the army quickly began reorganizing the regiment, with the RLG losing twenty-nine men sent to the

101st Military Police, fifty men transferred to the 104th Infantry, and fifteen sent to the 102d Infantry. All of those units belonged to the 51st and 52nd Brigades in the new 26th (Yankee) Division, which included National Guard units from most of New England. The Sixth Regiment had been rife with rumors of such a change since early August. The RLG had already become something quite different from its prewar version. With about 72 members in early 1917, it had recruited to 100 men around the time the United States declared war. In May the RLG recruited to 150 men. Now, in August, the army was whittling it down to a size where it could no longer be expected to function as a company. By the time the RLG left Camp Devens in early September 1917, it contained just over fifty men. Transfers and discharges continued throughout the war, although the lineage of the RLG continued in theory.[15] None of the men who had served in the prewar RLG would see overseas service with the company. Instead, the 26th Division, containing almost twenty-five thousand men from all over New England, was more the entity through which the town of Wakefield, as all New England, followed the war. The second volume of the history of the RLG, published in 1926, covered the war mostly by tracing the 26th Division, with emphasis on units such as the 101st Infantry and the Military Police that absorbed many of the men of the RLG.

On 4 September, the much-diminished Sixth Massachusetts Regiment left Camp Devens and travelled by train to Camp Bartlett in Westfield, Massachusetts. As the RLG was unloading from the train, new orders came transferring twenty-nine men to another unit, this time to the 101st Infantry Regiment, which was then in camp at Framingham. Unlike Camp Devens, Camp Bartlett was in the western part of the state and not convenient for day trips from Wakefield. Once in camp, the remaining men of the RLG cleared their assigned section and set up their company area. As soon as this was accomplished, the company was moved to a different part of the camp and had to perform the task all over again. Because of the small size of their company, the RLG combined with Company B for drill practice and mess.[16]

Captain Connelly was soon assigned the additional task of camp quartermaster. Instead of taking a detail composed of soldiers sent by all the units in the camp, he was allowed to use his remaining RLG men, now numbering in the teens, as his detail.[17] With their company-owned truck, the men were able to organize the mass of material flowing into the camp efficiently, but being federal soldiers in wartime was turning out to be very different than in the past. Captain Connelly himself received his transfer in late September, taking command of G Company in the 104th Regiment.

After having a few short-term company commanders, the remaining men in the RLG, numbering only seventeen, moved to Camp Greene, in Charlotte, North Carolina. Much to their surprise, North Carolina turned out to be colder than they expected, and the men passed an unpleasant winter. In early February 1918, the RLG and the entire Sixth Regiment moved to Camp Wadsworth, near Spartanburg. Here the rump of the RLG finally lost its old designation as Company A of the Sixth Massachusetts. The remains of the depleted regiment were incorporated into the new 4th Pioneer Regiment. After a few more transfers of men out, the remainder finally departed for France in September 1918. Once in France, the 4th Pioneers was broken up and the men scattered to various units.

The overwhelming majority of the men who had served in the prewar RLG, and even most of the men who joined in the spring and early summer of 1917, served overseas in other companies and regiments in the 26th Division. The division saw action in France at Chemin des Dames, the Toul sector, Aisne-Marne, St. Mihiel, and the Meuse-Argonne. People throughout New England followed the exploits of the Yankee Division carefully, seeing it as their own. As the commander of G Company in the 104th Regiment, Captain Connelly won much admiration for leading two platoons over the top to retake a trench from German storm troopers on the evening of 11 April 1918. Shortly afterward, Corporal Harry E. Nelson, who had joined the RLG in May 1917, was killed by a German shell. The postwar American Legion chapter in Wakefield would be named for him. He was the first man from Wakefield to be killed in the war, although he would not be the last. In the Pas Fini sector, near Chauteau Thierry, in late July 1918, four former members of the RLG were killed, and another twenty-five either gassed or otherwise wounded. In mid–October, another four former RLG members in the 104th Infantry were killed and four wounded in an operation in support of French tanks near Bois Haumont, in the Meuse-Argonne. Of the four, one had been in the RLG since 1912, and the others had joined in the late spring of 1917.

For many though, the war brought new opportunities to prove their martial worth. Fred H. Rogers, who had been in the RLG since 1901 and entered the war as a first lieutenant, left the army as a major. The former commander of the RLG, now Major Connelly, was wounded by German machine gun fire on the day before the Armistice, and left the army with a Silver Star medal, a French Croix de Guerre, and a reputation for courage and sound judgment.[18] As a whole, the men who had entered the army through the RLG could claim great things; not counting men discharged, all but four of the original men served in France before the Armistice, and one of those four did not go because he was then a cadet at West Point. The town tabulated the service of the men who entered the war with the RLG and found that fourteen died in the war, and another forty-eight were wounded. One the other hand, twenty-four were cited for gallantly, and seven were decorated by the U.S. or by France.[19] Nine men from the RLG who entered the war as NCOs eventually received commissions. But the men of the RLG, as with most men in the National Guard, had originally joined assuming that in war they would serve with their neighbors. While most of the men from the RLG served with some men they knew from the RLG, most of their fellow soldiers in their new companies and regiments were strangers in the late summer of 1917.

In all some nine-hundred men and women from Wakefield served in the armed forces during the Great War.[20] With such a large number serving, those who entered as part of the RLG represented a relatively small percentage of residents serving in the war. Those from the town not serving in the 26th Division were scattered throughout the army, navy, Marines, and Coast Guard, which made following the exploits of residents in uniform difficult. The experiences of the entire 26th Division made an easier focus, and thus came to represent the war effort to the town, as the army's constant movement of men out of the RLG made that company far less of an expression of the town in war.

The federal military took control over another Wakefield military institution during the war, the rifle range. Since 1902 the Massachusetts Rifle Association, to which most MVM officers belonged, owned the range at the northern end of the town.[21] Perhaps inevitably, the state and later federal governments had taken an interest in and control of the range within a few years of its establishment. The U.S. Navy, followed by the Marines, began using the range for training their teams for annual national rifle competitions in 1904. The connection between the rifle range and the Marines is perhaps even more significant. Retired Major General William A. Worton of the Marine Corps recalled in 1964 that as a member of the Marine Detachment, Massachusetts Naval Militia, between 1913

Sailors and Marines using the Bay State Rifle Range, possibly while it was designated as Camp Plunkett during World War I, but probably between the world wars, when the Navy used the facility each year to train its marksmanship teams (courtesy Lucius Beebe Memorial Library of Wakefield).

and 1917, he had "voluntarily spent many weekends at the Wakefield Rifle Range learning the rudiments of field service."[22] He believed this detachment, later a company, to be the origin of the U.S. Marine Corps Reserve. The Navy leased the whole 435 acre site for two years during the Great War and renamed it Camp Plunkett. As such, it was the second largest small arms range under Navy control.

With the federal government assuming control over the RLG and the rifle range, the town and commonwealth were effectively divorced from direct participation in the war. However, the scattered men formerly of the RLG serving as part of the 26th Division, and the rump of the RLG which became part of the 4th Pioneers, was not the only RLG in existence during the war. As in the Spanish American War, Wakefield created another RLG for local service during the war. This new RLG was one of the organizations escorting the National Guard RLG to the train station on 26 August. Symbolically, this new, home guard RLG was there to show the departing National Guardsmen and the residents of the town that Wakefield still had an organized militia when the RLG departed the town for distant battlefields.

The Massachusetts Home Guard had its origins more than two weeks before the United States declared war against Germany. On 22 March 1917, Governor Samuel McCall asked for legislation to create such a force. Governor McCall explained his concern for the "bridges, water powers, [and] factories" which would be likely targets in the event of war.[23] The age and conditions of enlistment in the Home Guard were set so as not to compete for recruits with the National Guard, which was then seeking new members.[24] The state law creating the Home Guard passed on 5 April 1917 — one day before the declaration of war.[25] The "Home Guard Law" authorized a militia force of volunteers who were inhabitants of the commonwealth over the age of 35 years, younger men whom through marriage or dependents were not liable to serve in the federal forces, or men physically unfit for federal service.[26] The Commonwealth pledged $200,000 for the force.[27] Although Home Guardsmen received no pay for weekly training, they were to receive pay at the same rate as National Guardsmen when called out on state duty.

The home guard version of the RLG resembled the departed National Guard version, although because of federal conscription members tended to be either older or younger than their National Guard counterparts. Congress had the authority to equip state forces,[28] but due to the vast expansion of the armed forces, the army had no equipment to issue home guards. Instead Massachusetts relied on its own resources, as well as the resources of towns and members, to maintain its new militia. The home guard RLG, like earlier home guard companies, gave former members of the RLG who had been refused for federal service due to health, job, or dependents, a way to show themselves and their community that they were not shirkers or slackers. Wakefield was far from alone in Massachusetts communities creating home guard companies to replace the departed National Guard. Towns did this with little direction from state officials. Communities like Wakefield had been creating their own militia companies for almost three centuries, and had done so as recently as the Spanish American War, less than two decades in the past.

The new home guard RLG of Wakefield had been authorized by the town to recruit up to seventy-five men, most of whom were exempt from federal service due to age or physical disability. Two months later, when the company petitioned to become part of the new Massachusetts Home Guard, fifteen men left the force rather than accept the new, commonwealth-wide responsibilities.[29] Although the Massachusetts Home Guard was not liable for service outside the commonwealth, all units were liable for service anywhere within it. Wakefield's new RLG became Company H, 12th Regiment, Massachusetts Home Guard. From early 1917 through late 1919, the company drilled in the town's armory at least one evening per week. On a few occasions, the men spent a weekend training in the field at a farm in north Wilmington. In addition to training as a company, Company H, following directions from the State Adjutant General's office, began a program of pre-induction training for local Selectees. Along with the entire Massachusetts Home Guard, the company spent 25 through 29 July 1918, training at the state camp in Framingham. The company and its commander won high praise from Lieutenant Governor Calvin Coolidge when he inspected the State Guardsmen.

The war passed without a serious breakdown of law and order in the town or commonwealth. Whether this relative peaceful situation could be credited to the existence of the home guard is doubtful, but the existence of the home guard did have perhaps some deterrence affect on potential dissidents. However the large numbers of young men away in the military and the full employment for those at home probably had more of an affect. In some areas of the nation locally organized groups took on a more ominous, vigilante air, but the Massachusetts Home Guard remained firmly under civic control.[30] The Home Guard company's major public event in Wakefield was marching in the annual Decoration Day parade. Early in the war, the company participated with the rest of the Home Guard in a parade through Boston. On 12 November 1918, the company again participated in a Boston parade with the entire 12th Regiment, this time in celebration of the Armistice. In the spring of 1919, the Home Guard was officially renamed the State Guard, to better reflect its commonwealth-wide obligation.³¹ The RLG returned to Boston, along with the entire State Guard, that April to welcome home the Yankee Division from Europe.³²

On 9 April 1919, almost two years after the National Guard RLG left Wakefield, was a day of celebration for the whole town. A committee of automobile owners drove to Camp Devens to pick up around fifty soldiers from the town being discharged from the army. Unlike similar celebrations during the Civil War or Spanish American War, however, these men came as individually discharged veterans, not as a single company. The returning soldiers were treated to a deafening new cacophony of sirens, bells, whistles, and anything else that made noise. Speeches of thanks were made by prominent citizens, patriotic songs were sung, the men were fed, their pockets filled with cigarettes and cigars, and finally the veterans were brought to their homes.³³ This small group, rather than the RLG, would stand in as the symbolic return of all Wakefield residents from the war. During the next several months, smaller ceremonies were held to pin medals on returning Wakefield men who had served in the war. The medals were created by the town and presented by the Lady Associates of Company A. Wounded veterans who had returned home earlier received their medals shortly after returning, while the bulk were pinned at a Welcome Home ceremony held on Monday, 13 October 1919. That ceremony was a town-wide celebration, featuring a parade through town in which some four-thousand people participated. Essentially, every club and organization from town and many from surrounding towns marched in the afternoon parade to show their support for the veterans. Colonel McMahon served as marshal. The parade began at the armory, marching through the Center and nearby upper-middle-class neighborhoods. The clear day helped bring out a crowd of around 15,000 spectators to watch the parade, said to be the largest Wakefield ever held and surpassing the celebrations at many larger towns and cities.³⁴ Although the Richardson house was again decorated for the event, Dr. Richardson himself did not attend.³⁵

Marching at the front were members of the local chapter of the GAR. The old veterans of the Civil War, bemedaled, bearded, and wearing their dark blue uniforms, formed a color guard to lead the parade.³⁶ The Civil War veterans were followed by the recently returned Great War veterans from all branches of the military. Behind them were the officers from the State Guard company. Also marching were the local chapter of the Red Cross, children's groups, fraternal organizations, and the fire department.³⁷ The parade ended on the Lower Common, at the bandstand, where a ceremony paying homage to the town's heroes was held. Following the ceremony, the veterans returned to the armory for a banquet, and later attended a free show for veterans at the local theater. At the theater, the veterans were entertained by six vaudeville acts and movies featuring Charlie Chaplin and Fatty Arbuckle. This celebration, in October 1919, for all the veterans of the war from Wakefield, filled the public notice of the end of the war that had in past wars been marked by the return of the RLG. With the discharge of the former members of the National Guard RLG from the federal military as individuals, the State Guard RLG became the only RLG embodied in any form.

Until the fall of 1919, the State Guard RLG performed few missions other than training and marching in parades. With the end of the war in Europe, the future of the State Guard became uncertain, but the National Guard would take several months or even years before it was reformed and again became a viable force for local service. National Guard units could be demobilized from federal service back to state service, but the individuals who comprised those units in the summer of 1917 had all been discharged as individuals or did not survive the war. Upon demobilization and discharge of the former members of the RLG from the Army of the United States, individuals were under no obligation to rejoin the RLG. Massachusetts was among a handful of states that felt secure enough with their state forces to adopt a wait-and-see attitude toward the eventual form the National Guard would take. The federal government, for its part, was uncertain of what the post-war military establishment would look like, and if the National Guard would even have a federal role at all in the future. Not until Congress passed the National Defense Act of 1920 would any certainty about the future of the National Guard exist.[38]

The summer encampment of the State Guard RLG in 1919, at Camp Robert Bancroft in Boxford, came after expiration of the original two-year enlistments. Keeping men in proved difficult after the Armistice. By the spring of 1919, the State Guard RLG counted only thirty-three members. The end of the draft in November 1918 meant that the State Guard could recruit from men previously liable for induction into the army. Vigorous recruiting brought the numbers back up, but left the company full of young men with little previous military experience. One of the new men was who entered at this time was Cyrus Wakefield, grand-nephew of the man for whom the town was named, who served in the State Guard RLG from 16 July to 27 October 1919. Unlike most other State Guard companies, the RLG came to their second summer camp with three officers and sixty-one men, but it was a far younger and less experienced company than it had been.

Unexpectedly, most serious unrest occurred after the Armistice. In 1919 Massachusetts faced a breakdown in law and order in Boston as a result of the police of that city going on strike over dismal pay and long hours. The Massachusetts State Police, which dated back to 1865, was still a small force, but the governor could also draw from the Metropolitan Police in an emergency. As the negotiations between the Boston police and the police commissioner reached an impasse, the commissioner began preparing for a strike. In August the city began advertising for volunteer policemen to serve during a potential strike. The commissioner believed that a few hundred volunteers would enough to augment the eight-hundred or so police he anticipated would refuse to strike.[39] However, of the 1,544 men on the Boston police force, 1,117 left their posts following the afternoon role call on 9 September 1919.[40] The Metropolitan Police supplied one hundred men for the emergency in Boston, while the State Police sent an additional 60.[41] The State Police was more professional than the Metropolitan Police but lacked the manpower to assume completely the police duties of Boston. The volunteer police, consisting mainly of middle-aged men from the upper-classes of the city, was not of the size or mettle for tedious and lengthy service. Shortly before midnight after the walkout, unrest begn to spread through the city. Numerous businesses and homes were broken into by rougher elements of the city intent on taking full advantage of the lack of police on the streets. The next day, the mayor took control of all units of the State Guard then in the city, and requested that the governor mobilize more State Guard regiments.[42]

On 10 September the State Guard RLG received orders to assemble for service in Boston.[43] By 1:30 P.M., the RLG assembled at the armory in Wakefield and departed for Cambridge to join the rest of the regiment. At Cambridge, one lieutenant and thirty-nine men from an under-strength company were attached to the RLG, bringing its strength up to four officers and ninety-three men. The company then went to the Brighton section of Boston where they occupied a police station. Shortly after they arrived, in consultation with the local police captain, the company moved out to disperse a mob blocking Market Square and the adjacent streets. After restoring quiet to the area, the RLG set up posts at strategic positions along the streets and began to enforce order in their district, which ran from Market Square to Oak Square to the Charles River. In addition, they patrolled the "more quiet residential sections adjoining Brookline."[44]

On September 13, the Adjutant General's Office increased the authorized strength of each company to one hundred men. The RLG detailed a sergeant and clerk to Wakefield to recruit thirty-seven new members. The first day eight men enlisted, most of whom were veterans recently discharged from the army after returning from Europe, while the rest were teenagers. However, only one of the recruits, Leon A. Blanchard, had originally entered federal service with the old RLG in 1917, having originally enlisted in the RLG on 29 March 1917. In the Great War he served with the 101st Military Police, and the 104th Infantry. He was seriously wounded in May, 1918, and discharged from federal service on 6 March 1919. He served as a corporal in the State Guard version of the RLG from 13 September 1919 until 31 January 1920. The following day the RLG moved from Brighton to the armory in Cambridge. The RLG later served a few weeks in Roxbury before returning to Cambridge.

The same day, the reorganization of the National Guard version of the RLG began, when Colonel Sweetser, who had commanded the old Sixth Regiment before the war, asked Major Connelly to begin recruiting the new company. The town newspaper assumed this new company would also be known as the Richardson Light Guard.[45] The status of this new version of the RLG was apparently uncertain, in that the War Department hoped that the National Guard would not be recreated in its pre-war form and would instead be replaced by a wholly federally-controlled force.[46] As a result, some confusion existed over whether the new RLG, indeed all the new companies, would be part of the National Guard, or the Massachusetts Volunteer Militia.[47] Major Connelly sought men who had previously served in the military for the new RLG, although that would put it in direct competition for recruits with the State Guard company. Within two days, thirty-five veterans had joined the new company, with drills scheduled to commence in January. Local opinion held that the State Guard was about to be disbanded, and thus having a National Guard version of the RLG was an urgent matter. About six thousand men joined the new National Guard across Massachusetts, and companies and regiments began forming, while the State Guard remained on active duty in Boston.

In mid–October, headquarters ordered the discharge of all State Guardsmen less than twenty-five years of age. By October 25, the situation in Boston had stabilized enough that half of the State Guardsmen were relieved from active duty and returned to their homes. From then until the final removal of all State Guard forces from the city, units underwent constant restructuring as reductions continued. The RLG allowed members with pressing business to return home first. By the end of the State Guard's involvement in Boston, only men who really wanted to remain on active duty were required to do so. On 18 December, the State Guard was demobilized. The remainder of the RLG returned to Wakefield and the company continued armory drills into 1920, but interest waned and attendance dwindled. During the winter of 1920, the new, National Guard version of the RLG began drilling. In May, the State Guard RLG participated in town ceremonies for the last time. The State Guard version of the RLG mustered out on 1 November 1920, when the entire State Guard was disbanded in a ceremony in Boston. The new Richardson Light Guard, part of the National Guard, took its place as the town's organized militia. Major Connelly had been serving as acting commander, but declined to accept the position permanently due to business concerns, and instead Captain James Brown became the commander. Only one man from the State Guard RLG joined the new National Guard version, but nineteen men who entered active federal service with the RLG in 1917 rejoined the postwar National Guard RLG.[48]

The economic base of the town was changing after the Great War. The merger of the Wakefield Rattan Company and the Heywood Brothers had turned out to be a bigger boon to the town of Gardner, home of the Heywood Brothers, than to Wakefield. The other rattan company in Wakefield, the A. D. Jenkin's and Company, evolved into the Jenkins-Phipps Company after Jenkins partnered with a salesman named Irvin E. Phipps, who left the Heywood Brothers and Wakefield Company around 1903. The new company sold out to Pierce and Foley in the late spring of 1918. Under new ownership the factory in Wakefield expanded, becoming the American Reed and Willow Company, with its main manufacturing plant at the southern end of Wakefield Center, near what had once been the southernmost corner of the Cyrus Wakefield estate. Although the American Reed and Willow Company, and the much larger Heywood Brothers and Wakefield Company ensured employment for rattan workers in the town of Wakefield, they were only local branches of large concerns now owned and run by distant businessmen, men who were not part of the community.

With the town's population expanding, the high school in the Layfayette building was no longer adequate for the number of students attending it. After much discussion, the townspeople decided to build a new high school. With the growth of businesses and residential neighborhoods around the center of the town, few remaining centrally located sites existed. The last sizable open area of land was the old Wakefield Estate, then in disrepair. The town purchased it in 1913 from George L. Wakefield, son of Cyrus Wakefield II, for $25,000. At the 1917 Town Meeting, the town voted to raze the buildings and build the new high school on a portion of the land, near the site of Cyrus Wakefield's mansion. The town voted over a quarter of a million dollars to create a modern high school. The project had to wait a few years due to the war, but the days of the Wakefield Estate were numbered. After the U.S. entrance into the war ended all major construction projects, the town of Wakefield put the old mansion to use for war-related activities. The town's insurance carrier wanted the building occupied anyway, so the town's Red Cross headquarters, Food Conservation, and Civilian Relief established themselves in the former home of Cyrus Wakefield. In 1919, when the town began to implement their long-term plans for the site, the Town Meeting raised the total amount of money to be spent building the new high school on the site to almost a half of a million dollars. On 17 October 1921, Cyrus Wakefield's mansion, less than sixty years old, fell under the hammer so that construction could begin on the desperately needed high school. In 1923, the new high school opened. Military training continued at the new High School, which bucked an anti-"guns in schools" movement that had been gathering in strength since 1911, while supporters of marksmanship training in schools argued that it helped make Americans out of immigrants.[49] In 1924, the cadets received new uniforms from the Junior Reserve Officers Training Corps, to which the cadet program was now aligned. The cadets showed off their new uniforms, which were similar in appearance to the uniform of the army prior to the Great War, at the town's 1924 Loyalty Day parade, held on May Day.

The general anti-militarism that swept the United States in the 1920s had little impact on the relationship between the town of Wakefield and the RLG. The town continued to support financially the RLG throughout the inter-war years, although increased federal support had made town contributions nice but not essential. Under the *National Defense Act of 1920*, the federal government continued paying National Guardsmen for armory drills, with privates still earning one dollar for each evening drill, to be paid every three months. Despite this important step in the assumption of more responsibility by the federal government, the town continued to give its company financial support. In 1920, for

Tents on the Bay State Rifle Range. One of the state-built structures, just behind the firing line, is visible on the left (courtesy Nancy Bertrand).

example, the town budget included $500 as "military aide."[50] Most of the town funds were used in practicing for and attending target shooting competitions. As with most National Guard units of the period, marksmanship received much attention between summer annual training periods. The close proximity of the Bay State Rifle Range gave the RLG an advantage in rifle training over most National Guard companies, and indeed the entire nation. It also provided a link with the federal military throughout the interwar years, indeed since the beginning of the century, as the Navy and Marines continued to use the site for training their rifle teams.

In 1926, the Commonwealth, fearing that a proposed golf course in the area would bring about the end of so fine and useful of a firing range, had the Armory Commissioners buy the Bay State Rifle Range for $64,000, and turn it over to the state Quarter Master to run it.[51] Governor Alvan T. Fuller renamed it by general order "Camp Curtis Guild" after the former brigadier general in the MVM and Republican governor.[52] The name "Camp Curtis Guild" had been used during the Great War for a mobilization camp of about four square miles just south of state Highway 133 in southwest Georgetown and Boxford, ten miles north of Wakefield.[53] Since the end of the war and the reversion of that camp to its prewar status in 1919, the name had been unused.[54] With its new name, and with its front gate now opening into the town of Reading, the former Wakefield Rifle Range lost much of its association with the town of Wakefield.[55] Its origins in the decision of the town of Wakefield in 1891 to support the RLG was largely forgotten.

During the years between the world wars, the town began building monuments to the recent wars. A ceremony on 23 May 1920 in front of the Greenwood School dedicated a boulder with a bronze plaque listing the men of the village of Greenwood who served in the Great War, with a special service for Robert Gibbs, who died in the war. Gibbs had not been a member of the RLG. Major Connelly, who was the first commander of the Corporal Harry E. Nelson Post 63 of the American Legion in Wakefield, conducted the unveiling and read the list of names on the tablet. Perhaps spurred into action by the activities of the smaller village of the town, Wakefield unveiled a boulder on the Common with a bronze

Wakefield World War Memorial, on the southern end of the Common (courtesy Michael Boucher).

plaque with the names of all from the entire town who died in the Great War. Again Major Connelly took a prominent role in the affair. Unlike the single war death recorded on the Greenwood boulder, the town boulder included the names of twenty-nine residents. Of those, ten had formerly been members of the RLG. The prewar National Guard made up only about 10 percent of the total wartime Army of the United States, but 34 percent of

VII. In the Army Now

Greenwood World War Memorial, dedicated May 23, 1920 (courtesy Michael Boucher).

Wakefield soldiers who died in the war entered the army as members of the Richardson Light Guard.

The Great War had been a terrific shock to the company, not just for the casualties suffered by former members but also for the way the company had been reorganized out of existence. The service of former members was spread over dozens of units in the army. To some extent, the experience of the State Guard version in the 1919 Boston Police Strike brought back a sense of unity, but even that company had trickled back to Wakefield, and received no town-wide celebration on its return. And further, although the State Guard company originally received many former members of the National Guard version of the RLG, only one former State Guard member joined the new National Guard version of the RLG. Still, following the demobilization of the massive wartime army in 1919 and 1920, and the reorganization of the National Guard, the RLG was again the military face of Wakefield.

Chapter VIII

The Oldest Regiment

In 1923, following a few years of shifting company designations and regimental affiliation,[1] the RLG became Company E of the 182d Regiment, the successor to the old Fifth Massachusetts Regiment, recognized by the Army in 1926 as the oldest military organization in the nation. The 182d Regiment was formed on 26 March 1923 and officially designated by the U.S. Army as the successor to the old Fifth Infantry Regiment of the MVM and later Massachusetts National Guard. The Fifth Massachusetts Regiment traced its lineage to the North Regiment created by Massachusetts Bay Colony in 1636, and therefore predated the Ancient and Honorable Artillery Company of Massachusetts and is the oldest military lineage in the United States. The old Reading militia company had belonged to the North Regiment when it was created in 1644, and so placing the RLG in the regiment had precedence.[2] Unlike earlier regimental designations, which were regiments in the state militia, the designation "182d" came from the U.S. Army. The newly christened 182d Regiment was one of the four infantry regiments in the reconfigured 26th "Yankee" Division in the National Guard. Most of the men who were mobilized as members of the RLG during the Great War served overseas in the Yankee Division. The RLG seemed to rest on as solid a foundation as any military organization in the United States. Lying ahead was the Great Depression followed by World War II. The latter, far more than the former, would bring the RLG to new heights of glory and at the same time break its bonds to the town of Wakefield.

More than the company designation came from outside the town. From settlement of the town until 1668, and again from 1774, militia companies raised in the town—enrolled militia, active militia, MVM, home guard, or whatever—chose officers by election. However, Captain Connelly's election of 1914 was the last time the men of the RLG elected their own commander.[3] In most states, enlisted Guardsmen seeking commissions were encouraged to take the Army's Series 10 correspondence courses in order to qualify, but Massachusetts took a more direct to approach to preparing future officers. In 1913, Massachusetts became the first state in the nation to establish a training school—what would later be called an Officers Candidate School—as a source of National Guard officers. The school had been suspended at the entrance of the United States into the Great War, but reopened in 1927.[4] Graduates were placed on the statewide list of those eligible for commissions in the National Guard, and were commissioned as vacancies occurred.[5] Even while the school was on hiatus following the Great War, records of the RLG refer to Thomas J. Quinn being "chosen," captain, although who or what was doing the choosing was not recorded.[6] However, Quinn, who had been in the RLG since 1914, had attended an officers training school while on active duty during the Great War and was commissioned in the infantry. Despite the increased professionalism in the selection of officers of the RLG, and concurrent loss of local control, 1926 marked three quarters of a century since the founding of the company, and again the town and company paused to celebrate the past.

On Columbus Day, 12 October 1926, the town held a dual celebration for the seventy-fifth anniversary of the RLG, and to dedicate one of the more conspicuous war monuments in the town, this one to the Spanish American War and other military campaigns from the turn of the century. The town again seemed to be a thoroughly military town, with throngs of generals, colonels, majors, and captains on the reviewing stand and on horseback near the stand. Most of the officers earned their ranks in the Volunteers or National Guard. The host of all this martial activity was again Colonel Gihon, who had commanded the RLG during the Spanish American War. He held memberships in the USWV, the Fine Members, and the Elks, as well as an honorary membership in the GAR. He had been appointed the deputy commissioner of aid and pensions for Massachusetts in 1918, and would keep that position for most of the remainder of his life.[7] He had become the most famous "old soldier" in Wakefield.

The anniversary celebration began with a target shoot at Camp Curtis Guild, with members of the RLG competing for medals such as the Gihon Medal and the Carpenter Medal, named for previous commanders. The Fine Members also turned out to compete with the active members. In the afternoon the town held a large military parade with four bands. The line of march included police, National Guard battalions, Marine Corps Reservists, the high school cadets, and local veterans groups, as well as the Girl Scouts and Boy Scouts.[8] The parade worked its way around the Center, past the reviewing stand erected on the Rockery, where the monument awaited unveiling. The Rockery contained a fountain with a small pool in front of the half-dome of boulders, with a walkway down the open area in front of the fountain. The site was an ironic one for a monument to the men of the town who served overseas at the turn of the century; just before the American declaration of war against Spain, some residents of the town attempted to have the Rockery destroyed as an impediment to traffic and a general eyesore.[9] But the Rockery continued to exist, as it still does, located across a short street from the southern point of the Common, where Main Street forks, with the eastern branch wending around the eastern shore of the lake, while the western branch becomes Common Street and intersects Church Street. There, in front of the fountain, the new monument was placed. The steps onto the Rockery list the names Porto Rico, Cuba, Guam, Philippines — the theaters of the Spanish American War.

The monument itself consisted of a statue of a soldier created by Theo Alice Ruggles Kitson. The statue, known as the *Hiker*, is a bronze man, slightly larger than life-sized, wearing the slouch hat and carrying a Krag-Jorgensen rifle of the war with Spain, yet reminiscent of the statue of Minuteman Captain John Parker on Lexington Common created by her husband, Henry Hudson Kitson.[10] The *Hiker* statue became one of the most recognizable symbols of the Spanish American War. A total of 52 similar statues were produced between 1921 and 1956, and others are displayed in Allentown, Pennsylvania, the Texas State capitol grounds, Arlington Memorial Bridge, the University of Minnesota, Capital Park in Sacramento, Bronson Park in Kalamazoo, and Penn Valley Park in Kansas City. The statue in Wakefield Center stands on a small boulder bearing a plaque stating that the monument served as a memorial to the "men of this town who served in the war with Spain, Philippine Insurrection, and China Relief Expedition of 1898–1902." Following the dedication of the memorial, the RLG hosted a large banquet at the armory to cap the events of the day and to mark the milestone in the life of the company. The host, Captain Thomas J. Quinn of the RLG, thanked the town for its "moral and financial support" given to the

Close-up of the Hiker stature by Theo Alice Ruggles Kitson on the Spanish American War Memorial, dedicated on October 12, 1926 (courtesy Michael Boucher).

RLG. The town had contributed $5,000 for the dedication ceremony, with the Fine Members raising another $1,000 for the RLG celebration, while the USWV raised even more funds for the day's activities.

Absent from the festivities were any members of the Richardson family. Solon O. Richardson, former member, Fine Member, and longtime booster, died at his home in Wakefield on the afternoon of 2 October 1922, after a long illness.[11] He was buried in Wakefield, with former officers of the RLG such as Colonel Woodward, Lieutenant Colonel Connelly, who had rejoined the National Guard, and others, serving as pallbearers, but with no direct involvement of the RLG. With his son and namesake, as well as his other son Dana, long since removed to Ohio, the Richardson family had no remaining immediate ties to the company or town.[12]

The glories of the RLG in the conquest of Puerto Rico were two and a half decades, and a world war, in the past. To commemorate the seventy-fifth anniversary of the RLG, the town published a second volume of the history of the company, covering the years from 1901 through 1926. The Great War fitted awkwardly into the history, but the RLG had recovered after the war. As with the 1901 volume, the company produced a handsome volume, hardbound and full of photographs. The end of the book, in summing up the close working relationship between Fine Members, the town monument committee, and the USWV, over the planning for the joint dedication of the monument and observance of the seventy-fifth anniversary of the RLG, mentions that "harmony prevailed" at the meetings because all saw that "the Richardson Light Guard was recognized as a town institution."[13]

In 1931, the past glories of the RLG were again brought back, this time by the Commonwealth of Massachusetts. The commonwealth had commissioned artist Richard Andrew to paint a mural in the state house in Boston commemorating the Sixth Massachusetts Regiment during the Spanish American War, as part of a series of murals depicting Massachusetts soldiers in various wars. The result, *Veterans of the Sixth Regiment Memorial: Landing in Porto Rico*, was unveiled in October 1931.[14] The painting featured Captain Gihon as the central figure leading the men of Company A — the RLG — from their ship. A local hero, and indeed the entire company, was thus featured as the best of Massachusetts men at war.

The RLG was more than seventy-five years old, and had served in three wars and had twice helped maintain order during urban strikes. Fundamental shifts in the Massachusetts National Guard were underway. Widespread ownership of automobiles allowed Wakefield men to travel to several surrounding towns to attend evening armory training, but it also allowed the RLG to draw from a larger geographic base.[15] Whereas in the decades before the Spanish American War the peacetime RLG had drawn almost all of its regular members from Wakefield, with the bulk of others from the neighboring town of Reading, the predominance of Wakefield and Reading men continued to decline between the world wars, and the trend became even more pronounced after World War II. Between 1920 and 1941, a total of one thousand, forty-three men served in the RLG. Of these, five hundred and seventy-eight, or just over half, listed Wakefield as their town of residence. Another eighty-seven came from Reading, and ten from North Reading. However three hundred and sixty-eight, more than a third, came from other towns, mostly from Malden or Melrose, but with some from all over Massachusetts.[16] Competition for members also came from within the town. In 1926, residents of Wakefield who wanted to attend drill in their hometown had a choice of two units to join, when the medical detachment of the 182d Regiment, which was about half the size of a company, began sharing the armory.[17] While the RLG normally held drill on Monday nights, and the Medical Detachment met on Tuesday evenings, National Guard Day was held on Saturday, 19 September 1930, with both units participating. Congress had declared National Guard Day to aid recruiting. As part of the observance, Company E and the Medical Detachment held a joint drill in the armory that was open to the public. Visitors got to see displays of equipment and to watch a demonstration of medics treating simulated casualties on the infantrymen.[18] Perhaps significantly, the local newspaper coverage of the event did not use the name "Richardson Light Guard" for the infantry company, calling instead by its official Army designation, "E of the 182d Infantry."[19]

Despite this sign that the RLG was losing its identity as a Wakefield institution, the National Guardsmen proved to be a local asset a few days later when a hurricane hit the town. Following several days of rain and wind, downed trees blocked roads and the town was left without power. Schools closed for several days, and the Methodist Church was damaged beyond repair. The selectmen took the unusual step of calling up the Guardsmen — both Company E and the medical detachment — to town service. The Guardsmen were activated on the evening of Monday, the 21st, via the town fire whistle using the militia call of "5–5." The Guardsmen worked until dawn chopping wood to clear roads of downed trees.[20]

In 1931, the War Department reclassified Camp Devens as a fort, which indicated that the Army considered the federal post as permanent and not simply a wartime installation.[21] With little chance that Fort Devens would be given to Massachusetts as a state training area, the commonwealth in 1935 established its own large permanent training area for the Massachusetts National Guard. Previously, annual training had normally been conducted using

"LANDING IN PORTO RICO IN 1898 CARRYING THE FLAG OF A UNITED COUNTRY EXTENDING THE JURISDICTION OF THE UNITED STATES OVER A NEW EMPIRE AND BRINGING MANY MILLIONS OF OPPRESSED PEOPLE WITHIN THE DOMAIN OF LIBERTY."

Sixth Regiment Memorial: Marching Through Porto Rico, mural by Richard Andrew, 1932, unveiled at the Massachusetts State House in October 1931, showing Company A of the Sixth Massachusetts landing in Puerto Rico in 1898. The man in the khaki shirt with raised pistol represents Captain Edward Gihon. The mural was part of a series representing Massachusetts soldiers in each war (courtesy Commonwealth of Massachusetts).

rented farmlands or fairgrounds, with training in large tactical problems often carried out along public roads. But the increased lethality of modern weapons and the need for Guardsmen to train off-road led to the establishment of the Massachusetts Military Reservation on the scrublands of what was then the relatively sparsely populated western end of Cape Cod. The cantonment area was designated as "Camp Edwards," after Major General Clarence R. Edwards, the Regular Army officer who commanded the Yankee Division during the Great War, and who has died that year. Throughout the rest of the 1930s, the federal and state governments constructed a variety of permanent buildings as well as two landing strips at the camp, mostly through the Works Progress Administration (WPA).[22] The much smaller Camp Curtis Guild continued to be an important training center for the Massachusetts National Guard, especially its small arms ranges, but it had definitely slipped in importance as a training area.

Still, the proximity of the rifle ranges on Camp Curtis Guild did assist the RLG — or "Company E," as even members usually referred to it — in maintaining a high level of proficiency with rifles. The company's competition rifle teams consistently scored high in regional and national competitions since the founding of the company, despite cutbacks in federal support to the National Guard. Because of the Depression, the federal government cut the number of drills for which it would pay National Guardsmen from forty-eight to thirty-six annually.[23] In 1935, the number of paid drills was returned to forty-eight per year, which Company E continued to hold on Monday evenings. Throughout the years of reduced federal support, the town continued to support rifle training. The rifle teams gave the town a way to show its continued support for the RLG after the state and federal governments had absorbed most of the burden for supporting the company. In 1931, at the height of the Depression, the town still gave the RLG $550, as it had for several years, to support target practice.[24] The town's continual support, closeness to the rifle ranges, and a tradition of marksmanship paid dividends in competitions. In 1934, the team won the team match sponsored by the National Rifle Association that August. The following October, it placed second, losing by one point, in automatic rifle competition in the regiment, and came in first in rifle competition.[25] Such honors, while common for the RLG, nevertheless were a source of pride for the RLG and the town of Wakefield.

More subtly, the National Guard as a whole underwent a further change in its relationship with the federal government in 1933. The *National Defense Act of 1933* changed the name of the Militia Bureau to the National Guard Bureau, finally ending the use of the term "militia," which had long ago become discredited in many parts of the nation. More important, the 1933 Act stipulated that National Guardsmen belonged to both the National Guard of their state, and also to the National Guard of the United States. The new act eliminated any ambiguity over the dual oath created in 1916. In theory, this new change meant that the federal government would call the National Guard into federal service as units, with membership intact. It also meant that National Guard officers already had federal standing, and thus no longer needed to simultaneously belong to the federal Officers Reserve Corps in order to ensure their standing as officers when called into federal service.

Despite federal changes to the relationship between the federal government and the National Guard, the viability of the Guard remained dependant on local enthusiasm for bringing in members. The RLG lost an important means of ensuring that the next generation of young residents from Wakefield became interested in serving in the RLG in 1931 when Wakefield high school ended military training. Major McMahon would long regret the end of military instruction at the high school, noting that all the cadets he had selected for leadership later distinguished themselves during the Great War.[26] The dissolution of the cadet companies came not from high cost, lack of interest on the part of students, or opposition to militarism, but because the continued growth in the student population had led to crowding at the high school, and simply no room or time was available to continue the corps of cadets. In addition to those from Wakefield, the high school also served students from the neighboring town of Lynnfield, although the population of Wakefield far exceeded that of relatively rural Lynnfield. The population of Wakefield had been growing for almost a century, until a sort of natural limit on growth had been reached at more than sixteen thousand in 1930.[27] The town was in danger of becoming choked in its own success, as was much of eastern Massachusetts.[28]

The look of the town was continuing to change. Ice houses, once common along the shore of area lakes and ponds, became increasingly obsolete. The Philpot Icehouse on Crystal Lake burned in 1911, before the Great War, and was not rebuilt. In the decade after the Great War, electric refrigerators became increasingly common in American households, and the typical middle-class family that had previously taken daily ice deliveries was the same type of family that bought the new electric ice boxes. A massive fire on 26 September 1929 burned the Porter-Milton Ice House, on the southwestern shores of Lake Quannapowitt. This ice house was likewise not rebuilt, and instead the town purchased the land and turned it into athletic fields, named Veterans' Field, using WPA funds.

But even more so than fire or the Depression, long-term changes in national tastes altered the economic base of the town. Wicker furniture, once so fashionable, had become bargain basement material. The market was flooded with shoddy wicker products, increasingly of man-made materials, which often began unraveling before the consumer got the furniture home. With the increased association with lower-end goods, the public demand for quality wicker declined to a point where the town of Wakefield had no remaining wicker factories. In the late 1920s, the Heywood Brothers and Wakefield Company was the last manufacturer producing wicker furniture by hand, but this practice was abandoned in 1929. A year later, the company ended almost all operations at its Wakefield complex,[29] while maintaining decreased operations in the central Massachusetts town of Gardner. Only the jute mat manufacturing part of the business remained in Wakefield. Most of the large brick buildings that had once housed the Wakefield Rattan Company works had long been rented or sold and now were filled with a variety of smaller industrial concerns. Where once local workers made nationally famous wicker furniture, now smaller companies made valves, electrical switches, and shoe related items, among other products. A few service businesses such as a lunch counter, a laundry, and an auto repair center also operated at the site. By the late 1930s, the American Reed and Willow Company building stood empty, joining other large brick buildings around town that were increasingly seen as blights on the land. But some of the other industries in the town thrived at the end of the 1930s. The large shoe factory of the L.B. Evans' Sons Company survived the Depression and began operating at full capacity as Europe again erupted in war in the late 1930s. Likewise, the Winship-Boit Company, making underwear, prospered during the world wars, and survived the Depression intact, becoming for a while the town's largest employer.

Widespread ownership of automobiles brought a more fundamental change to Wakefield, indeed all of eastern Massachusetts. The greatest physical impact of the car came with the building of a road that would alter the landscape in ways never dreamed of by all but a few. The new road would bring changes to Wakefield on a scale that rivaled the opening of the railroad in 1845. The new road was state highway Route 128. The genesis of the new road came in the mid–1920s, when Massachusetts copied the new federal practice of assigning numbers to existing roads to create longer routes. Route 128 circumscribed Boston roughly eight miles from downtown, running through the centers of suburbs, including Wakefield. But initially the route was little more than an exceedingly crooked line on maps, and certainly not practical for high speed travel.[30]

In the early 1930s, William F. Callahan, the new commissioner of the Massachusetts Department of Public Works (MassDPW) unveiled his plans for relocating Route 128 a few miles further out from Boston, away from built-up areas, building a new roadway through what was then mostly farmland, forests, and abandoned railroad beds. The new road would be a limited access-divided highway around Boston, the first "beltway" to be built in the United States. Construction began piecemeal in 1936, during the Great Depression, and by the entry of the United States into World War II, a short completed section began at the

Wakefield-Lynnfield border and ran east to Danvers. The declaration of war by the United States in December 1941 halted further construction until after the war. However the plans for the new highway, as well as the completed section to the east of Wakefield, remained during the war years as harbingers of things to come.

The 182d Regiment spent the year 1936 concerned more with its past. In that year the status of the RLG as a member of the oldest regiment in the United States kept the company busy with ceremonial duties as well as state emergency service. On Monday, 16 March 1936, Company E underwent its annual federal inspection. Captain James Brown, the commander, invited the Fine Members to the armory for the event. Former commander, Edward Connelly, who at the time served in the Massachusetts legislature as the representative for the 19th Middlesex District, which included Wakefield, also attended.[31] However, rain had been falling for the past several days, and coupled with melting snow, some areas of eastern Massachusetts were starting to flood. On the evening of the 20th, Company E and the medical detachment were called to state active duty, along with the entire 182d, for flooding in the Merrimack River Valley. Again the old militia call was struck on the fire whistle, but due to the greater area from which the company drew its members, the police from nearby towns of Reading, Stoneham, Andover, and Melrose had to assist in locating members and informing them of the mobilization.[32] Except for two on sick leave, and three members in the WPA, all members reported. Captain Brown was in the hospital for a minor ailment when the call came, so the company was under the command of First Lieutenant Richard B. Dolbeare, whose family owned the local press and newspaper.[33]

The newspaper reported that "war-time scenes [were] re-enacted at armory — military call brings out big crowds," estimated in the hundreds.[34] This time the paper referred to the infantry company as "Company E, 182d Infantry (Richardson Light Guard)," again emphasizing the changing identity of the company.[35] The 182d Regiment spent almost fourteen days on state active service, with Company E performing duty for ten days in Lowell as part of the effort, most of it protecting a hospital.[36] Company E returned to Wakefield on 30 March. The next day, the *Daily Item* published on the front page a letter signed by members of the company thanking the town for its assistance and support.[37] The medical detachment returned a few days later.

More pleasant duties filled the remainder of 1936. The year marked the three-hundredth anniversary of the regiment, and the occasion was observed by a regimental banquet, followed by ceremonies in Concord on 19 April to commemorate the opening battles of the Revolution. On 30 May, the regiment attended a ceremony on Boston Common to honor the four living veterans of the regiment from the Civil War, none of whom had served in the RLG. The last local veteran of the Civil War, Benjamin Day, a musician who served with the RLG in the Port Hudson campaign and for a few years after the war, had died shortly before.[38] At the ceremony, Company E wore its nonstandard ceremonial uniforms, which harkened back to the eighteenth century, and "attracted much favorable comments in the Boston newspapers."[39]

The unique dress uniforms were one of the last vestiges of Company E's separate identity in addition to its status in the National Guard. During 1939 and 1940, the federal government published a commemorative book for the National Guard of each state. As in all volumes, the Massachusetts book included pictures of all armories, training sites, and most important, units. Page after page shows units assembled for formal group photos, and individual photos of the officers, all dressed in nearly identical standard U.S. Army dress uniforms, with minor differences such as riding boots and pants for cavalry and artillery units.

Company photograph of the RLG in the 1939 National Guard commemorative volume. The company was referred to only as Company E of the 182nd. They are dressed in their unique ceremonial uniforms that had received such favorable notice in the Boston newspapers in 1936. All other companies in the Massachusetts National Guard were photographed wearing standard U.S. Army uniforms.

That is until a reader turns to the page with Company E of the 182d Infantry of Wakefield.[40] Unlike every other company, detachment, and headquarters in the entire Massachusetts National Guard, the Wakefield company—the Richardson Light Guard—chose to wear its unique uniform for the official photos. After more than three decades in the National Guard, Company E still retained its distinct double breasted cadet-style dress uniforms with white straps across the chest, and a rigid and high hat with chin strap and topped with a plume. Very few militia units in the United States still maintained separate ceremonial uniforms. Most, such as the Ancient and Honorable Artillery Company of Massachusetts, the Governors Foot and Horse Guard of Connecticut, the Cleveland Grays, and a few other organizations, remained outside of the National Guard. Company E clung to its dual identity of prominent social militia company and National Guard company. However, nowhere in the volume is any mention of the name "Richardson Light Guard." Instead, the company is identified only by its letter designation, and its battalion, regiment, and division affiliation.

On 11 November 1936, Wakefield held a large Armistice Day parade, with Colonel McMahon as marshal. Colonel Connelly attended as a member of the USWV. Also marching was the local American Legion post, and the Company A Veterans' Association. Colonel Gihon made short speech at a ceremony following the parade. The newspaper made no mention of the Fine Members Association. Conspicuously absent from the town's activities was Company E, which was at that time was marching with its regiment in a parade in nearby Malden.[41] Company E belonged to the regiment, not to Wakefield. The year-long festivities ended with a military ball at the armory in Charlestown.[42] As part of the 182d Regiment, the Company E saw further state active duty in September 1938, in response to a hurricane. But other matters seemed to preoccupy state military leaders. A data sheet sent by the adjutant general's office to all armories in Massachusetts in 1938 asked for the particulars on the armory, such as the size, material, and other physical characteristics. The questionnaire was quite standard, but included one question, however, that was perhaps unnecessary in a town such as Wakefield, which asked whether the armory was "[d]efensible against mobs?" Although the Wakefield armory did not have the gunports of the castellated armories, and looked as defensible as an average high school, the question was answered in the affirmative.[43] But in 1938 the danger faced by Company E came not from the workers of Wakefield, from whom it had never come, but from overseas.

Events in Europe and the Far East loomed much more heavily on the horizon during the 1930s than the fear of local workers or plans for new highways. The Japanese war in China outraged many Americans. The German invasion of Poland on 1 September 1939 signaled the start of World War II in Europe. Britain and France declared war on Germany in the days following the German invasion. The United States did not mobilize for war immediately after Germany invaded Poland, but the nation did begin taking serious steps toward preparing militarily. For the National Guard, the outbreak of war in Europe meant additional training time. The federal government normally paid the National Guard for weekly drill, a period of usually two hours in the evening training at an armory. During two weeks in the summer, the National Guard trained at state-run camps in battalions or larger units for two weeks. On 8 September 1939, as a response to the outbreak of war in Europe, Pres. Franklin D. Roosevelt declared a limited national emergency. The next month the federal government increased to sixty the number of drills for which it would pay the National Guard per year.[44] The Chief of the National Guard Bureau recommended that in the following years a week be added to the summer training period, giving National Guardsmen three weeks of field training. His proposal was adopted for 1941.[45] The planned extended period of annual training for the National Guard never occurred, and instead the president on 31 August ordered the entire National Guard, around 300,000 men, onto active federal service for twelve months of full-time training, with the first units to begin their year-long tour on 16 September 1940.[46] During the Great War, mobilization of the enormous army had begun in the spring, which allowed the War Department some time as far as building suitable shelters for the large influx of men. The situation was different in the fall of 1940. Facilities did not exist to house the entire National Guard over the winter, so the activation of the National Guard occurred over a period of four months, with some units not scheduled to enter federal service until early 1941. The 182d, still part of the 26th Division, was scheduled to begin its year-long term relatively late, and was not to enter active duty until January 1941. The added time gave the RLG time to get affairs in order and recruit new members.

More signs of the increased instability of the world soon came to Wakefield. The uneasiness felt in the nation at the Japanese conquests in China and revived German militarism in Europe was reflected in the creation of the nation's first peacetime draft in 1940. On 15 October 1940 the town's Selective Service Board held its first meeting. Colonel McMahon headed this local draft board, which had jurisdiction over Reading as well as Wakefield.[47] Colonel McMahon had served in the RLG in many of its incarnations for half a century. He first enlisted in June 1893 and remained until June 1915, seeing combat during the Spanish American War with the RLG in Guanica. He became an officer shortly after the war when he was appointed a second lieutenant. He retired from the Massachusetts National Guard as a lieutenant colonel shortly before the U.S. entered the Great War, but was back in uniform in June, 1917 as a captain commanding Wakefield's State Guard company. That August he was elevated to the rank of major and became a regimental officer in the State Guard. In 1923 he accepted a commission as a major in the U.S. Army Reserve Quartermaster Corps. He continued to serve as the military instructor for Wakefield, Stoneham, and Woburn high schools. He was a man of long experience with local military affairs and one of the most respected military men in the town. The activities of the draft board did not pass without notice, but in the tightly knit town, such activities were seen more as a sign of the times rather than something to oppose. The occasional gathering of young men at the Lafayette building to make the trip via train to Boston for induction brought mixed feelings of pride in the youths going off to do their part, and the regret that such measures were required. Local ministers came to these musters to offer prayers and solace, and town officials joined family and friends who came to send off their loved ones. The Red Cross also sent members to make refreshments available to the young men leaving Wakefield. On at least one occasion, the Reading High School band came to play for the departing selectees.[48] Other young men began joining the National Guard, not in hopes of avoiding active service, because all knew the National Guard was being called to active duty, but for the appeal of serving with men they already knew, men from their own area. In fact, company leaders used this as a recruiting tool. After obtaining the names of residents at the top of the Selective Service list, they would visit those men and attempt to persuade them to join the Guard in order to serve their year on active service with men they knew.

As the selectees from Wakefield began entering active service, Company E took in new members and prepared to enter active service for its year of training. At the same time, the company began purging minors and men physically unfit.[49] The 182d Infantry was lucky in that members had a few months to adjust to their coming year of active service, and it entered active service four months after the first National Guard units had done so. For Company E and the medical detachment, and the entire 182d, activation came at eight o'clock on the morning on 16 January 1941. Company E mustered with seventy-seven enlisted men and four officers. Of those, only sixteen had been in the company two years earlier.[50] The company commander, Captain James G. Brown, forty-three years old, from Melrose, had originally joined the RLG in 1914, and had been commissioned as a second lieutenant during the Great War. First Lieutenant Frank Marchetti, forty-two, who lived in Winchester but originally from Wakefield, had also served in the RLG during the Great War.[51] Of the two second lieutenants, one was from Reading and the other from Saugus.[52] Of the enlisted men, thirty-five, or just less than half, called Wakefield home. Another seven hailed from Reading. The front page of the *Reading Chronicle* proudly listed the Reading men in the company, as well as another five serving in the medical detachment.[53] Twelve enlisted men came from Melrose, a neighboring city which hosted a battalion headquarters company but not a line company in the 182d Regiment. Men from outside Wakefield were more common in the higher ranks. Of twenty men with the rank of corporal or higher, only five came from Wakefield, with another one from Reading. Of the sixty-one men with the rank of private first class or below, thirty-one came from Wakefield, and seven from Reading. The predominance of local men in the lower ranks was increasingly the norm in the National Guard. Men originally joined their local unit. As they increased in rank, they often had to change units to fill an opening elsewhere in the regiment.[54]

For the men in Company E, the new emergency began in a similar manner to the first week after the U.S. entered the Great War. For several days following the activation, the men of Company E assembled each morning at the armory and spent the day packing equipment, straightening records, and training in basic tasks. The men were released each evening to return to their homes for supper and sleep if they lived close enough to do so. But on 26 January, Company E of the 182d departed Wakefield on its journey to Camp Edwards, on Cape Cod.[55] That morning, Company E marched out of the armory, up Armory Street to the upper station. There they boarded a train that took them to Boston. From North Station, in Boston, the men were taken across downtown Boston to South Station, where they boarded another train that took them to the railhead on Camp Edwards. The cooks left Wakefield earlier, taking the company truck through the clogged streets of Boston, in order to arrive earlier and begin preparing the meal to welcome the rest of the company.

Since the United States was not at war, and the federal government planned to release each National Guard unit after it completed its year of training, the departure of Company E from Wakefield lacked much of the emotion or holiday atmosphere of earlier sendoffs. After all, the men of Company E would only be a few hours away, on the Cape, training, rather than fighting in a war. Members would be able to come home on passes and leaves by taking trains for a few hours. However, many towns, including Wakefield, did not want to gamble that the year their National Guard company would be away in federal service would be a time of peace and calm, or that no natural disasters would occur. Additionally, some state politicians believed that those communities that showed their desire for a local company by creating a replacement militia would be more likely to continue to have a National Guard company based in the town after it was released from federal training.[56] As Wakefield had done for centuries, it again began creating a new local militia company to replace Company E. Although some Reading men served in the new State Guard company in Wakefield, for the first time in a century Reading also maintained a small militia force of its own, the Reading Home Defense Corps. This body operated under the town's Committee of Public Safety, and performed missions such as watching for enemy aircraft, and guarding critical infrastructure after Pearl Harbor. It was not officially connected to the State Guard.[57]

Just after Christmas in 1940, the *Daily Item* ran an article stating that the local armory would have two companies of the new State Guard based in it. The new units were to be Company H of the 23rd Infantry Regiment, and the Headquarters Company of the 2nd Battalion, also of the 23rd Infantry.[58] The commander of Company H was First Lieutenant Gray B. Brockbank,[59] who had originally joined the RLG in 1915, and been commissioned in the Officers Reserve Corps during the Great War. Company H, as all line companies, was intended to draw most of its members from its hometown, while the Headquarters company drew from the area of the entire battalion. The new regimental commander was Colonel Connelly, the man who had commanded the RLG at the start of the First World War. Regimental officers met in Malden to organize the new State Guard regiment. Leaders sought men between 21 and 50 years of age to fill the units.[60] If the potential recruit was less than 35 years of age, he could not be in Class 1 for Selective Service. Only men deferred from the draft could be accepted. Prior military service was mandatory for officers, but only desirable for enlisted men. Recruits who could pass the physical exam signed three year enlistments. After serving the full enlistment, they could reenlist for periods of one year.[61]

The newspaper article really announced the creation of a State Guard out of home guard units that had already formed in many towns. This was exactly the same situation as has occurred in 1917. Wakefield, as with many other Massachusetts towns, began the creation of a home guard company almost as soon at the call-up of the National Guard had been announced. Massachusetts held a week-long refresher course for three-hundred potential State Guard officers in the summer of 1940 on Camp Edwards, almost as soon as the president announced the call up of the National Guard.[62] The new Wakefield home guard company held drill on Monday evenings, the traditional evening the RLG drilled.[63] The company had been drilling for several weeks prior to the creation of the State Guard structure, although the newspaper mentioned that drill would not be held that week or the next due to the holidays.[64] However, with Company E of the 182d Infantry Regiment in the Massachusetts National Guard carrying the legacy of the RLG, the new home guard company could not call itself the RLG. During the nineteenth century, and for the RLG as recently as the Great War, having more than one company bearing the same name, one at home and one at war, was not unusual. However, the U.S. Army took lineages quite seriously and would deny the Wakefield home guard company the right to call itself the RLG while the original RLG remained embodied as Company E of the 182d.

During this period of emergency between the outbreak of war in Europe in September 1939, and the entrance of the United States in December 1941, the supremacy of the federal military took precedence over all local efforts. The commonwealth lost control of Camp Curtis Guild to the federal government for the duration in 1941. Although it was owned by the commonwealth since 1926, the U.S. Navy had continued to lease the camp for part of each year until 1941. In that year, the U.S. Army decided it needed the camp, and especially its ranges, to support small arms training for Fort Devens, which was thirty miles to the west. The camp also served as a staging area for the Port of Boston and allowed soldiers stationed in that city to train in small weapons and field duties.[65] The increased federal presence at the camp, as well as soldiers from it seen in town and using the small beach at the northern end of Lake Quannapowitt, became everyday reminders that, as United States was pulled into World War II, Wakefield remained in some ways a military town.

For the men of Company E, the months on Camp Edwards were an exciting time. The company drew three barracks, a mess hall, supply room, and recreation room. At first, the men had far more room in the barracks than one company really needed, but that situation soon began to change. The Organized Reserve divisions that were called to active service along with the National Guard absorbed the bulk of the selectees—the term "draftees" was not allowed to be used—because at the time Reserve units were cadre units. They had the structure and leadership down to the battalion level, but no enlisted soldiers until filled with selectees during mobilization. However, the National Guard divisions also received selectees to bring the strength of each company to over two-hundred men.[66] The first group of forty-three selectees was assigned to Company E on 6 March. This group came from all over the Greater Boston area. None were from Wakefield or Reading. Of the forty-three men in this group, thirty-seven would deploy with the company in January 1942.[67] For the next group, however, Captain Brown went to the selectee reception center on Camp Edwards to claim local men for his company. Of this second group of selectees, fifty-six men in all, who arrived on 21 March, twenty-six were from Wakefield, and another sixteen came from Reading. Of the remaining fourteen, four came from Stoneham, which bordered Wakefield on the west. All but nine of this second group of selectees would eventually go overseas with the company.[68] In this way Wakefield and Reading men continued to shape the identity of the company while on Camp Edwards, although men who had served in the pre-mobilization Company E were soon a minority.[69] Still, with the influx of local selectees into the company, it remained the main vehicle through which Wakefield and Reading men served in the military.

Captain Brown proved to be an indulgent commander on active duty, in line with regimental and divisional practices. Soldiers not performing duty on the weekend were usually allowed to go home, provided they were able to be back for duty when required. The train did not leave Camp Edwards until about noon on Saturday, after inspection, and most men would then have to walk or hitch a ride between South Station and North Station in Boston, for the last leg of their trip. They would not arrive in Wakefield until late in the afternoon on Saturday. The last train left Wakefield at 9:00 in the evening on Sunday, so weekends away from Edwards did not mean a lot of time spent in Wakefield. However, some soldiers owned cars, and began taking them back to Camp Edwards with them. Soon, carloads of soldiers were leaving Edwards Saturday mornings, and returning in the early hours of Mondays.

Camp Edwards had undergone a massive construction program beginning in mid–1940, with federal dollars pouring in to build the standard two-story wooden barracks that were going up in military camps from Alaska to Puerto Rico and all places in between. Still, the camp seemed like a sea of mud and snow in the early months of active service.[70] But while completing most of the massive structural changes on state military camps, the U.S. Army delayed implementing major structural changes to the National Guard itself. Some changes were made, such as the creation of regimental combat teams,[71] but much of the structure of the National Guard divisions was left as it had been in the Great War. When the original twelve-month period of active federal training for the National Guard was only about half complete, Congress, after a heated political battle, authorized the president to extend to eighteen months the term of active service for all men inducted under the earlier act — selectees, reservists, and National Guardsmen.[72] For the men in Company E in the summer of 1941, this meant that they would not be returning to Wakefield and civilian pursuits until the middle of July 1942. News of the extension of service brought some griping, but nothing serious. In some of the divisions that had been called up early, the extension came as they were anticipating their release in late September or early October. As a result, some wag coined the acronym "OHIO," meaning "Over the Hill In October," to describe their sentiments about the extension. The official history of the 26th Infantry Division claimed that resentment against the extension was stronger among selectees than Guardsmen.[73] On 19 August 1941, the 182d went to Fort Devens for three weeks of training. While there, the "OHIO" protest came to the 182d. One former member recalled an officer telling the Guardsmen that anyone caught saying "OHIO" would be promptly tried by court martial. Of course, as soon as lights were out, the term could be heard echoing around the bivouac site.[74] But the incident seems to have been one more of good humor rather than a serious challenge to military discipline, and nothing more was heard of the incident. After leaving Fort Devens, the regiment returned to Camp Edwards for more maneuver training.

Their time back on Edwards would be short, as they soon started preparations for movement to Fort Bragg, North Carolina, to participate in much larger maneuvers. While getting ready to go, the Army declared Captain Brown, Lieutenant Marchetti, and Second Lieutenant Rollin Foster too old for their ranks, and transferred them out. Brown was forty-four, Marchetti was a month away from his forty-third birthday, and Foster was thirty-six. Their ages were not unusual for National Guard officers of the period. Approximately 22 percent of first lieutenants were at least forty years old. These men were not discharged, but reassigned to other duties.[75] Only George Gromlie, newly promoted to first lieutenant, who had just celebrated his twenty-sixth birthday, remained of the original officers. However the new commander was Captain Dolbeare of Wakefield, who had commanded the company in the late 1930s and had since been serving as a regimental officer.

On 29 September 1941, the entire Yankee Division went to the Fort Bragg area in North Carolina to participate in the Carolina Maneuvers.[76] After a long three-day ride on trucks, the company was assigned a peanut field near the town of Ellerbe as a bivouac site. For the men of Company E, the Carolina maneuvers were a series of hot days and cool night, long dusty marches, and the novelty of red clay everywhere. Very often the men in Company E, as with other enlisted men throughout the maneuvers, had little idea of the larger issues of the exercises. The 26th Infantry Division was assigned to the "Blue Army" under the command of Lieutenant General Hugh Drum, along with Regular Army divisions such as the 1st, 8th, and 9th, and National Guard divisions such as the 28th, 29th, 30th, and 44th. They faced the "Red Army," which had a less traditional structure, fighting a series of mock battles that tested men, equipment, doctrine, and leaders. For two months the divisions trained, but by early December, the entire 26th Infantry Division was packed and heading back to Camp Edwards.[77] The company clerk and a few other men had returned early, in order to out-process all men over the age of twenty-eight who wanted a discharge, in accordance with a recently passed law. But even for the younger men, the hardest training was now in the past, and they anticipated that the remainder of their time on active duty would be filled with normal duties on Camp Edwards, with a generous leave and pass policy. The hard part of their federal service was behind them, and soon the company would be detraining at Wakefield's upper station and the men returning to civilian pursuits.

Chapter IX

In the Americal Division

The Yankee Division returned from the Carolina Maneuvers on Saturday, 6 December 1941.[1] The men, worn out from long days riding in backs of trucks, dragged themselves into the barracks and fell asleep. The next morning, a Sunday, began like a normal duty day, with the soldiers assuming the routine duties of life at Camp Edwards. Many family members were arriving on post to welcome the returning Guardsmen. But late in the day rumors of the Japanese attack on Pearl Harbor began to spread around the camp. Commercial radio stations began broadcasting sketchy reports of the attack.[2] The next day rumors were confirmed and headquarters cancelled all leaves and passes. In the day following the Japanese attack on Pearl Harbor on 7 December 1941, the United States declared war on Japan, which was followed by a German declaration of war against the United States, which brought about a reciprocal U.S. declaration against Germany on 11 December. Italy followed Germany's lead and declared war on the United States. The United States was now fully committed to fighting World War II and Company E would not be returning to Wakefield for the duration. Men on pass or leave scrambled to get back to Camp Edwards, where the tempo of operations had suddenly and drastically increased. The men in Company E understood that there would be no discharge back to civilian life for a long time.

As in the Great War, the first impulse during World War II was to use soldiers for local security. The War Department had contingency plans to use some of the divisions to provide increased security along the coasts, and the 26th Infantry Division was one of the divisions so selected. Soldiers from Camp Edwards found themselves guarding canals, bridges, and other points deemed vulnerable along the New England coast armed with old .03 bolt-action rifles, although some began to receive the new M-1 rifles. Most of the men of Company E were soon on Maine beaches walking sentry.[3] The times were so uncertain, and the men so jittery, that some men in Company L, based in Malden, shot at some fishing boats thinking they might be German or Japanese warships.[4] Back in Wakefield, the entrance of the United States into the war brought the local State Guard company to alert. Members of Company H of the 23rd Massachusetts spent the days following Pearl Harbor awaiting the call to state active duty to perhaps watch the coast or to patrol bridges, tunnels, and reservoirs to protect them from Axis saboteurs.[5] Wakefield was proud of its National Guard company as well as its local militiamen, whom it considered "the first line of defense within the State of Massachusetts."[6] Although the people of Wakefield believed in local defense, the federal war effort took precedence over everything.

By late December, the initial shock of Pearl Harbor had subsided and the military set itself to the task of defeating the enemies of the nation. Despite the increased seriousness of military service, some of the informal aspects of life in Company E continued. Captain Brown had been a lenient commander, and Captain Dolbeare found imposing strict army discipline a challenge. For Private James C. Buckle, December had been a mostly enjoyable month. On the return trip from Fort Bragg, he had jumped off the truck in Rhode Island and hitchhiked back to Wakefield. He then returned to Camp Edwards with his car, and afterward was able to go home, usually with a carload of fellow soldiers, when not on duty. While most of the company patrolled Maine beaches, he remained on Camp Edwards stoking the coal furnaces. The stokers were supposed to work an "eight hours on, sixteen off" schedule, but rearranged it so they worked sixteen hours on, and then had over a full day off. During his off time, Private Buckle drove home. Eventually, his lack of sleep caught up with him, and on the day after Christmas, 1941, he totaled his car in the town of Bridgewater, half way between Wakefield and Camp Edwards.[7]

Units from the 26th Infantry Division, especially the 181st Regiment, would continue to patrol the eastern seaboard into November 1943, but the Army soon had other plans for the 182d Regiment.[8] In early January 1942, the 182d Regiment was recalled from coast watching duty and returned to Camp Edwards to begin integrating a new group of fourteen selectees. This new group contained only one Bay Stater, and so the Wakefield, Massachusetts, and even National Guard, character of the company was diluted even more. The new men were originally been at Camp Croft, South Carolina, and had been sent to Company E, 182d, in order to bring it closer to war strength, an indication that the Army was already envisioning an early deployment for the regiment. Nine men in this group came from Indiana, two from New Jersey, and one each from Pennsylvania and Tennessee. All these new men eventually deployed with Company E.[9] Passes were again available, which many soldiers took advantage of to go home to say goodbye to their family. Major structural changes in the National Guard divisions were underway. In the Great War, a U.S. Army division contained four infantry regiments. In the late 1930s, the U.S. Army adopted a new model for its division based on three infantry regiments. The new, lighter "triangular" divisions were considered to be more mobile, while the older "square" divisions with four infantry regiments were deemed more suitable for defense. For this reason, the Army had delayed converting the National Guard divisions in the belief that the National Guard would play a defensive role, rather than fight as part of any expeditionary force. The scale of World War II and the nature of America's entrance into it destroyed any hope among the leadership of the Regular Army of implementing any plan to use the National Guard solely for defense, and so shortly after the entry of the United States into the war, the Army implemented its new divisional structure on the National Guard divisions. As a result of this reorganization, the 182d regiment, to which Company E still belonged, was shorn from the 26th "Yankee" Division on 14 January 1942.[10] For the enlisted men of the 182d, the relief from the Yankee Division came without ceremony or emotion, just an announcement that they were to prepare to move out as a regiment, although their eventual destination was unknown. While the 182d Regiment was absorbing this disturbing turn of events, a situation in the South Pacific was about to have a lasting impact on the regiment.

The French island of New Caledonia, located about 850 miles northeast of Brisbane, Australia, concerned American, Australian, New Zealand, and of course Free French defense planners. The surrender of France to Nazi Germany made the island vulnerable to exploitation by the Japanese, as had happened in French Indochina. Control of the islands, in the southern Coral Sea, by Japan would cut Australia off from the United States, and thus American and Australian defense planners scrambled to get a force on the island using whatever available units could be scraped together. Many of the available units were made surplus from the conversion of the 26th Infantry Division, and the 33rd Infantry Division from Illinois, into triangular divisions. Two infantry regiments—the 182d and the 132d—plus an odd assortment of surplus supporting units, were to be sent to buttress the defenses of the island. More men were transferred into Company E from other units of the 26th Infantry Division just before the company left Camp Edwards,[11] bringing it up to a total of five officers and 218 enlisted men. Of the fifty-one enlisted men reassigned to Co. E. on 17 January 1942, none were from Wakefield or Reading, although one man came from North Reading, and all were from Massachusetts.[12] Company E and the 182d Regiment were reorganized according to the current U.S. Army infantry tables. The company now included three rifle platoons and a weapons platoon. Each platoon had about forty men, normally commanded by a second lieutenant. Riflemen were armed with either the M1 rifle or the Browning Automatic Rifle (BAR). The weapons platoon was armed with 60 mm mortars and .30 caliber light machine guns. The company headquarters contained the commander, usually a captain, and the executive officer, usually a first lieutenant. Headquarters also contained the cooks, clerks, and supply soldiers.

The pattern of company organization was repeated on a larger scale at the battalion level. Each battalion had a headquarters company, and three rifle companies and a weapons company. The companies were lettered sequentially. Thus, the first battalion had companies A, B, C, as rifle companies and D as the weapons company. The weapons companies were equipped with heavy machine guns and 81 mm mortars. In the second battalion, Company H was the weapons company. The regiment included three such battalions, plus a regimental headquarters. The 182d left Camp Edwards by train on 20 January to go to its Port of Embarkation at the Brooklyn Navy Yard. After arriving, the men of Company E were assigned to the starboard side of the promenade deck on the U.S.S. *Santa Elena* and settled in. On board the *Santa Elena*, the men tried to make themselves comfortable in their makeshift quarters. Their assigned area was a deck that had been enclosed. Pipe-frame bunks five-berths high served as sleeping area, each man was issued a piece of canvas to lace into the frame to make his bunk. The Company E that boarded the *Santa Elena* was considerably different from the one that had entered active duty only a year earlier. Of the original four officers and seventy-seven enlisted men, one officer and sixty-six enlisted men remained. The eleven enlisted men no longer in the company had left for a variety of reasons. Two were commissioned and destined to serve in Europe, three were transferred to G Company, one transferred into the Army Air Corps, three were assigned to other units, and two were dropped from the company rolls and eventually discharged.[13] Fifty-four men in the company were from Wakefield, and another twenty-three came from Reading.[14] Because the company carried more men than was standard, 218 enlisted men, the soldiers joked that they were carrying their casualty replacements with them.[15]

In the early hours of 23 January 1942, Task Force 6814, as this hastily assembled force was called, packed onto seven transports, with escorting destroyers, heavy-cruisers, and blimps, began the voyage that would eventually take them to New Caledonia.[16] The task force, also known as the "Poppy Task Force," contained about 20,000 officers, men, and a few women nurses. Through what began as this ad hoc task force, Company E would fight in World War II. To command the task force, what the official history later described as a "wartime military stew of men and equipment,"[17] the Army selected Major General Alexander M. Patch, Jr. General Patch had been a colonel commanding a regiment in the 9th Infantry Division that past August, promoted to brigadier general, and put in command of the Infantry Replacement Training Center on Camp Croft, South Carolina. He was relieved of that assignment and promoted again just prior to assuming command of Task

Force 6814.[18] He had earlier served as the senior Regular Army advisor to the Alabama National Guard, and thus understood something of the strengths and weaknesses of the National Guard.[19]

The voyage to New Caledonia took Company E through the Panama Canal, which they cleared on 31 January 1942, and into the Pacific. At that time, their initial destination of Australia became known, although most did not know their final destination.[20] Daily routine on the transports consisted of whatever training could be accomplished, endless lifeboat drills and the occasional scare of a Japanese submarine. Meals were served twice a day, and the men spent an inordinate amount of time standing in the chow lines that wrapped around the ship. The lack of fresh water on the transports limited each man to one canteen per day, while showers were taken with salt water.[21] General Patch, who remained in the United States gathering information and battling pneumonia while the task force steamed toward the South Pacific, and the local men in charge, were unaware of the immediate situation on New Caledonia. The island had earlier been the scene of a small struggle between Free French and the Vichyites, and was protected only by a company of Australian commandos and some French militia. As a result of the uncertain condition on the island, the task force first docked in Melbourne, Australia, on the evening of 26 February. The men of the task force unloaded and began to sort the equipment that had been so hastily and haphazardly loaded in Brooklyn, and combat loaded it back on the ships, ensuring that most urgently needed items would be available first.

In part as protection against air raids, most of the task force was dispersed throughout the suburbs of Melbourne, with Company E marching through streets filled with welcoming residents to spend its first night in Australia in a park. The next day Company E boarded a train for a trip to the small town of Ballarat, about seventy-five miles northwest of Melbourne. There the men of Company E were billeted in private homes, for which the homeowners were compensated by the government. Other men from the regiment stayed in tents, but overall found the experience like a vacation.[22] Relations between the homeowners and their American boarders were good, and members of Company E often kept in contact with their Australian hosts long afterward. One former member recalls that in Australia, they did no training; all they were required to do was present themselves at an athletic field in town for roll-call once in the morning. Commanders were quite liberal with passes.[23]

The Americans arrived during the Australian summer wearing their winter uniforms, and since their fatigue uniforms were still packed away on the transports, the men had to wear their khaki shirts and pants. Some men still remembered the heat of the Australian summer sixty years later.[24] On passes into Melbourne the troops drank warm beer and flirted with local girls, whom the Americans found pretty, except for their teeth. The girls, for their part, greeted the Americans with flowers and kisses. All in all, the men found the Australians pleasant, and reminded some of the men of New Englanders. But for some members of Task Force 6814, their time in Australia was filled with more pressing business. In addition to sorting through the equipment that had been loaded so quickly in New York and trying to make sense of it, some men trained under tutelage of Australian and American soldiers for the type of fighting they might face on New Caledonia should the Japanese invade.[25] After eight days in Australia, the soldiers were told they had to return to Melbourne and reboard the ships immediately.[26] When Private Joseph J. D'Alessandro and another soldier returned to the home in which they were billeted, they found the homeowners were out for the day. After climbing in through a window, they gathered their belongings and left a note thanking their hosts for their hospitality and left.[27] On 4 March the men of Company E were back at the docks of Melbourne. The task force left Melbourne for the final leg of its journey to New Caledonia on 6 March on the seven transports, this time with Company E on the U.S.S. *Argentina*.[28]

The task force arrived in Noumea Harbor on the southwest shore of New Caledonia on 12 March. Noumea, normally a sleepy outpost of French colonialism, was the capital of the island and its largest town. Because its harbor was small, and not designed for large transports, the men of Company E had to disembark over the side of the ship, climb down cargo nets, and be taken to shore on lighters. When all were on shore, they began a seven-mile hike to their initial bivouac site, where they soon became acquainted with the local mosquitoes. However, the next day they were taken by truck back to the harbor.[29] For the next month, the company bivouacked on a nearby hillside in pup tents, while providing harbor security, with an additional detail to guard the local hospital in Noumea, which was "off-limits" to American servicemen. The town of Noumea was overrun with soldiers and sailors, many of them drunk, and combined with the requirement that all military personnel carry loaded weapons, incidents were perhaps inevitable. On 16 March, Corporal Parker C. Kimball of Reading was killed while on duty guarding the hospital. He had responded to an incident where a private was attempting to subdue a pair of drunken sailors, when one of the sailors fatally shot Corporal Kimball.[30]

Added to the mission of the task force was the defense of the New Hebrides, 250 miles to the northeast. To defend the islands, Force A, consisting of Companies L and M of 182d, plus a medical and a service detachment and engineer platoon, was sent. The remainder of the men on the 182d, and the entire task force, spent next few months preparing to defend New Caledonia against a Japanese invasion that never came. The 182d garrisoned the southern half of the island, a sector containing the capital of Noumea and its precious harbor, plus the only operational airfield on the island. To assist them, the 182d had some 1,400 or so Free French militia attached. One former member of Company E remembered the duty as consisting mainly of standing sentry at bridges, the hospital, and the airfield.[31] The Allied victory at the naval Battle of the Coral Sea from 3 to 8 May greatly diminished likelihood of a Japanese invasion of New Caledonia. Manning defenses became less urgent, while training became more routine. Off duty, few opportunities for recreation existed aside from deer hunting and fishing. The larger war remained mostly unknown to the men, and they suspected, similarly unknown to their officers. They started to feel cut off on what was again quickly returning to its state as a forgotten backwater. The men began to wonder if they would spend the entire war garrisoning New Caledonia.

In April, the 164th Infantry Regiment, an orphan of the North Dakota National Guard recently separated from the 34th Infantry Division, arrived to join the defense of the island.[32] With the addition of the 164th, Task Force 6814 began to resemble a division and so the War Department decided to turn this assortment of leftovers into a division. Although it had the required three infantry regiments, its other organic and attached units hardly came close to a standard U.S. Army division. In part because of the unusual complement of units that made up the new division, the Army broke with standard practice and gave the division an official name rather than a numerical designation.[33] It was also the only division in the U.S. Army created outside of the United States. After an earlier War Department suggestion of "Necal," or General Patch's suggestion of "Bush" failed to generate any enthusiasm among the men who served in the division, an unofficial contest was held gather suggestions. Private First Class David Fonesca of Roxbury, Massachusetts, who served in the signal company of the 182d, submitted the name eventually chosen: "Americal." The name combined "*Ameri*can" with "New *Cal*edonia."[34] The existence of the new division was officially announced on 27 May 1942 without ceremony.[35] For the men of Company E, as well as the rest of the new division, the change in status from task force to division had little noticeable immediate effect. Instead the men began a program of training and reorganizing to function like a proper division. Company E moved several miles to a new bivouac site near the Saint Louis Mission and began patrolling and mapping the area. In July, the 182d was mostly reassembled and began more training in tactics and maneuvers.[36] Life on the island became even more routine. According to a member of Company L, from Malden, the chief complaint was the food. The men were fed something they called "ug mutton" more often than the troops would have liked.[37]

Throughout most of the remainder of 1942, the Americal Division attempted to remain above the squabbles between native New Caledonians, the colonists, and the recently arrived metropolitan French, although General Patch became involved whenever he believed their struggles threatened the security of the island. General Patch was later awarded the Distinguished Service Medal for his success in dealing with the difficult political situation on the island. Largely as a result of his fairness, tact, and good sense, the island remained under civilian control.[38] Thanks largely to the deft handling of the situation by General Patch, rumors and the occasional alert were the sole impact the political intrigues had on Company E. However, the Americal Division would not be spending the entire war garrisoning this half-forgotten outpost of French colonialism. Instead the War Department would soon be turning defense of the island over to elements of the U.S. 43rd Infantry Division and the Kiwi Division from New Zealand, and the Americal began preparing for a new mission.

One chronic problem for the Americal was the shortage of commissioned officers to serve as platoon leaders. For this reason, the division held its own officers candidate school on Camp Stevens, in Noumea, after the threat of invasion passed. Company E lost three members to this school, all of whom received commissions and were assigned to other companies in the Americal. The three men were First Sergeant Charles Willis, Sergeant William J. Blanchard, and Sergeant Lewis W. Doyle. Normal army practice was to avoid assigning officers who had previously served as enlisted men in the same company or even battalion. The school lasted six weeks, after which the officer candidate was assigned to a unit in a probationary status, before receiving his commission. In all about 385 men in the Americal received commissions through this program.[39] Other soldiers left Company E for a variety of reasons. One soldier, at forty-five years of age, was found unable to bear the physical strains of service and was reassigned to a stateside hospital and later discharged.[40] Another soldier transferred to the Mobile Reconnaissance Squadron, or as the men of the 182d called them, the "Peep Troops," while still another transferred to the service command.[41] One soldier lost a couple of fingers in a weapon accident, and was reassigned to the battalion headquarters as a cook. A final man was sent home and discharged. While popular in the company, most men agreed that he never should have been accepted by the Army, and the other soldiers had long been covering for him.[42] On 23 September, the first replacements reached the Americal, and were assigned to below-strength units.[43]

After the relatively safe and dull duty on New Caledonia, the Americal Division became involved in heavy combat. In early August 1942, about one thousand miles northwest of New Caledonia, the U.S. Navy and Marines began operations in the southern Solomon Islands, particularly on the island of Guadalcanal, where the Japanese were building an airfield. On 13 October, the first elements of the American began to arrive from New Caledonia, when the 164th Regiment landed. They were quickly placed under control of the First Marine Division, which was overseeing operations on the island. Shortly after arriving, the Americal suffered its first combat casualty of the war when a corporal was killed during a noontime air raid.

Company E of the 182d boarded the USNT *McCawley* for the trip. The 182d Regiment was delayed about a month in its arrival by the fierce Japanese attacks on the U.S. naval forces around the island. It arrived on Guadalcanal, without its third battalion, in the early morning of 12 November 1942, and thus missed a month of furious Japanese attempts to dislodge the Americans. However, their arrival was exciting enough, as an artillery shell hit just off the stern of the *McCawley* as it arrived at the island. The company went over the side of the transport and boarded Higgins boats for the final leg of their journey to Guadalcanal. Once on the beach, they began to dig in, and soon watched as group of Japanese warplanes flew overhead. However, the Japanese pilots were after the American fleet and took no notice of the infantry men on the beach. The "wall of steel" fired at the airplanes from the ships brought down all but one aircraft, and it flew back over the infantrymen trailing smoke, which heartened the soldiers on the beach.[44]

Due to transfers, discharges, and the commissioning of three soldiers, Company E of the 182d arrived on Guadalcanal with fifty-nine of its members remaining from the group that mobilized in January 1941.[45] The 182d first went into action shortly after arriving, under Brigadier General Edmund B. Sebree, along with the 164th Regiment and the 8th Marines. The first battalion of the 182d took the brunt of the early clashes with the Japanese, but the second battalion, to which Company E belonged, did not stay out of the fighting for long. The presence of the Americal on the island made both necessary and practical the enlargement of the salient the Marines had been defending. Company E was sent out to establish an ambuscade on the west bank of the Matanikau River, two kilometers inland, on Ridge 13. From their position, the company sent out small patrols daily, which reported only sporadic contact with Japanese.

On 23 November 1942, a patrol from the 164th infantry, operating in front of the position of Company E, came under attack and requested assistance retrieving two of their dead. The commander of Company E asked for volunteers, and seven men, Sergeant Americo A. DeFeo, Sergeant Michael F. Doyle, Sergeant Harry E. Mohla, Corporal Samuel DeMarco, Corporal Joseph M. Conway, Private First Class Roland L. Hatch, and Private First Class Francis T. O'Brien, stepped forward. While on that mission, the Company E men ran into a Japanese ambuscade, and had to fight their way back to friendly lines. All made it back and received the Silver Star Medal for their actions, although Private O'Brien was wounded and also received a Purple Heart for his wound. A few days later, on 26 November, another Company E soldier, Corporal Edmund W. Torrance, from Saugus, earned a Purple Heart when he was wounded. That night, Private First Class Charles E. Parry, from Reading, was killed as the company spent the night in their foxholes. He was the first Company E soldier listed as Killed in Action. There would be more.

Around 1:00 in the afternoon of 4 December, members of the company saw what they thought was a white man approaching their position. A small patrol led by Sergeant Harry E. Mohla went forward to contact the man. After establishing that he was indeed a white man, and apparently an American, he was brought in. The man turned out to be Seaman Dale Land, an eighteen-year-old survivor from the U.S.S. *Walke*, a destroyer sunk during the naval battle of Guadalcanal. After spending three days in the water, the man had come ashore and made his way through Japanese positions to reach friendly forces. A week after finding the sailor, Company E rotated off the front line and back to their base camp. In the rear, the men were able to bathe, and food was a little better, sometimes as a result of "midnight requisitioning" from the supply dumps. Occasionally, they attended movies shown at the Air Forces Reconnaissance Squadron base. Duty in the rear was an odd assortment of the fatiguing and the dangerous. Often they had to pull nighttime security at the airfield. Later they were tasked to lay the Matson mats, a heavy matting that provided an instant runway, and were strafed by a Japanese airplane while doing so, although none of the men were hit. Another time, they were tasked to patrol up the Lunga River, scene of a recent Japanese defeat, on a mopping up mission. However, all they found were dead Japanese soldiers and abandoned artillery pieces.[46]

Company E rotated back to the front lines, but by late January the battle for the island was definitely slackening, and on 9 February, the U.S. military declared the campaign officially over, and an American victory. In May 1943, the U.S. Navy submitted the First Marine Division, reinforced, to receive the Presidential Unit Citation for its actions on Guadalcanal from 7 August 1942 through 8 December 1942.[47] Since several elements of the Americal, including the 1st and 2nd battalions of the 182d Regiment, had been attached to the First Marine Division during this period, Company E received this honor.[48]

After the fighting on Guadalcanal subsided, the Americal moved to the Fiji island group for rest and refit. The first elements left Guadalcanal on 1 March 1943, with the 182d following on 24 and 25 March.[49] The 182d arrived on the island of Viti Levu, in the British colony of Fiji, a week later, and took over a sector from Momi Bay south and east to Suva Harbor. Shortly afterward, on 1 May, the War Department reorganized the Americal to make it more in line with the standard infantry division, although the term "infantry" would not be officially attached to the name of the division until late in the war. Life in Fiji passed without too much excitement, other than the scourge of malaria, which swept through the Americal. By the time the epidemic ended, medical officers estimated that about 85 percent of Guadalcanal veterans had been hospitalized at one time for the virus.[50] Men continued to trickle out of the company. Three men from Company E volunteered for combat duty, and ended up in Southeast Asia, serving in Merrell's Marauders.[51] All the while, replacement soldiers, raised through conscription and trained partially in the United States, began to fill units depleted over the past year. During the summer of 1943, the 182d Regiment lost much of what was left of its prewar identity as replacement officers and men joined the unit. As the United States had lowered the draft age to eighteen, the average age of the men in the company dropped. The new selectees came from all over the United States. After August, none of the three battalion commanders had served in the prewar 182d. Instead, officers from the Regular Army increasingly filled those positions.[52] Major Dolbeare, who served as commander of Company E in peacetime and for a while after mobilization, and later commanded the second battalion for about six weeks on Fiji, was rotated back to a stateside assignment that August.[53] By 10 December 1943, when five lieutenants joined Company E, none of the officers in the company had been in since January 1942.[54] Except for one brief two-week period in June 1945, all the regimental commanders after May 1943 were men from outside the regiment.[55]

The rebuilt 182d Regiment, as with the entire Americal, settled into a routine of training in jungle warfare, sports, and in general enjoying being away from Guadalcanal. Horses were rather common on Fiji, and many of the soldiers acquired one.[56] By late 1943, the rainy season made life in Fiji less pleasant, but by then rumors of deployment to the northern

Solomon Islands began to spread through the Americal. After a nine-month stay on Viti Levu, a detachment from the 182d left Fiji on 19 December 1943 with the advance party from the Americal. The remainder of the 182d left on 22 December, in the second echelon. Their destination was the island of Bougainville, at the northern end of the Solomons, which the Japanese had controlled since early 1942 and on which the U.S. Marines had been engaged since 1 November 1943.

Bougainville was a substantial island, between twenty-five and forty miles across, and about 120 miles long, southeast to northwest. Thick jungle covered much of the island. Running through the middle of the island were the Emperor Range and the Crown Prince Range, reaching heights of more than five thousand feet, with two active volcanoes. Allied military planners wanted the island as a site for an air base. From Bougainville, Allied airpower could be brought to bear on the massive Japanese base at Rabaul, on New Britain, some three hundred miles to the northwest. The Japanese understood the importance of Bougainville to Allied plans, and fought desperately to dislodge the Americans from the island to prevent the construction of any Allied airfield. On 28 December 1943, the main body of the 182d landed on Bougainville, at Empress Augusta Bay, in the middle of the west coast, and began relieving elements of the Third Marine Division. On 2 January 1944, the 182d moved to take up a position on the perimeter with the 164th Infantry on its left flank and the 2nd Marine Raider Battalion on the right. The Marines had the beach on their right flank. The 132d Infantry soon began relieving the Marines, and by early March, the three infantry regiments of the Americal stood on line. The 37th Infantry Division held the line to the left of the Americal.

After a few months of sporadic contact, the Americans had strong evidence of a coming Japanese offensive. Companies throughout the Americal sent out squads on patrols to their front searching for signs of Japanese activity. Divisional policy stated that an officer would lead each patrol, meaning that platoon leaders were going out three times as much as their enlisted soldiers. On 31 January 1944, the Company E executive officer, First Lieutenant Gerald B. Lyons, gave a platoon leader a break when he volunteered to lead one of the patrols. Company E lost its Executive Officer that day when he was Killed in Action.[57] For a brief period, the battalions took turns acting as regimental reserve, but fearing the Japanese might sense the vulnerability of the line when the battalions were rotating, the scheme ended. Instead, the 1st and 3rd Battalions of the 182d held the division center, while most of the 2nd battalion, including Company E, remained behind the lines as a regimental reserve. Some 800 yards east of the defensive line stood Hill 260, a hill soon to become notorious to the men of Company E.

Hill 260 was an hourglass shaped hill running northeast to southwest. It had two knobs, about 150 yards apart, connected by a small trail that ran through the saddle between the knobs. Each knob was about half the size of a football field. The slopes on the east and west of the hill were very steep.[58] A small stream, which the Americans dubbed "Eagle River," ran between the defensive line and the hill. The hill stood outside of the defensive line but held a vital artillery observation post in a tall banyan tree on the southern knob, which the division headquarters believed had to be held if the entire defensive line were to remain tenable. While the importance of the hill was never explicitly explained to the men of the 182d, they were experienced enough by this time to understand that the Japanese would need to take the hill before launching a general assault on the main defensive line. The hill was defended by a rifle platoon augmented with a section from the heavy weapons company, about eighty men in all. The defenders were dug in around the banyan tree. The responsibility to provide a rifle platoon to defend the hill rotated between Companies E, F, and G. The soldiers of the Americal continued to improve the defenses, while the daily rains turned much of the road network into mud as fast as the engineers could repair it. By 7 March, the Americal was as ready as they were going to be for the coming Japanese onslaught.

The Japanese also understood the tactical importance of Hill 260 and began a concerted effort to take it. Possession of Hill 260 would give the Japanese a commanding tactical advantage over the Americal defensive lines, and deny the use of the observation post to the Americans. The Japanese assigned a battalion reinforced by two infantry companies to take the hill. Despite constant patrolling by the 182d, the Japanese were able to infiltrate the area in front of the hill almost without detection. On the night of 9–10 March, a small unit of Japanese infiltrated between the hill and main defensive line, while a larger force assembled east of the hill for the main attack. In the early morning of 8 March, the normal American harassing howitzer fire was suddenly drowned out by an enormous Japanese artillery barrage hitting it seemed every airfield, supply dump, road junction, and artillery position. The American artillery eventually got the upper hand, but Japanese infantry patrols became more aggressive as they searched the defensive line for a weak spot.

A platoon from Company G, based in Woburn, held Hill 260 that morning. On 9 March, a patrol from the 164th encountered a group of about fifty Japanese soldiers just north of Hill 260. At dusk that same day, the Japanese subjected the Company G platoon to murderous fire from machine guns, rifles, and mortar, but were soon quieted by American artillery. At 6:30 on the morning of 10 March 1944, heavy Japanese machine gun and mortar fire poured in from the east. Five minutes later the ground attack began. At least one reinforced Japanese company came up the steepest part of the hill and overran the platoon from Company G. At 6:38 A.M., the artillery observer in the banyan tree reported that the Japanese had overrun the defenders, and that Japanese soldiers were all around the base of the tree. He was not heard from again.[59] The remaining defenders retreated to the north knob and continued their defense despite being surrounded. The Japanese forces, estimated by the Americans to be about a company, but which were actually a battalion,[60] began to reinforce their position, fully expecting an American counterattack. Major General Oscar W. Griswald, commander of XIV Corps, which had taken over operations on Bougainville from the 3rd Marine Division, ordered Major General John R. Hodge, the commander of the Americal since the previous May, to hold Hill 260 at all costs. By the time this order was issued, the hill was mostly in Japanese hands. General Hodge ordered his assistant division commander, Brigadier General William A. McCulloch, to oversee the campaign to retake the hill. He established his command post at the 182d Regiment's command post.[61] Surprised by the order to hold the hill, Colonel William D. Long, the Regular Army commander of the 182d, released his battalion reserve under Lieutenant Colonel Dexter A. Lowry.

Colonel Lowry originally planned to use Company F, from Waltham, and the remaining elements of Company G to reclaim the hill, relieve the survivors of the Company G platoon, and perhaps reestablish contact with the observation post in the banyan tree. However, Company G was not ready to attack, and so Colonel Lowry sent in Company E — the unit bearing the heritage of the Richardson Light Guard. Company E, then under the command of First Lieutenant Melvin E. Carlson, had about 128 men present for duty. Companies E and F began to attempt to retake the hill before the Japanese could consolidate their hold. The plan called for Company F to maneuver to the northern side of the hill, for an attack on the northern knob, while Company E was to attack from the south. At 8:45 in the morning, Company E received the order to attack the south knob from the southwest. Two hours later, the third platoon of Company E, led by Second Lieutenant Leonard C. Hurley, started climbing the southwest slope, while Company F attacked from the north, where it soon made contact with the survivors from the Company G platoon. After advancing about forty meters, the advance of the third platoon of Company E was halted by entrenched Japanese defenders. Soon the other platoons from E were committed, as one went north to assist Company F while another sought a more southerly route up the hill.

The first and second platoons had been attempting a double envelopment of the Japanese position, but enemy machine guns, mortars, hand grenades, and other firepower drove the first

The observation post in the banyan tree atop Hill 260 on Bougainville before the battle.

and second platoons back with heavy casualties. The communications sergeant, James C. Buckle, who was with the commander, volunteered to check the wire running to the observation post for breaks. A squad was sent with him. At one point, they had a big scare when they spotted another squad nearby, but it turned out to be another group of Americans. They continued following the wire and eventually found a break. After repairing the break, Sergeant Buckle could still not contact the observation post. Further movement up the hill was impossible owing to the large number of Japanese soldiers on the crest. Eventually the Americans learned that the men in the observation post had all been killed by mortar fire. By the time Sergeant Buckle and his squad returned, Company E had suffered its first wounded in the battle, including the commander, Lieutenant Carlson, who had to be evacuated off the hill. Lieutenant Carlson had been assigned to the company while it was on Fiji, and had taken command on 10 December 1943. At the time of the fight on Hill 260, Sergeant Buckle was the company communications sergeant. His duty was to remain with the company commander and provide radio communications for him. After First Lieutenant Carlson was evacuated, Carlson spent time in the hospital. Upon his discharge from the hospital, he left the regiment and had no further contact with Company E.[62] The third platoon also took heavy casualties by this point. The second platoon again attempted to take the Japanese position with a charge and suffered even more casualties. With Lieutenant Carlson evacuated, Colonel Lowry took direct control of Company E, and Sergeant Buckle, as the communications sergeant, remained by his side for the remainder of the time Company E remained in the battle.

Colonel Lowry, originally from Florida, was a 1929 graduate of the Military Academy at West Point. He stood more than six feet tall, and never showed any fear to his men. At one point, he was walking forward on the hill, along with the commo sergeant and a runner, when a burst of Japanese machine gun fire opened up on them. Sergeant Buckle and the runner instinctively dropped flat on the ground. Colonel Lowry, still standing, looked down at them and calmly asked "what is the matter boys, are you scared?" His unwavering leadership, always from the front, brought him great respect from the men of Company E. Throughout the day, as Company E continued to lose men, Colonel Lowry repeatedly requested — begged — General McCulloch to release the remainder of Company G. His requests were denied.

As darkness fell, a seriously weakened Company E formed a perimeter on the southeast slope and dug in for the night.[63] During the night the Japanese made a bayonet charge in a failed attempt to dislodge Company F from the northern knob of the hill, while at the same time bringing in more reinforcements for their own defensive positions. The remaining Company E soldiers managed to hold their ground throughout the night. The morning of the 11th brought new assaults from the Japanese, as they began an all-out attempt to drive Company E from the southeast slope. Despite the fury of the assault, Company E held its position, killing twenty Japanese, but with their own strength diminished to only thirty-five men combat effective. By this time, all of the platoon leaders either were dead or wounded. First Lieutenant Fred H. Willard of first platoon was wounded 11 March. After his evacuation, he remained hospitalized until 20 March. First Lieutenant George A. Karl of second platoon was wounded on 10 March and evacuated. He remained hospitalized until 15 March. Second Lieutenant Hurley of third platoon was declared Missing in Action on 11 March, and declared Killed in Action 2 April 1944. After a short break in the attack, the Japanese, perhaps realizing the small number of American defenders still fighting, launched another company-sized assault, and again Company E, down to about a quarter of its original strength, refused to give ground. After the fight, only twenty-four enlisted men remained still able to perform duty, and many of them were wounded.

By midmorning of the 11th, the once lush jungle had been cleared by battle, and the men found little cover from Japanese eyes or the blistering sun. Again Colonel Lowry asked for Company G to be sent in. Following the second attack of the day, Company E began to withdraw behind the perimeter of the 182d rather than risk encirclement.[64] At last Company G was allowed to move forward, through the withdrawing remains of Company E, but failed to make progress against the Japanese, who were then launching a new attack. Colonel Lowry remained on the hill with Company G, although the next day tripped on one of the booby traps that dotted the hill, and was wounded enough to require medical

South knob on Hill 260 after the battle. Company E of the 182nd became the only unit to earn the Distinguished Unit Citation in the Americal Division during World War II for its gallant stand here. A most expensive piece of land in terms of blood expended to possess it.

evacuation. It was the last the men of Company E would see of the man who led them through their worst fight.[65] In all, Company E lost approximately one officer and fifteen men Killed In Action, and another three officers and ninety-three men Wounded In Action, of whom two later died of their wounds.[66]

The battle for Hill 260 dragged on for a few more days, with an attack each morning petering out after suffering several casualties. The 182d had become so weakened by the battle that companies from the 132d had to become actively involved in the fight. New attacks on 18 March by two companies of the 132nd failed to dislodge the Japanese. The Japanese were dug in around the thick roots of the tree. Fearing that the Japanese would use the observation post in the banyan tree themselves, the Americans began a concerted effort to bring down the tree. Mortars, flamethrowers, and tunnels were all employed to fell the tree. The top of the banyan tree finally toppled, leaving a charred stump about fifteen feet high. Regarding the degeneration of the battle into a struggle to take down the tree, the historian Jerry Cooper said "[p]erhaps never in the annals of warfare had soldiers spent so much effort, ammunition, and blood for the possession of one tree."[67] The flame thrower platoon of the 164th Regiment, then attached to the 182d Regiment, suffered about one third of its strength as casualties in the battle.[68] After the failure of the 18 March attack, the commander of the Americal decided that taking Hill 260 would not be worth the further casualties it would cause, and instead decided to neutralize the Japanese on it by constant artillery and mortar fire, while patrols around the hill effectively isolated the Japanese forces on it.

Someone had to take the blame for the failure to retake the hill despite the high casualty rate, and Lieutenant Colonel Lowry was relieved on 20 March, and returned to the

United States.⁶⁹ The men of Company E believed he had been made a scapegoat. Throughout the battle, Sergeant Buckle remained with Lowry, serving as radioman. Buckle believed that the main problem during the battle was a faulty chain-of-command. The regimental commander, Colonel William D. Long, had previously served mainly as a staff officer. Buckle believed he was assigned command of an infantry regiment in order to make him eligible for promotion to general. Brigadier General McCulloch was an artillery officer. Although McCulloch was supposed to advise Long, he effectively directed the regiment in battle. Neither Long nor McCulloch listened to Lowry's suggestions, although only Lowry was actually in the battle and thus knew what was going on.⁷⁰ Colonel Long was relieved of command of the 182d Regiment the next day.⁷¹ On the morning of 28 March, a composite company manned by the remaining men from the second battalion of the 182d began the daily attack, when an artillery spotter reported that he had observed two Japanese soldiers leaving the hill. When the company arrived on the top of the south knob, they found Japanese had all left, and Hill 260 was again in American hands.⁷² The next morning, the 182d turned control of the hill over to the first battalion of the 24th Infantry, the first unit of black infantry soldiers committed in the Pacific.

For their actions on Hill 260 on 10 and 11 March 1944, Company E of the 182d — the Richardson Light Guard — received the Distinguished Unit Citation.⁷³ The granting of the Distinguished Unit Citation to an individual company was rare; indeed, Company E was the only unit in the entire Americal Division to receive that honor during World War II. However, by the time of the battle for Hill 260, Company E contained only thirty-eight men who had served in the company since it left the U.S. in January 1942, and only eight men who had been in from its induction in January 1941, all of who had since become NCOs. These men were SSgt Thomas J. Burbine, SSgt Angelo Calcagno, SSgt Herbert W. Dillon, SSgt Harry E. Mohla, Sgt James C. Buckle, Sgt Anthony T. Cincotta, Sgt Joseph M. Conway, and Sgt James J. Sullivan. In addition, T/5 Louis F. Burbine, who had served in Company E. until he transferred to the service company after the Guadalcanal campaign, fought on Hill 260 as part of a flame thrower team. Of these nine men, seven were from Wakefield. Sgt Buckle received the Silver Star for his actions on Hill 260. Six of these men were wounded in the battle, although only SSGT Calcagno was wounded seriously enough to warrant his return to the United States.⁷⁴

On 21 November 1944, the Americal turned over operations on Bougainville to Australian II Corps and returned to training. This time, great emphasis was placed on amphibious operations. Rumors as to next assignment again spread through the division. The Americal left Bougainville in several waves between 8 January and 27 January 1945, its destination the island of Leyte in the center of the Philippines. The U.S. Sixth Army had been fighting on Leyte since late October 1944, against stubborn Japanese defenders. The first part of the Americal, the 164th Infantry Regiment, arrived on Leyte on 21 January 1945, and began the division's main role in the struggle, searching for Japanese forces on the island and killing them. Although General Douglas MacArthur termed this a mopping up operation, in reality it was a difficult and bloody operation lasting almost four months. Despite the earlier amphibious training, the actual landings were performed dockside. The 182d Regiment arrived a couple of weeks later, and by 18 February, the second battalion of the 182d was moving into action. By this time, twenty men who had been in since January 1942, remained, with three from the original company. The Americal fought pockets of Japanese defenders into March. On 23 December 1944, while training on Leyte, Company E got a new first sergeant. This man, Thomas J. Burbine, from Wakefield, was one of the last remaining men who had served in the prewar Company E, having first enlisted in 1937. Although he had been sent to an officers candidate school that past September, he had not completed the course and returned to Company E about a week before his promotion to first sergeant.[75]

Following relief from the mopping up mission on Leyte, the Americal made an amphibious assault at Talisay on the shores of Cebu on 26 March. The Americal splashed ashore at 8:30 in the morning, with the 182d having the mission of securing the railroad and highway crossings of the Mananga River. A total of nine men from the January 1942 company made it Cebu, with only three who had been in since January 1941.[76] From there, the Americal began a struggle lasting into the summer for control of the island. Although the liberation of Cebu City came within days, the Japanese defenders were scattered through the island and the Americal had to wage a long and bloody campaign against them. For their actions, the Americal received the Philippine Presidential Unit Citation. By June, the situation on the island had stabilized enough so that most of the Americal began more amphibious training by regiment. Despite the training and the continued mopping up operations on Cebu, the men were allowed some rest and recreation. But all understood that this period was the calm before the storm, and that storm was the invasion of the Japanese home islands.

Throughout the war, the town of Wakefield remained interested in Company E, but the scale of the participation by local residents diluted that focus on Company E considerably. By 1944, the town counted almost two thousand men and women in the federal military. W. Eaton's *History of Wakefield*, published while the war was still being waged, mentioned that the local company was "now part of the regular army, performing valiant service in the far-away islands of the South Pacific," but also noted that "through enlistments, drafts and selective service almost two-thousand men and women have joined the armed forces, serving in Africa, Italy, India, South Pacific Islands, Iceland, Alaska, England and other foreign countries."[77] Those from the town who entered as part of Company E back in January 1941, amounted to about 5 percent of the total residents who eventually served during the war, and the constant reassignment of men out of Company E due to dependents, illness, wounds, and general needs of the Army, not to mention death, meant that the percentage of Wakefield men in Company E — the RLG — dwindled to the point that by 1944, it was negligible as an expression of the town in war. Instead the town took pride in the breadth of places and units in which its sons and daughters served the nation in war.

Meanwhile, the local State Guard infantry company provided the town with a visual reminder of the existence of war, and carried on many of the traditions of the RLG. At a ceremony hosted by the State Guard company on 27 October 1944, Major Dolbeare was presented his Presidential Unit Citation, which the 1st and 2nd Battalion of the 182d had received for their service on Guadalcanal. Major Dolbeare had since returned from the South Pacific and was serving locally as an instructor in the First Service Command, headquartered in Boston.[78] By 1944, the Wakefield State Guard company had maintained its allotted strength of sixty-one enlisted men and three officers, but only through constant recruiting of new members. One hundred men had already been discharged from the company since its creation, about half for entry into federal service either through voluntary enlistment or conscription.[79] Two became officers, while many of the remainder became NCOs in the federal military.[80] The company held its weekly armory drills on Monday night, as had the RLG and Company E for generations. The Wakefield State Guard company held a party for Major Brockbank of the State Guard on the evening of 28 January 1945, at the Wakefield armory.[81] A magician followed a catered dinner, after which the men sang. The speaker, from the First Services Command, praised the State Guard and told them to continue their service to their town and to Massachusetts.[82]

A few months later, on 11 June 1945, the man who led the RLG during the Spanish American War, Colonel Gihon, died quietly at the Soldiers Home and Hospital in Chelsea, and was buried in a small Catholic cemetery after a memorial service at Saint Joseph's Church.[83] Childless, he had been living with a niece in Wakefield for several years. The local newspaper carried an extended obituary of Colonel Gihon, describing him as "one of Wakefield's best known citizens"[84] for more than fifty years, but his exploits were forty-seven years in the past. His death near the end of a long, bloody war that affected Wakefield and the RLG far more profoundly than had the Spanish American War muted the impact of his death. The descendent of the company he had commanded, Company E of the 182d, had been chopped to pieces just a year before on Hill 260. Wakefield residents by the thousands were scattered over the globe fighting on several fronts, and the experience of the RLG in 1898, with its one day of actual fighting and minimal casualties, must have seemed quaint. But its descendant in the 182d faced a potentially grim future in 1945.

The Americal Division was slated to participate in the invasion of the Japanese home islands, dubbed Operation Olympic, in early November 1945. The Americal, as part of General Walter Krueger's Sixth Army, had the tentative mission of securing beachheads on the southern shore of the main island of Kyushu. The Americal Division was grouped with the 42nd Infantry Division and the 1st Cavalry Division to make IX Corps, under Lieutenant General C.P. Hall.[85] Along with planning and training for the invasion, as well as making the acquaintance of the young women of Cebu City, the men of the Americal continued to send out patrols against the Japanese still holding the northern areas of the island. However, on 7 August an official press release told the men of the Americal Division about the destruction of the Japanese city of Hiroshima the day before. Rumors of Japanese capitulation began to spread through the ranks. However, the division leadership continued the planning and preparation for the invasion in case peace did not come. Finally, on 15 August, all offensive operations against the Japanese ended. Classes in fire support scheduled for that day were cancelled. In fact, all preparations for Operation Olympic ended, and instead, the Americal Division began training in riot control and working on their personal appearance in support of Operation Blacklist — the occupation of postwar Japan.[86]

Convincing their Japanese adversaries on Cebu of the reality of the surrender took some time, but on 28 August the largest Japanese command on the island surrendered to Major General William H. Arnold, the commander of the Americal. By 30 August, the Americal had already disarmed and processed 9,800 Japanese soldiers, a figure far larger than anticipated.[87] But the Americal had to begin movement to its assigned sector in Japan for occupation duty and turned operations on Cebu over to the 77th Infantry Division to complete the disarming and processing of the surrendered Japanese.

While the official surrender of the Japanese Empire to the United Nations was underway on the battleship USS *Missouri* in Tokyo Bay on 2 September, the Americal prepared to move to Japan for its new assignment. Although the Japanese Emperor had ordered his subjects to offer no resistance, American military planners worried rogue elements in the Japanese military might attempt to oppose American landings. For this reason, two plans for getting the Americal on Japanese soil were drafted; one for a beach assault landing and one for a peaceful dockside landing. In the event an assault was necessary, the 182d was to land abreast of the 132nd somewhere in Sagami Bay-Tokyo Bay area. However, the landing was peaceful and performed dockside in Yokohama. The Americal Division began disembarking in Yokohama on the morning of 8 September. The Americal had an overall area of responsibility to the immediate south and west of Tokyo, mostly in Kanagawa Prefecture, consisting of 1650 square miles, which would grow even larger. The 182d Infantry was assigned a central sector, with the regiment's area spreading about forty miles east to west.[88] Once established, the Americal began carrying out their occupation duties of destroying Japanese war material, arresting war criminals, making surveys of their areas, and in general seeing that local government continued to function, albeit in a non-militaristic manner. The Americal tried, beginning in October, to begin formal training again, but the burden of normal occupation duties and the lack of interest in men hoping to be discharged soon made such training poorly implemented.

The end of the war had brought a natural desire on the part of most soldiers to return to the United States and civilian life. Rumors spread through the Americal, with some saying the division would remain in Japan for the foreseeable future, where others had the division returning to New Caledonia for deactivation. In order to make the process of demobilizing soldiers as fair as possible, the U.S. military created a point system, whereby all soldiers were awarded points based on several factors such as time in service, time in combat zones, and other pertinent aspects of a soldier's time in the Army.[89] By the end of September, the point system began to affect the Americal. As the 43rd Infantry Division was due to move back to the United States, the Americal began rotating its men with higher point totals into the 43rd, while the 43rd sent its soldiers with lower point totals to other divisions, including the Americal. Thus the Americal lost its more experienced men, and began receiving men whose time in the Army and overseas was more limited.[90] On 14 October, the Americal learned that it had officially been classified as Category IV by the Army, meaning that it would be sent back to the United States and inactivated as soon as possible. The 1st Cavalry Division was to relieve the Americal Division of occupation duty, and shipment of the Americal back to the United States was slated to begin around 1 November. On 22 October, the 82nd Field Artillery of the 1st Cavalry Division relieved the 182d from its assigned areas around Hachioji and Hara-Machida. For Company E of the 182d, the war was really over. The regiment would ship back to the United States and the men in it would be out processed. Some of the men used their brief time remaining in Japan without official responsibilities to visit Tokyo and see some of the area, but most had little opportunity for sightseeing. Despite the relief by the 1st Cavalry, work remained. Equipment had to be turned in, gear had to be packed, and paperwork had to be completed, but the end of the time in the Army of the United States for the 182d Regiment was in sight.

As Company E of the 182d Infantry Regiment, part of the Americal Division, the Richardson Light Guard protected New Caledonia, fought on Guadalcanal, Bougainville, and in the Philippines. It earned the Navy Presidential Unit Citation, the Distinguished Unit Citation, and the Philippine Presidential Unit Citation. The company ended the war on occupation duty in Japan. But throughout the war, due to wounds, transfers, and reorganizations, Company E lost its Wakefield and even National Guard character. Selectees filled most of the ranks by the end of the war. As such, it hardly could carry the name "Richardson Light Guard," and most of the soldiers in the company at war's end knew little of the name or the town that had created the company a century earlier. But until the 182d Regiment, and with it Company E, was released from the Army of the United States back into the National Guard, the Army kept the unit lineage of the Richardson Light Guard.

During the second half of November 1945, the Army shipped the Americal to the United States to discharge the soldiers in it. The Army deactivated the 2nd Battalion, which included Company E, on 29 November 1945 at Fort Lewis, Washington. The remainder of the 182d was deactivated on 2 December 1945 at the Fort Lawton Staging Area.[91] Despite being deactivated, the regiment, albeit without any soldiers, remained part of the now deactivated Americal Division. The Army did not separate the 182d Regiment from the Americal Division and return it to the Massachusetts National Guard until 8 July 1946.[92] Neither the deactivation nor the separation from the Americal meant that the men of Company E of the 182d would soon be arriving in Wakefield's north station for a glorious homecoming. Few soldiers in the Americal as it arrived in Washington state for its demobilization had been in since its origins as Task Force 6814 in January 1942. General William Arnold, the last divisional commander, estimated that less than two-hundred soldiers in the Americal at the end had been in since it was formed on New Caledonia.[93] The 182d Infantry, and its Company E that carried the lineage of the Richardson Light Guard, had by this time no one who before the war had been a National Guardsman going to weekly drills at the state armory in Wakefield. Apparently only a single man who had belonged to the prewar Company E made it to Japan as part of the 182d Regiment, and he had been commissioned during the war and assigned to another company. That man was Captain William Blanchard, who had been a private in 1939. Blanchard had been commissioned through the Officers Candidate School the Americal held on New Caledonia, and later took command of the service company of the 182d Regiment. In all, six men who entered active duty with Company E in January 1941, received commissions during the war. Two men served as officers in the European theater, one of whom was killed in action. The other four served with various units in the 182d Regiment, one of whom eventually retired as a lieutenant colonel, and another as a chief warrant officer. [94] Company E of the 182d, carrying the lineage of the RLG, had evolved into something else, it had not been replaced. The lineage was not meaningless myth, as the continuity had been preserved. But Company E in the fall of 1945 was no longer the face of the town of Wakefield, Massachusetts, at war. Instead, the company that the town recognized as its own at the end of the war was the Sixth Company of the 23rd Regiment, Massachusetts State Guard.

While Company E of the 182d Regiment, the unit that officially held the linage of the Richardson Light Guard, served in the Americal, and thousands of Wakefield's sons and some of its daughters served throughout the world in all branches of the service, the State Guard company occupying the state armory in Wakefield continued to perform many of the local homefront roles of militia during wartime. Because the Sixth Company occupied the armory in Wakefield, and most of its leaders had long served in the prewar RLG, townspeople began to think of it as *the* Richardson Light Guard. However, not until 30 April 1946, five months after Company E of the 182d Regiment was deactivated, did the Wakefield State Guard company received permission from the state adjutant general to bear the name "Richardson Light Guard" in addition to its designation as the Sixth Company of the 23rd Regiment.[95] The name, which had been chosen by the founders of the town's militia company in 1851, now could only be bestowed by the state adjutant general with the permission of the U.S. Army. In a sense the granting of the lineage of the RLG to the Wakefield State Guard company marked the true return of the Richardson Light Guard to its hometown. The men who had left with Company E for a year of training in January 1941, either returned as individuals, or would not be returning at all. Of the five officers and 218 enlisted men who deployed with the company in January 1942, one officer and five enlisted men were Killed in Action, twenty-nine enlisted men were Wounded in Action, and most spent at least some time being treated for malaria, dysentery, and other maladies. The war had been too long, too disruptive. The sons and daughters from the town had served in countless units, and thus the RLG would receive no tumultuous welcome on its return.

Chapter X

Bedroom Suburb

Wakefield's State Guard company survived only a year after receiving the official designation as the Richardson Light Guard. During World War II, plans had been laid for the separation of the Air Force from the Army, although implementation of such a radical restructuring of the military was delayed until the end of the war. In 1947, the newly-created Department of Defense replaced the former War Department and Navy Department as a cabinet-level department. The War Department divided into the Department of the Army and the Department of the Air Force, and joined the Department of the Navy at the sub-cabinet level. Following a year and a half of uncertainty over the future of a state-based military reserve, the Department of the Army began organizing the postwar National Guard. On 16 December 1946, Wakefield learned that the town's postwar National Guard company would also be designated as Company E of the 182d, although the 182d was much enlarged[1] and would now be part of a Regimental Combat Team (RCT) and not part of a division.[2] Regimental combat teams contained an engineer company and a field artillery battalion in addition to the infantry regiment, although company E would still be an infantry company.[3] Despite an erroneous report in the local newspaper indicating the town would only have the infantry company, the armory would also continue to host a medical detachment.[4] The reformed company had none of the men who deployed with Company E in January 1942.[5] Still, the town was pleased to have a National Guard infantry company again, and the new Company E began forming and drilling at the armory. Company E received federal recognition the following April, meaning it had been accepted by Army inspectors and was then considered part of the National Guard.

Massachusetts maintained a large National Guard structure, with a total of more than 26,000 positions, which made it the seventh largest National Guard in 1948. Only California, Texas, New York, Pennsylvania, Illinois, and New Jersey had larger forces. Massachusetts contributed around $900,000 annually to its National Guard's budget between the end of World War II and the start of the Korean War, although this amount was increasingly dwarfed by federal support.[6] The federal government handed the National Guard an important tool for recruiting World War II veterans in June 1948, when it authorized retirement pay at sixty years of age for former Guardsmen who had served at least twenty years, five of which were on federal active service.[7]

However, Massachusetts, like most states, found recruiting its National Guard to its authorized strength difficult in the years immediately after World War II. By the end of June 1949, Massachusetts had only 11,479 men in its Army National Guard. It ranked 38th out of 52 states and territories in having its units federally recognized, with just under 80 percent. However, all twenty-seven units of the 182d RCT had received federal recognition by that time.[8] While Wakefield men served in the local company, the predominance of local men in Company E had ended, and Wakefield men more often than not served in units based in other towns.⁹

The National Guard, indeed the whole reserve structure of the United States military, had become too complex and specialized for a strictly town-based organization. Widespread car ownership allowed National Guardsmen to travel conveniently to surrounding towns for evening drill. The development of fleshed out units in the Army Reserve, as well as the enlarged reserves of the Navy, Marines, Coast Guard, and the newly separate Air Force, gave Wakefield residents a variety of organizations through which to serve part-time in the military.¹⁰ Service in the reserves became more attractive in 1948 when the federal government began to pay reservists for drill at rates comparable to those paid to Guardsmen. The National Guard itself divided in 1948 into the Army National Guard (ARNG) and the Air National Guard.¹¹ This tendency of men to serve in units away from their hometown became more practical when the National Guard and other reserve components began to hold drill assemblies one weekend a month rather than one weekday evening in the late 1950s and early 1960s. This development occurred over several years as National Guard units began using more of their 48 allotted training periods in combined weekend sessions. The Air National Guard began the practice earlier, but during 1957, the Army National Guard reported that 11 percent of its training periods were held on weekends. That figure had risen to almost 25 percent two years later. Using weekends rather than week-nights potentially doubled the training time of the Guard, as weekend training sessions required four hours to qualify for a pay period, while evening sessions required only two hours. The change in part allowed reserve component units to conduct more in-depth training during drills away from their armories, and break away from the idea of just meeting for a few hours in the evening. However, the practice of using two four-hour paid training periods per day on the weekend brought the additional requirement to provide a noon meal to the Guardsmen, adding to the cost of training.¹²

The fall of 1947 was an exceptionally dry season, and brush and forest fires were increasingly common throughout New England. On 23 October, a huge fire swept through the woodlands on the western side of North Reading, near the state sanitarium and threatening many cottages scattered on the small dirt roads around Martin's Pond.¹³ Fire departments from Boston to Nashua, New Hampshire, sent crews to help battle the blaze, reaching a total of around 1,000 men fighting the four-thousand acre blaze. Late in the afternoon of 23 October, a Thursday, Company E and two other companies from the 182d were called to active duty to assist. Under the command of Lieutenant Thomas H. Freeman, 38 men from Company E—close to its actual strength—fought the fire. The men worked overnight near the sanitarium, and returned to Wakefield at 8 o'clock the next morning. The *Daily Item* referred to the fire as the "worst forest conflagration ever known in this vicinity," although damage was mainly confined to forests and marshes.¹⁴ The service of the National Guard in the fire helped advertise that the Guard was back, and spurred sorely needed enlistments.¹⁵

The centennial of the RLG occurred in 1951, but little recognition came from the town. The nation was involved in the Korean War at the time, and about one-third of Army National Guard had been mobilized in support of the war.¹⁶ However, the 182d Regimental Combat Team remained at home, unmobilized throughout the Korean War. The town newspaper briefly mentioned on 10 October that the National Guard company was hosting a banquet at the armory the next evening, a Thursday, in celebration of the centennial.¹⁷ All former members, whether service had been in a Volunteer Militia, National Guard, or

State Guard incarnation, were invited to attend. Some 225 former members came to enjoy a banquet together and reminisce over the past.[18] Colonel Connelly served as toastmaster for the event, at which Colonel Otis M. Whitney, commander of the 182d, presented a "special unit presidential citation" streamer to the company guidon in recognition of the company's sacrifice at Hill 260 on Bougainville during World War II.[19] Dancing followed the banquet, presentation, and the speeches, and the event did not end until after 11 P.M. However the anniversary was a eulogy for the Richardson Light Guard, a remembrance of things past rather than a celebration of the present or future. On the following Monday, the newspaper carried a story about the event, and included a few columns giving a brief history of the company. Unlike milestones in 1901 and 1926, the town appointed no committee to oversee the event, and published no volume on the history of the company since 1926. The centennial was a company event, and not a town event. Gone were the distinctive uniforms and the designation as the "Richardson Light Guard." Postwar town budgets did not include even a token payment toward the company.[20] Although the new company was hardly an institution of the town of Wakefield, Company E at least bore the lineage of the Richardson Light Guard and had the state armory in the town as its home, so many of the older residents referred to it as the "Richardson Light Guard" out of habit, even if the unit never officially received that designation. A National Guard company existed in Wakefield, E of the 182d, but that substantial bond between town and militia company had been relegated to history and did not exist in the present or future.

The town of Wakefield in the first couple of decades after World War II had little interest in its past and instead looked toward the future. Many of the once proud factories that made the town of Wakefield a center for furniture, shoes, and textiles, had become ugly and run-down buildings, occupied by low-rent businesses that generated little wealth for the town. On 13 December 1950, a small fire broke out in the Town Hall, the one Cyrus Wakefield had built for the town. The fire spread through several offices in the building before the fire department extinguished it. Although the building itself survived, the damage was extensive in parts. Many of the town's offices were moved to the old high school building, the Lafayette Building, while the town entered into a long debate over the fate of the Town Hall. The debate dragged on almost eight years, centering on whether the old-fashioned building was worth the money that would be needed to restore it to continue to serve as the seat of town government. All the while, weather continued to damage the structure. Eventually, the March 1958, Town Meeting judged the estimated cost of $95,000 to restore the building a luxury they could not afford and voted to tear the building down and permanently move the town government to the Lafayette Building. The cost to move the town government and alter the Lafayette building to serve as Town Hall cost about $95,000, an irony not lost on later generations who would long regret the loss of Mr. Wakefield's most tangible legacy to the town.[21] In its place the town created a parking lot, which went a small way to ease downtown parking concerns but lacked the panache of the former edifice. At least the marble tablets bearing the names of the town's Civil War dead that once stood in the old Town Hall, which provided the justification for its construction, were moved to the Lafayette Building. The larger-than-life portrait of Cyrus Wakefield, the one that with a portrait of George Washington flanked the stage in the auditorium, also survived the fire and was salvaged from the building before it fell to the wrecking-ball. The portrait was moved to the high school, the one that had been built on the grounds of the former Cyrus Wakefield estate, so the school provided an appropriate place for the portrait of the man whose name the town bore.

In the early 1950s, the town of Wakefield, as well as most of eastern Massachusetts, was undergoing yet another profound change. Boston had become increasingly choked and traffic bound.[22] The prewar plans for a new state Route 128 as a limited access divided highway that would wend through the suburbs around Boston were given new emphasis. The new highway was named the Yankee Division Highway, after the 26th Infantry Division in which most of the National Guard of New England — although not the RLG — fought in the Great War. Construction resumed in 1949. The new Route 128 was one of the first limited access divided highways to loop around an American city. In essence, Route 128 connected Boston's North Shore to its South Shore by a circular route laying about ten miles from the center of Boston. Although critics derided it as the "road to nowhere," it quickly began the transformation of eastern Massachusetts as technology-based firms moved to Route 128 from Boston and Cambridge. The general sweep of the road around Boston would have placed it right across the southern end of Lake Quannapowitt, near Wakefield Common, the old burial ground, and the Congregation Church, so instead engineers planned the route with a sharp bump in its generally circular route as it bends north of the lake, staying just inside the Wakefield side of the Wakefield-Lynnfield and Wakefield-Reading town lines. This placement effectively separated Camp Curtis Guild from Wakefield. The camp had expanded by eminent domain in 1948,[23] and would continue to do so, but that expansion was away from Wakefield. Its front gate opened into an area that belonged to the town of Reading, and the camp's mailing address listed Reading, and thus its final link to the town of Wakefield was severed.[24] A small triangle of land at the southern end of the camp, containing part of the motor pool and a maintenance building, is on the northernmost spit of land belonging to Wakefield. At the end of the twentieth century, the camp contained 681.7 acres, of which 23.9 were in Wakefield, 285.1 in Reading, 332.7 in Lynnfield, and 40 in North Reading. However, as the entire camp is north of Route 128, few people realize that any of the camp is inside Wakefield. But Wakefield itself was firmly inside what would become the "inner loop[25]" of Boston, and the town became increasingly popular for both automobile commuters and high-tech related businesses.[26] The rural areas to the north of Wakefield Center quickly filled with houses, condominiums, and office parks. In 1959, Pleasure Island, an amusement park that sought to rival Disneyland, opened just north of Route 128, although it lasted only a decade. The high development density prevented the building of a true bypass, which spelled the decline and sometimes death of so many Main Streets throughout the United States, around Wakefield Center. Instead, Main Street in Wakefield continued to have a small-town feel, as if suspended in time and oblivious to the replacement of surrounding factories and farmland by office parks and subdivisions.[27]

Like the town of Wakefield, the United States Army was also in one of its periods of transformation during the 1950s and early 1960s. Trying to incorporate the lessons from the Korean War, and more importantly, to orient itself toward fighting in a war where each side might employ tactical nuclear weapons, the Army seemed constantly to reorganize. The National Guard began shifting units in and out of the Wakefield armory, further destroying whatever bonds still existed between the town and the local unit of the National Guard. In 1950, the Medical Company, as the unit sharing the Wakefield armory had been designated for two years, moved from Wakefield to Charleston, and the tank company of the 182d came from Charleston to Wakefield.[28] While the infantry company drilled on Monday evenings, the tank company drilled on Thursday evenings. The tank company was smaller than the infantry company, with only about 50 men authorized and less actually in the unit, and few came from Wakefield.[29] The company kept none of its tanks at the local armory, and only one at nearby Camp Curtis Guild.[30] For a few years in the early 1950s, the company brought the tank from the camp to the town to use in local parades, and its presence on Main Street during the Fourth of July parade was becoming a familiar sight.[31] However, the tank company only remained in Wakefield for five years, and was never designated as the successor to the Richardson Light Guard. Instead, heritage of the RLG remained with the infantry company. However, even the infantry company, E, was hardly suited to carry the moniker. The name had become outdated and was not applied to the company in any official connection. The Richardson family had no connections with the unit. No longer were the commanders prominent men from Wakefield. When a man from Greenwood, one of the smaller villages within the bounds of Wakefield, became commander in 1958, the novelty of a local man as commander was worthy of favorable comment by the surviving Fine Members Association.[32]

The Americal Division, now numbered as the 23d Division, was reactivated in December 1954, for service in the Canal Zone, but as a Regular Army division and with no connection to the 182d RCT. Instead it was filled with a combination of Regulars and draftees. The new Americal lasted until April 1956, when it was again deactivated. The Army reactivated it again for the Vietnam War from September 1967 through November 1971. During the Vietnam War it was again a Regular Army division fleshed out with draftees.[33] A more drastic restructuring of the Army began in 1956, which would have serious repercussions on the 182d. This was the adoption of the "pentomic" division structure for all infantry divisions—Regular Army, Army Reserve, and National Guard. Whereas the old infantry divisions had been built around three infantry regiments, in the new structure divisions were built around five "battle groups," which were smaller than regiments but larger than battalions. Each battle group contained four companies of infantry plus supporting units, whereas formerly a regiment had twelve infantry companies.[34]

For the 182d, indeed almost all regiments in the Army of the United States, it meant the end of the regiment, which had been the basic building block of American combat power since settlement, as a tactical unit. Instead, most "regiments" became simply lineages shared by two or three battalions.[35] In 1959, the 182d transformed from a Regimental Combat Team, with three infantry battalions plus supporting elements, into the First Battle Group, with four infantry companies. The armory in Wakefield was assigned only the Combat Support Company,[36] which inherited the lineage of the RLG. The designation of "Company E" was given to the company based in Everett.[37] The 182d again became part of the 26th Infantry Division, as it had been before January 1942.

In spite of the need to train for its federal mission, the National Guard companies in Wakefield, as well as the National Guard as a whole, would see little employment in their war fighting role between the end of the Korean War and the 1991 war against Iraq. For domestic political and global strategic reasons, the National Guard saw very limited involvement in the Vietnam War.[38] At the same time, the involvement of the National Guard in domestic law-enforcement was also in decline. Massachusetts, with a professional and large state police force, would seldom use its National Guard for responding to strikes or riots after the 1920s. Only the occasional major snow storm, flood, or fire caused Massachusetts to call up National Guard units for state missions. In 1954 and again in 1956, the Wakefield company spent a few days on state active duty in response to hurricanes,[39] but such missions were increasingly rare. The entire regiment was activated on 12 September 1960 in response to Hurricane Donna, and was activated piecemeal in response to the Chelsea Fire, with the Wakefield company serving on State Active Duty from 22 to 26 October 1973.[40]

The company's rifle team continued to hold competitions at the indoor range at the armory with .22 caliber rifles, with .30 caliber training held at Camp Curtis Guild.[41] In some ways, rifle training became more important as many of the basics of military training for new National Guardsmen that had been performed during evening and weekend drills were increasingly provided by the Army. In 1949, for example, the National Guard trained on a three year cycle based on 48 evenings a year, which would take a new recruit to a level of familiarity with the military. Recruits were trained by the National Guard at their armory, not by the Army.[42] However, the Army required beginning 1 October 1956 that at least 20 percent of enlistees in the National Guard either have prior military service or attend six months of active training with the Army. This requirement was increased a year later, when all non-prior service recruits were required to spend six months on active duty attending basic and advanced training with the Army.[43]

Although the absorption by the Army of the responsibility of providing basic training for new Guardsmen allowed the National Guard to concentrate on higher-level unit training,[44] it also broke much of the fierce identity with the National Guard rather than the Army that had earlier characterized the National Guardsman. It also made recruiting difficult. The Chief of the National Guard Bureau blamed the inability of the National Guard to recruit to strength in 1959 on several factors, the primary reason being the lack of enough slots for National Guard recruits at Army training centers. He also blamed the disinterest in the National Guard from men who had served their required two years of federal military service, and the decreased numbers of men drafted into the Army.[45] That negative motivation to join the National Guard provided by the draft would lead to an overabundance of potential recruits attempting to get into the National Guard as the Vietnam War led to increasingly high draft calls as the 1960s progressed.

During 1961, many units of the National Guard and Reserve were called to federal active duty in response to the Berlin Crisis, but the 182d did not deploy.[46] Still, increased tensions brought additional training time to the Wakefield company. In July 1962, it had a practice alert. The local newspaper reported on the exercise, noting that the local National Guard company was "formerly company 'E' and before that the Richardson Light Guard."[47] When the town newspaper mentioned the local National Guard company at all, it usually had to explain what the unit was, a scenario unthinkable in the late nineteenth century. A National Guard company was based in the state armory in the town, but was not the town's militia.

The pentomic divisional structure did not perform as well as had been expected during large exercises and was replaced in the early 1960s by the Reorganization Objectives Army Division (ROAD), which was based around three brigades. The adoption of ROAD did not, however, bring a return of the infantry regiment as a tactical unit. Under ROAD, infantry battalions fell directly under brigade commanders. Wakefield's Combat Support Company was converted back to an infantry company, designated as Company C.[48] It joined those of the nearby towns of Melrose, Malden, and Everett in making up the 1st (and only) battalion of the 182d in March 1963.[49] For National Guard and Army Reserve units, such reorganizations could wreak havoc on morale and readiness. Although members of the reserve components in heavily populated areas could travel to several units within an hour's drive from their homes, the redesignation of units usually meant many members had to learn new skills or attend drill at a different armory.[50] Ominously, the Army also decided that it could maintain ancient lineages of regiments as long as one of the battalions claiming that lineage remained in existence. For residents of Wakefield, this constant changing of the identity of the local National Guard company underscored its standing as part of the national military structure and not an institution of the town. The federal government decided what type of unit a National Guard company would be, not the members of the company or even the Commonwealth of Massachusetts. The Fine Members Association had largely faded away. With the constant shifting of unit identifications within the rump of the regiment, few former National Guardsmen by the 1970s identified strongly with a single company. In 1971, the "182d Infantry Association" formed as a club for former members of the regiment, which by then contained a single battalion.[51] The mission of the National Guard company in Wakefield changed several times during the decades after World War II in part because the Army itself was in a period of almost constant flux as it tried to orient itself toward modern warfare, and find again the proper role for the reserve components in war. This period of transition lasted into the 1970s, when the Department of Defense adopted the Total Force policy, which structured the military in such a way that fighting an extended war, such as the Vietnam War, would not be possible without using large amounts of the reserve components.

In the early 1970s, the Massachusetts legislature created the Special Commission on Military Affairs to study the state's National Guard and recommend any changes to its size, distribution, and any other matters over which the state still had any influence. Coming under scrutiny were the older armories, many of which were in need of massive spending to repair, maintain, and bring up to modern standards. In January 1973, rumors hinted that the state armory in Wakefield would be one of the armories closed as a result of the Special Commission's recommendations. The commonwealth's Quartermaster, Colonel William Quigley, dismissed such "erroneous" rumors, because, as he explained, Massachusetts wanted to close armories in need of repair, and Wakefield's was in good shape. Instead he said the armories in Salem and Springfield were most likely targets of closure. However, that August, the Special Commission concluded its two-year study, and recommended the closing of thirty armories, including Wakefield's.[52] The town was officially informed of the Commission's decision to close the armory in Wakefield in January 1974.

The town also learned that Massachusetts had no intention of replacing the armory with a new, modern structure to house a National Guard company in the town. Instead, the closing was part of a general realignment of the Massachusetts Army National Guard (MA-ARNG). Armories in nearby towns such as Stoneham, Everett, and Gloucester were also closing. The MA-ARNG had positions for 14,588 soldiers, but was cutting down to 12,473 soldiers. Federal law authorizing conscription of individuals into the federal military expired in 1973, although the military stopped conscripting in late 1972. After the draft ended in 1972, maintaining fully manned units in the National Guard had become problematic. While the National Guard did not draft men to serve, membership in the National Guard or Reserve made a man ineligible for individual conscription and thus the National Guard was usually filled to a level above 100 percent of its authorized strength while the draft was in force. Each armory closed would save an estimated fifteen thousand dollars. The *Daily Item* reported that "[t]he closing of the Armories, the spokesmen say, is in the interests of economy and efficiency, and they feel it will also 'improve manpower.'" National Guardsmen who belonged to units housed in the soon-to-be closed armories would not be involuntarily discharged, but were instead reassigned to below-strength units.

News of the imminent closure of the armory, and thus the end of a National Guard company based in Wakefield, had little impact on the town. Newspaper stories did mention that the grandson of Dr. Richardson had attended the opening of the armory in 1913, making the journey from Toledo to do so, and that his grandfather had donated money to start a militia company in 1851, but except for the one brief reference, made no further mention of the deep connections previously shared by the town and the local company. The story in the local newspaper did note that "the regionalization of the Guard is a departure from the traditional community-based character of the state military force, which predates colonial times," but voiced no regret at this loss to Wakefield.[53] Instead, the town looked to the closing of the state armory as an opportunity. Rather than hold ties to Company C of the 182d, the town valued the armory itself—the one built by the commonwealth and ever since maintained by the commonwealth—far more than it valued the National Guard company that occupied it. The end of a National Guard company based in Wakefield was relatively unimportant as long as the town would continue to be able to use the armory. All armories scheduled to be closed were built before World War II, and Wakefield's armory was considered to be in better shape than most of the others. Colonel Quigley told the town that sometimes the commonwealth traded old armories to towns in exchange for property on which to build a new armory, or sold them to the highest bidder. However, occasionally armories were sold to towns for one dollar for civic uses.[54] Armories often served other purposes besides housing a National Guard unit, with their large drill halls often serving as a municipal gymnasium or auditorium. Wakefield public schools had long used the drill shed for athletics, and the town wanted to continue using the building for this and other civic purposes.

Town officials created a subcommittee on Armory Acquisition, which sought to find if local clubs and organizations wanted space in the armory should the town acquire it.[55] The town entered into negotiations with the commonwealth adjutant general's office to purchase the armory for one dollar. With the reorganization of the National Guard scheduled to be complete by 1 April, the commonwealth hoped to have all the targeted armories vacated in June.[56] The Town Meeting voted to buy the armory for one dollar by a vote of 165 to 11.[57] Although the town voted overwhelming to acquire the armory, the dissenters worried the costs of maintaining the building would outweigh any benefits to the town. Massachusetts turned the armory over to the town, which in honor of the division in which the town's National Guard company had served in World War II, named it the Americal Civic Center.

And the final, by then tenuous, relationship between the town and the Richardson Light Guard came to an end. The Army preserved the lineage carried by the former Wakefield National Guard company, designated as Company C since 1963,[58] by combining it with the Support Company in Melrose to make a new Company B,[59] but for all practical purposes, the Richardson Light Guard, and any Wakefield militia company, was dead.[60] Except for a few years before the creation of the RLG in 1851, the town had a militia company since its founding. Now the town would be without a militia company, yet the only interest the town had in this end of a legacy was on the fate of the armory, not on the unit the armory was built to house. While the National Guard company was being reorganized out of existence, the daily newspaper in Wakefield covered the progress of the town in acquiring the armory, and the conquest of South Vietnam by North Vietnam. The town itself seemed to take only a passing interest in the final end of the town militia in Wakefield.[61]

During the spring of 1975, with the final chapter of the Vietnam War being played out largely without direct United States involvement, the prestige of the American military was at low ebb. However, the celebration of the American Revolution Bicentennial was at the same time gaining momentum. While the American Revolution was at heart a political

Looking north up Main Street next to the Rockery in 2012 (courtesy Michael Boucher).

struggle, the accompanying war for independence was of course a military struggle. Celebrating an essentially military victory at the end of the humiliating and divisive Vietnam War presented image problems for the Bicentennial. One of the ways this uncomfortably military aspect of the American Revolution was dealt with was by making "militia" a static and quaint institution of the distant past, and largely neglecting the role the Continental Army played in the victory.

The front page of the *Daily Item* on 8 April 1975, less than two weeks before the two-hundredth anniversary of the Battle of Lexington and Concord, contained the headline "Redding First Parish Militia and Minute Companies Reactivated." The accompanying story explained that the town's selectmen had authorized the creation of this new "militia" the past November. It was not a new Richardson Light Guard nor an inheritor of the town's National Guard company lineage. Instead the new "militia" company was a town-sanctioned reenactor group dressing in knickers and tri-cornered hats to help celebrate the Bicentennial. This town militia was purely ceremonial, and carried flintlocks loaded only with black powder. Towns throughout east coast states were creating similar organizations. The name chosen for this new "militia" company demonstrates the new static image for militia, existing in a time vaguely defined as the Colonial and Revolutionary period. The spelling of the town's name as "Redding" had been used only briefly, in 1643, while the spelling as "Reading" had been in common use for well over a century by the time of the Revolution. Similarly, a decade earlier, the "Linn Village Drum Band" formed in the town, likewise attired in clothing from the Revolutionary era, although taking a name for the town briefly used 130 years before the Revolution. These militia and minuteman companies, really groups of local reenactors pretending to be militia companies of the 1770s, did offer members a few of the benefits that the Richardson Light Guard had once provided men from the town; a chance to put on clothing different from that worn by other residents, meet together for a purpose, and occasionally parade through town. Of course unlike the RLG, participating in parades was the main focus of this new "militia," whereas in the old RLG, fighting wars was the reason for existence. But at the end of the Vietnam War, the concept of the town militia was something to be celebrated as existing safely in the past, rather than a vital part of the contemporary community. The newspaper claimed that this new group was the "very same militia company [that] marched to Concord on that fateful day of April 19, 1775."[62] That claim had far less validity than the lineage of the local National Guard company in the spring of 1975 to the Richardson Light Guard in 1851 but the irony seems to have occurred to no one. The newspaper would give regular coverage to the "Redding First Parish Militia" as it made appearances in Wakefield and surrounding towns. This was the acceptable military face of a leafy suburb of middle-class commuters.

* * *

The modern town of Wakefield remains a desirable suburb in the early years of the twenty-first century.[63] Its Common, with adjacent park at the southern end of Lake Quannapowitt, often throngs with residents and visitors, as well as flocks of sometimes disagreeable Canada geese. Many people make a daily four-mile walk around the lake for exercise. Wicker furniture still adorns many front porches around town, albeit no longer made in local factories. Wakefield's picturesque Main Street still draws shoppers and diners. Many of the central institutions of the town remain in the Center: a part of the former high school, now the junior high, sitting on what was once the Wakefield estate, the Beebe Memorial Library, the banks— now branches of much larger institutions— and the Baptist, Unitarian, Episcopalian, and of course, the Congregational churches. The Lafayette Building, stripped of it architectural grandeur and renamed, serves as the current Town Hall. But one town institution faded away and its passing is hardly noticed by most residents— the militia company.

The origins of the Americal Civic Center, and the adjoining Armory Street that runs next to it, are fodder for local history buffs and little more. The close proximity of Camp Curtis Guild, with its mailing address in Reading, allows some residents to assume a local National Guard company still exists, and even a few older residents assume that the RLG still exists somewhere, perhaps on the camp. While some National Guard units do use the camp as a home station,[64] none have deep roots in either Wakefield or Reading. The modern Massachusetts National Guard, like that of all states and territories, is largely a federal institution, with around 95 percent of its funding coming from the federal government, and the remainder from the commonwealth.[65] Prominent residents and towns contribute nothing to the budget.

The town never created a memorial specifically for its former militia company. Memorials exist for all residents who served in the Civil War, Spanish American War, and the wars of the twentieth century, but none for the RLG itself or Company E. Actually, the Hiker monument on the Rockery commemorates the Philippine War (formerly called the Philippine Insurrection) and the Boxer Rebellion — events in which the RLG did not participate — as well as the Spanish American War. Two bronze tablets engraved with the names of the major theaters of the Spanish American War formerly adorned the sides of the Rockery, near the fountain, but they have been missing for several years, leaving only their out-

The 1912 State Armory — the Americal Civic Center since 1975 (courtesy Michael Boucher).

line on the stones on which they were formerly mounted. The former armory, now the Americal Civic Center, honors the memory of the division in which the RLG served in World War II, but not really the company itself. In the basement, a room is designated as the "Light Guard Room," with a plaque over the door marking it as such — a small truncated memorial to what had once been one of the finest militia companies in the nation, and one that strips away the most personal bond between the militia company and the town, that of Dr. Richardson.

Although Cyrus Wakefield got the biggest honor in the nineteenth century, his legacy was more transitory than anyone could have suspected on that day in 1868 when his name became attached to the town. His Town Hall is gone, his mansion is gone, his former factory complex, long since renamed the Robie Industrial Park and converted to other businesses, was gutted by fire in the winter of 1972. Even his portrait, the one commissioned by Dr. Richardson that once shared the spotlight with that of George Washington in the Town Hall auditorium, disappeared from its last known location at the high school about the time when fire swept through that building in December 1971. The fate of the portrait remains unknown. The fire had not so totally destroyed the building that it would have been reduced to ashes, yet after the fire the portrait was not in its usual hanging place. It possibly was moved prior to the fire, or during the fire by a fire fighter and later unknowingly destroyed during fire fighting operations, or even stolen. The two redbrick buildings that survived the 1972 conflagration that destroyed most of the buildings of the former rattan factory, buildings that had long been eyesores, were torn down in early 2005 to make way for a supermarket. After fire gutted the Taylor building in the early 1970s, it was rebuilt without its distinctive fourth floor. In its new form, it was renamed the Wakefield Building. The destruction of so much of Wakefield by fire in the early 1970s was not coincidental. Throughout much of 1971 until 29 April 1972, Wakefield suffered through a series of major fires that ended with the arrest of a police officer, who later confessed to starting most of the fires. During the preceding 13 months, Wakefield had developed the reputation as the training ground for fire crews throughout the northern Greater Boston Area.[66]

The former armory, the wooden one built by the town in 1895 and damaged by fire in 1911, still stands, although it too is reduced to a nondescript commercial building, with few if any folks in the town aware of its former glory.[67] The name "Wakefield" remains on the town, but except for residents with an interest in local history, many are unaware of the connection. Massachusetts abounds with towns with "field" in their names, such as neighboring Lynnfield.[68] Folks unfamiliar with the history of Wakefield, or without an interest in rattan furniture, could easily assume that the town's name dates back to settlement and was perhaps chosen in honor of Wakefield, England.[69]

Light Guard Room in the basement of the Americal Civil Center, a humble and truncated memorial to what had once been one of the finest militia companies in the nation (courtesy Michael Boucher).

The Richardson family is also absent from the town. Solon O. Richardson III and his brother moved to the Mid-West generations ago. The militia company that bore his family name is long gone, memorialized on a forgotten basement door without the family name. Two streets in Wakefield bear the name "Richardson," as well as a commercial building on Main Street at the site of his former homestead. However, as Cyrus Wakefield's ghost has perhaps discovered, buildings are transitory monuments and not lasting claims to memory. The former Richardson house, the one moved back from Main Street at the beginning of the twentieth century to make way for the Richardson Building, burned while being torn down in 1971, erasing yet another tangible relic of what had once been one of the most prominent families in the town.[70]

The town of South Reading, with a mixed agricultural and light industrial economy in the mid-nineteenth century, evolved into the town of Wakefield, a bedroom suburb of Boston. During the same period, the Richardson Light Guard, part of the Massachusetts Volunteer Militia, evolved into Company C, of the 1st Battalion, 182d Regiment, Massachusetts Army National Guard, and was reorganized out of existence by forces beyond the town. The changes in names reflected changes in the community, commonwealth, and nation, and the way the land forces of the United States organized for war. The lineage between the company in 1851 and 1975 was as solid as the lineage of any military unit, but by the latter date the name "Richardson Light Guard" was a local curiosity and not the identity of the company. While the federal government did pull militia companies into a more centralized and nationally controlled structure, communities did not concurrently push their militia companies away. In the case of the Richardson Light Guard and its hometown, both town and militia company developed in separate directions during the twentieth century. Like former couples no longer in love, they simply drifted apart. Key events such as the requirement to take the dual oath in 1916, the opening of the new armory by the commonwealth in 1913, the ending of the election of officers in World War I, were all pivotal events in the dissolution of the ties between the RLG and the town of Wakefield, but the process was overall more evolutionary. Men such as Major McMahon and Colonel Connelly must have been aware of the increased federal involvement in the RLG, but registered no protest and instead probably saw the benefits that came with that federal involvement as outweighing the drawbacks. The massive impact of World War II on the

Rockery with the Spanish American War Memorial, and the Baptist Church behind it (courtesy Michael Boucher).

relationship between the RLG and Wakefield, with the local National Guard company representing only a small, and diminishing percentage of residents serving in the town, only accelerated a trend already well established. The post-war National Guard company never reestablished close ties with the town; both had become far more complex institutions.

Documenting the modernization of the town and militia company until they were quite separate institutions is in no way offered as an indictment of the National Guard, much less the modern town of Wakefield. As the Richardson Light Guard evolved during the twentieth century, so too did the town of Wakefield. Town and militia company each became less autonomous, more professional, and less susceptible to control by local prominent men. The nineteenth-century norm of a small group of wealthy men supporting and

The 1894 town armory in 2012 (courtesy Michael Boucher).

controlling the town government, businesses, banks, fraternal organizations, fire companies, and the militia company, had given way to a less personal, more professional managerial system. For the town in the 1970s to have held onto the local National Guard company as tightly as it held the RLG in the 1870s would have been anachronistic.

The story of the loss of a town militia is not simply an exercise in nostalgia, for it emphasizes the transformation of war away from an event that directly involved communities and local elites. The incorporation of most organized militia companies into the National Guard had the positive effect of increasing professionalism and competence. With the RLG, increased state and later federal control broke a critical bond between town and company. The story of the RLG demonstrates what has been lost with the disappearance of the town militia. The RLG was as much an institution of the town of Wakefield as were the schools, library, and fire departments. The RLG was created by a town, maintained by that town, later with the assistance of the commonwealth, and served as an institution of the town in war and peace. As the local National Guard company, which bore the heritage of the Richardson Light Guard, became more of an institution of the state and later the federal government, the bond with the town was broken, and the town militia ceased to exist.

Chapter Notes

Introduction

1. See David Blight, *Race and Reunion: The Civil War in American Memory* (Cambridge: Harvard University Press, 2001) on the changing way the Civil War was remembered and memorialized in the decades afterward.
2. *Wakefield Citizen and Banner*, 6 May 1898.
3. Susan G. Davis's *Parades and Power: Street Theatre in Nineteenth Century Philadelphia* (Philadelphia: Temple University Press, 1986) explores the use of the public parade and the role of the militia company in staging such parades as a means to bond the community together as much as to solidify the militia.
4. *Wakefield Citizen and Banner*, 6 May 1898.
5. Ibid., 29 April 1898.
6. Ibid.
7. Ibid., 6 May 1898.
8. Ibid.
9. Ibid.
10. Ibid.
11. Ibid.
12. William E. Eaton, *History of the Richardson Light Guard of Wakefield, Mass. 1851–1901* (Wakefield: Citizen and Banner, 1901), p. 127.
13. *Wakefield Citizen and Banner*, 6 May 1898.
14. Ibid.
15. Ibid.
16. W. Eaton, *History of the RLG, 1851–1901*, p. 129.
17. Photo of the event in the possession of the Wakefield Historical Society.
18. The literature on the development of the suburb and the inclusion of towns into metropolitan areas tends to stress the importance of railroads, street cars, automobiles, and, finally, the Interstate Highway System in drawing surrounding towns closer to central cities Sam Bass Warner, Jr.'s *Streetcar Suburbs: The Process of Growth in Boston, 1870–1900*, 2d ed. (Cambridge: Harvard University Press, 1979) was one of the first modern studies of the phenomenon, stressing, as its name implies, the role that streetcars—passenger street railways—had on this process. A more general study was Kenneth T. Jackson's *Crabgrass Frontier: Suburbanization of the United States* (New York: Oxford University Press, 1985), an unsympathetic look at the growth of the suburb as the dominant demographic development in the twentieth century. Robert Fishman also took a critical look at suburb development in his *Bourgeois Utopias: The Rise and Fall of Suburbia* (New York: Basic Books, 1987), while Henry C. Binford, like Bass before him, looked at the drawing in of towns immediately around Boston proper in his *The First Suburbs: Residential Communities on the Boston Periphery, 1815–1860* (Chicago: University of Chicago Press, 1988). Dolores Hayden, in her *Building Suburbia: Green Fields and Urban Growth, 1820–2000* (New York: Vintage Books, 2004), also followed Bass's work, stressing the role that technological advances, especially in mobility, had on suburban development. Alison Isenberg's *Downtown America: A History of the Place and the People Who Made It* (Chicago: University of Chicago Press, 2004) argued that town centers were more mythical than real, and while post–World War II developments of malls and box stores did cause decline to many downtown areas, the ideal was more imagined than real. And finally, Robert Bruegmann, in his *Sprawl: A Compact History* (Chicago: University of Chicago Press, 2005), takes a largely celebratory look at the development of the heavily developed large metropolitan areas, arguing that they arose because they fulfilled the needs of many people.
19. Michael E. Weaver's *Guard Wars: The 28th Infantry Division in World War II* (Bloomington: Indiana University Press, 2010) documents well the profound impact that several years of active federal service had on breaking the National Guard character and connections to original home communities of a division during World War II.
20. William M. Donnelly *Under Army Orders: The Army National Guard during the Korean War* (College Station: Texas A&M Press, 2001), ch. 5, "The Thunderbirds in Korea," on the deployment of one of the two Army National Guard divisions to serve in Korea during the war. While the 45th Infantry Division, from Oklahoma, deployed to war as a unit, its members were rotated home and discharged piecemeal, while the division itself remained in Korea, increasingly filled with draftees and Regulars. See also William Berebitsky, *A Very Long Weekend: The Army National Guard in Korea, 1950–1951* (Shippensburg, PA: White Mane, 1996).
21. John A. Logan, and Cornelius A. Logan, *The Volunteer Soldier of America with Memoir of the Author and Military Reminiscences from General Logan's Private Journal* (Chicago: R. S. Peale, 1887).
22. Although Emory Upton's *The Military Policy of the United States* (Washington: GPO, 1903) would not be published until 1904, copies of the manuscript passed through the War Department during the decades before the turn of the century. See also Stephen Ambrose, *Upton and the Army* (Baton Rouge: Louisiana State University Press, 1964).

23. One of the few such histories was John H. Sherburne, Jr., *Battery A Field Artillery M.V.M. 1895–1905* (Boston: Sparrell, 1908). While company histories are not unheard of, very few detail the life of a militia unit during peacetime. Sherburne was a lieutenant in the battery, and the battery paid for the book's publication. In a similar vein was *Squadron A: A History of Its First Fifty Years* (New York: Ex-Members of Squadron A, 1939). Rather a different sort of book is George N. Vourlojianis's *The Cleveland Grays: An Urban Militia Company, 1837–1919* (Kent, OH: Kent State University Press, 2002), one of the few modern studies of a single company over an extended period. Unlike the RLG, the Grays remained aloof from the National Guard and evolved into a private military organization. Bruce Allan Olson's Ph.D. dissertation at the University of Houston, "The Houston Light Guards: Elite Cohesion and Social Order in the New South, 1873–1940{in} (1989), showed that the lack of outside control in the nineteenth century allowed the Houston Light Guards to remain an elite institution, but increased federal control and responsibilities broke that autonomy in the twentieth century. One of the few academic works to focus on the relationship between a militia company and its hometown is G. Ward Hubbs's *Guarding Greensboro: A Confederate Company in the Making of a Southern Community* (Athens: University of Georgia Press, 2003). Hubbs found that the town of Greensboro, Alabama, formed around the militia company, the Greensboro Guards, in the decades before the Civil War. The Confederate loss of that war ended the existence of the militia company.

24. One of the first post-World War II works focusing on the colonial Massachusetts militia was Jack S. Radabaugh, "The Militia of Colonial Massachusetts," *Military Affairs* 18 (January 1954): 1–18. The approaching Bicentennial of the American Revolution and concurrent revival of popular interest in the Massachusetts Bay militia in part inspired Robert A. Gross, *Minutemen and Their World*, and Ronald L. Boucher, "The Colonial Militia as a Social Institution: Salem, Massachusetts 1764–1775," *Military Affairs* 37 (December 1973): 125–30. Other works from the past few decades focusing in the colonial Massachusetts Bay militia include Fred Anderson *A People's Army: Massachusetts Soldier and Society in the Seven Years' War*, and Richard P. Gildrie "Defiance, Division, and the Exercise of Arms: The Several Meanings of Colonial Training Days in Colonial Massachusetts," *Military Affairs* 52 (April 1988): 53–55. More recent is Kyle F. Zelner, *A Rabble in Arms: Massachusetts Towns and Militiamen during King Philip's War* (New York: New York University Press, 2009) which describes how Essex County raised expeditionary forces from the militia. One of the very few works to focus on the origins of militia in Virginia is William L. Shea, "The First American Militia," *Military Affairs* 46 (February 1982): 15–18], and *The Virginia Militia in the Seventeenth Century* (Baton Rouge: Louisiana State University Press, 1983). Shea demonstrates that militia in early Virginia was more ad hoc, with little or no existence during peacetime, and gradually evolved into a more exclusive institution more similar to militia in England..

25. Steven C. Eames, *Rustic Warriors: Warfare and the Provincial Soldier on the New England Frontier, 1689–1748* (New York: New York University Press, 2011).

26. Howard H. Peckham, *The Colonial Wars 1689–1762* (Chicago: University of Chicago Press, 1965).

27. Harold E. Selesky, *War and Society in Colonial Connecticut* (New Haven: Yale University Press, 1990).

28. Fred Anderson, *A People's Army: Massachusetts Soldiers and Society in the Seven Years War* (Chapel Hill: University of North Carolina Press, 1984).

29. Robert A. Gross, *The Minutemen and Their World* (New York: Hill & Wang, 1976). See also John R. Galvin, *The Minutemen, The First to Fight: Myths and Realities of the American Revolution*, 2d rev. ed. (McLean, VA: Pergamon-Brassey's, 1989). Galvin stresses that the minutemen at Concord were hardly an ad-hoc or spontaneously formed group of enraged farmers, but the product of a century and a half of institutionalized militia practices in New England.

30. Lawrence Dilbert Cress, *Citizens in Arms: The Army and Militia in American Society to the War of 1812* (Chapel Hill: University of North Carolina Press, 1982).

31. Marcus Cunliffe, *Soldiers and Civilians: The Martial Spirit in America, 1775–1865* (New York: Free Press, 1973).

32. National Guard Association of the United States, *The Nation's National Guard* (Buffalo: Baker, Jones, Hausaur, 1954).

33. R. Ernest Dupuy, *The National Guard: A Compact History* (New York: Hawthorn Books, 1971).

34. Jim Dan Hill, *The Minute Man in War and Peace: A History of the National Guard* (Harrisburg: Stackpole, 1964).

35. William H. Riker, *Soldiers of the States: The Role of the National Guard in American Democracy* (Washington: Public Affairs Press, 1957).

36. Martha Derthick, *The National Guard in Politics* (Cambridge: Harvard University Press, 1965).

37. John K. Mahon, *History of the Militia and National Guard* (New York: Macmillan, 1983).

38. Jerry M. Cooper, *The Rise of the National Guard: The Evolution of the American Militia, 1865–1920* (Lincoln: University of Nebraska Press, 1997).

39. Michael D. Doubler, *I Am the Guard: A History of the Army National Guard, 1636–2000* (Washington: GPO, 2001), pp. 142–49.

40. Jerry Cooper with Glenn Smith, *Citizens as Soldiers: A History of the North Dakota National Guard* (Fargo: North Dakota Institute for Regional Studies, 1986).

41. Eleanor L. Hannah, *Manhood, Citizenship, and the National Guard: Illinois, 1870–1917* (Columbus: Ohio State University Press, 2007).

42. The historiography of the early National Guard as a largely middle-class organization under the control of business leaders serving as an anti-labor force is well established. See, for examples of this view, Joseph John Holmes, "The National Guard of Pennsylvania, Policeman of Industry, 1865–1905," Ph.D. diss., University of Connecticut, 1971; Patrick Henry McLatchy, "The Development of the National Guard of Washington as an Instrument of Social Control, 1854–1916," Ph.D. diss., University of Washington, 1973; Robert Reinders, "Militia and Public Order in Nineteenth Century America," *Journal of American Studies* 2 (1977): 81–102; George G. Suggs, Jr., *Colorado's War on Militant Unionism: James H. Peabody and the Western Federation of Miners*, (Detroit: Wayne State University Press, 1972); Harry

Krenek, *The Power Vested: The Use of Martial Law and the National Guard in Texas, 1919–1932* (Austin: Presidial Press, 1980); Alan M. Osur, "The Role of the Colorado National Guard in Civil Disturbances," *Military Affairs* 46 (Feb. 1982): 19–24; and Brian Linn, "Pretty Scaly Times: The Ohio National Guard and the Railroad Strike of 1877," *Ohio History* 94 (Summer/Autumn 1985): 171–181. One work that breaks with the paradigm is Andrew Birtle, "Governor George Hoadly's Use of the Ohio National Guard in the Hocking Valley Coal Strike of 1884," *Ohio History* 91 (Annual 1982): 37–57. Birtle argues that the National Guard played a neutral role in enforcing the law during the labor unrest.

43. Hannah, *Manhood, Citizenship, and the National Guard*, pp. 4, 78–102.

44. The breakdown of local communities is probably best explained by Thomas Bender in his *Community and Social Change in America*. (Piscataway, N.J.: Rutgers University Press, 1978). Bender's study, which borrows from Ferdinand Toennies's Gemeinschaft-Gesellschaft concept of community, forms much of the underpinning of the current book.

Chapter I

1. "Records of Massachusetts Bay," vol. I, p. 272. Uncertainty over the proper spelling of the name of the town continued for several decades. A report submitted 10 September 1653 on the best route for a highway linking the town to Winnesimett (modern Chelsea) spells the name as "Reddinge." A letter of agreement written in 1708 by the town clerk regarding the establishment of a school uses the spelling "Readding," while a petition of 7 May 1719 on behalf of the Second Parish to use some of the land set aside by the town for the support of a minister employs the spelling "Reeding." On a map entitled "An Exact Mapp of New England and New York," in C. Mather's *Magnalia Christi Americana* (London, 1702), the town is labeled as "Reding."

2. Lilley Eaton, *Genealogical History of the Town of Reading, Mass. Including the Present Towns of Wakefield, Reading, and North Reading with Chronological and Historical Sketches, from 1639 to 1874* (Boston: Alfred Mudge & Son, 1874), p. 1.

3. Mass. Bay Records, Vol. II, p. 73. Quoted in Paul Guzzi, Secretary of the Commonwealth, *Historical Data Relating to Counties, Cities, and Towns in Massachusetts* (Boston: Commonwealth of Massachusetts, 1975), p. 55.

4. See William P. Cumming's *British Maps of Colonial America* (Chicago: University of Chicago Press, 1974), p. 33, for a detail of the Mather map showing eastern Massachusetts Bay, Plymouth, and Rhode Island colonies.

5. Loea Parker Howard, *Ancient Redding in Massachusetts Bay Colony: Its Planting as a Puritan Village and Sketches of Its Early Settlers from 1639–1652* (Boston: Thomas Todd, 1944), p. 25.

6. The name of the struggle that occurred between 1642 and 1649, primarily in England but also involving Wales, Scotland and Ireland, has changed often in the past and is currently in flux. Originally called the Great Rebellion, it later was known as the Puritan Revolution. The name English Civil War lasted the longest, but this name came under attack, and the struggle had been increasingly known as the British Civil Wars. By the end of the twentieth century, the rejection of strictly Anglocentric interpretations of the struggle led to the increased use of the name War of the Three Kingdoms, but this name has also had its detractors and the issue is probably far from settled.

7. George R. Stewart, *Names on the Land* (New York: Random House, 1945), p. 49. The town of Hull, on a peninsular south of Boston, received its name in this manner in 1644.

8. The social and economic development of New England towns during settlement and the colonial era is well-documented in Philip J. Greven, Jr.'s *Four Generations: Population, Land, and Family in Colonial Andover, Massachusetts* (Ithaca: Cornell University Press, 1970). Andover was the town immediately north of Reading. Kenneth A. Lockridge shows much of the same development in his *A New England Town: The First Hundred Years* (New York: W.W. Norton, 1970), which focuses on Dedham, about nine miles southwest of Boston. However, neither author includes the town militia as a subject of his study. An older study, Sumner Chilton Powell's *Puritan Village: The Formation of a New England Town* (Middletown, CT: Wesleyan University Press, 1963), shows what the earliest settlers took from England, and what they created after they settled.

9. The original warrant is held by the North Reading Historical and Antiquarian Society.

10. For more on the developing nature of Puritanism, see Edmund S. Morgan's *Visible Saints: The History of a Puritan Idea* (Ithaca: Cornell University Press, 1963), an older but still accurate explanation of Puritanism in New England. Perry Miller's *Errand into the Wilderness* (Cambridge: Harvard University Press, 1956) and *The New England Mind in the Seventeenth Century* (Cambridge: Harvard University Press, 1983) also provide useful context of the Puritan world view. Stephen Foster's *The Long Argument: English Puritanism and the Shaping of New England Culture, 1570–1700* (Chapel Hill: University of North Carolina Press, 1991) stressed the dynamism of Puritanism in the New World, as it adapted to new situations in New England.

11. See Alden T. Vaughan, *New England Frontier, Puritans and Indians 1620–1675*, 3d ed. (Norman: University of Oklahoma Press, 1995), for more on the complex and evolving relationship between Puritan settlers in New England and the Natives.

12. L. Eaton, *Genealogical History*, p. 11.

13. Jack S. Radabaugh, "The Militia of Colonial Massachusetts," *Military Affairs* 18 (1954), 2. See also his "The Military System of Colonial Massachusetts, 1690–1740," Ph.D. dissertation, University of Southern California, 1965, which focuses on the critical years when wars against the French led to more British involvement in the colonial militia.

14. The English militia can be traced to an institution known as the "Fyrd" in Saxon areas of Britain. The Fyrd enrolled free men for training in the use of weapons for local and kingdom defense. Over the years, the Fyrd developed into the General Levy, which, like the Fyrd, was based on the obligation of all able-bodied free men to perform military service within their own districts, and in the event of invasion or rebellion, march to any part of the kingdom for service. See C. Warren Hollister, *Anglo-Saxon Military Institutions on the Eve of the*

Norman Conquest (Oxford: Oxford University Press, 1962).

15. See Radabaugh, "The Militia of Colonial Massachusetts," p. 14. For more on New England militia during the seventeenth century, see Doubler, *I Am the Guard*, pp. 14–21, and Mahon, *History of the Militia and National Guard*, pp. 14–26. Harold E. Selesky's *War and Society in Colonial Connecticut* (New Haven: Yale University Press, 1990), pp. 3–47, shows a similar development in a sister New England colony.

16. Nathaniel B. Shurtleff, ed., *Records of the Governor and Company of the Massachusetts Bay in New England* (Boston: W. White, 1854), vol. 5, p. 30.

17. Doubler, *I Am the Guard*, p. 16. See also Richard P. Gildrie, "Defiance, Diversion, and the Exercise of Arms": The Several Meanings of Colonial Training Days in Colonial Massachusetts," *Military Affairs* 52 (April 1988): 53–55, and Ronald L. Bucher, "The Colonial Militia as a Social Institution: Salem, Massachusetts, 1764–1775," *Military Affairs* 37 (December 1973): 125–30.

18. L. Eaton, *Genealogical History*, p. 276. Chelsea, established as a separate town in 1739, had earlier been known as Winnissimet, Rumney Marshe, Pullen Poynte, and Number Thirteen.

19. See Jill Lepore's *The Name of War: King Philip's War and the Origins of American Identity* (New York: Alfred A. Knopf, 1998), pp. xv-xvii, on the problem of giving a name to the struggle, as well as the larger implications of the war. The standard military history of the conflict remains Douglas Edward Leach's *Flintlock and Tomahawk: New England in King Phillip's War* (New York: Macmillian, 1958), while Kyle F. Zelner, *A Rabble in Arms: Massachusetts Towns and Militiamen During King Philip's War* (New York: New York University Press, 2009), provides insight into how towns raised expeditionary forces for the war from the militia.

20. L. Eaton, *Genealogical History*, p. 24.

21. *Ibid*., p. 129. See Richard I. Melvoin's *New England Outpost: War and Society in Colonial Deerfield* (New York: W. W. Norton, 1989) for an account of the dangers faced by English settlements further west during King William's War and Queen Anne's War.

22. Chester E. Eaton and Warren E. Eaton, eds., *Proceedings of the 250th Anniversary of the Ancient Town of Redding Once Including the Territory Now Comprising the Towns of Reading, Wakefield, and North Reading, with Historical Chapters* (Reading: Loring & Twombly, 1896), p. 287.

23. More specifically, the colonial wars were known by the names King William's War (1689–1697), Queen Anne's War (1702–1713), King George's War (1744–1748), and the French and Indian War (1754–1763). In England, the names for the same wars were, respectively, the War of the League of Augsburg, War of Spanish Succession, War of Jenkin's Ear or Austrian Succession, and the Great War for Empire or Seven Years War. New England also fought Dummers War, also called Father Rale's War, in 1722–1725, but that war, fought mainly against French-allied Wabanaki Confederacy of Indian tribes in Maine and Acadia, was not part of a larger war.

24. C. Eaton and W. Eaton, *Ancient Town of Redding*, p. 287.

25. *Ibid*.

26. James A. Sawicki, *Infantry Regiments of the U.S. Army* (Dumfries, VA: Wyvern, 1981), pp. 355–57.

27. Lawrence Delbert Cress, *Citizens in Arms: The Army and Militia in Americans Society to the War of 1812* (Chapel Hill: University of North Carolina, 1982) covers this persistence of the colonial militia throughout the eighteenth century.

28. France did retain two small islands off the south coast of Newfoundland, St. Pierre and Miquelon, as well as several islands in the Caribbean.

29. Cress, *Citizens in Arms*, 48–49.

30. *Ibid*. See also T.H. Breen's *American Insurgents, American Patriots* (New York: Hill & Wang, 2010), 83–86, on the purging of the Massachusetts Bay militia of royalist officers.

31. Bruce N. Morans, *A Town that Went to War: A Chronicle of Our Bicentennial* (Reading: Bruce N. Morans, 1975), p. 18.

32. Morans, *A Town that Went to War*, p. 18.

33. For more on the social status of the minutemen, see Robert A. Gross, *The Minutemen and Their World* (New York: Hill & Wang, 1976). Gross studied the minutemen of Concord, Massachusetts Bay Colony, between 1763 and 1776. Concord was about fifteen miles west by southwest of Reading. In topography, demography, and economy, it was similar to Reading. See also Allen French's older but still useful *The First Year of the American Revolution* (Boston: Houghton Mifflin, 1934).

34. Doubler, *I Am the Guard*, pp. 28–29.

35. David Workman, "Revolutionary Years," in Nancy Bertrand, ed., *Wakefield: 350 Years by the Lake, an Anniversary History* (Wakefield: Wakefield Item Press, 1994), p. 29.

36. Hiram Barrus, "Reminiscences of the Revolution," reprinted as "Appendix HH" in L. Eaton's *Genealogical History*; also Gross, *The Minutemen and Their World*, p. 128.

37. William E. Eaton and History Committee, comps., *History of Wakefield, Massachusetts* (Wakefield: Wakefield Item Press, 1944), p. 39.

38. L. Eaton, *Genealogical History*, p. 176. For more on the role of the colonial militia in maintaining revolutionary discipline, see John Shy's "The Military Conflict Considered as a Revolutionary War," in *A People Numerous and Armed: Reflections on the Military Struggle for American Independence* (New York: Oxford University Press, 1975), pp. 193–224.

39. Morans, *A Town that Went to War*, p. 74.

40. L. Eaton, *Genealogical History*, pp. 5, 11.

41. W. Eaton, *History of Wakefield*, p. 124.

42. *The First Fifty Years History of the Golden Rule Lodge, A.F. & A.M, Wakefield, Massachusetts, 1888–1938* (Wakefield: Wakefield Item Press, 1938), pp. 144–45.

43. Jasper Ridley, *The Freemasons: A History of the World's Most Powerful Secret Society* (New York: Arcade, 1999), pp. 176–81.

44. *Ibid*., p. 186.

45. *First Fifty Years*, p. 145.

46. *Ibid*.

47. C. Edward Skeen, *Citizen Soldiers in the War of 1812* (Lexington: University Press of Kentucky, 1999), 4–61. See also Marcus Cunliffe's *Soldiers and Civilians: The Martial Spirit in America, 1775–1865* (New York: Free Press, 1973) for an in-depth study on the ideology that drove the political struggle over the role and organization of the militia in the years between the Revolution and the War of 1812.

48. "Report of the Committee on change of the Town's Name," 1868.

49. Guzzi, *Historical Data*, pp. 55, 68, 89.

50. W. Eaton, *History of the RLG, 1851–1901*, p. 3.

51. L. Eaton, *Genealogical History*, p. 392.

52. George N. Briggs, "Senate ... No. 15. Annual Report of the Adjutant General, 1847," p. 25.

53. Brig. Gen. Gardner W. Pearson, the Adjutant General of the Commonwealth of Massachusetts, *Records of the Massachusetts Volunteer Militia Called Out by the Governor of Massachusetts to Suppress a Threatened Invasion During the War of 1812–1814* (Boston: Wright & Potter Printing, State Printers, 1913), p. 96. The volume refers to the company as "Capt. Hay's Company," and notes that it was part of "Maj. W. Wood's Battalion of Rifles." The use of the commanding officer's name to identify military units was a common practice of the time. A note in the book mentions that the term "Massachusetts Volunteer Militia" in the title of the work was an anachronism, and instead the term "Massachusetts Militia" would have been accurate.

54. W. Eaton, *History of the RLG, 1851–1901*, p. 3.

55. Claire R. Snyder, *Citizen-Soldiers and Manly Warriors: Military Service and Gender in the Civic Republican Tradition* (New York: Rowman & Littlefield, 1999) ch. 4, "Civic Rituals of the American Citizen Soldier," more fully explores the role militia played in binding communities in the decades after the Revolution, while at the same time helping define the role of the masculine republican citizen.

56. Marcus Cunliffe *Soldiers and Civilians: The Martial Spirit in America, 1775–1865* (New York: Free Press, 1973), pp. 217–18, 230–41.

57. *Militia Act of 1792: An Act more effectually to provide for the National Defence by establishing a Uniform Militia throughout the United States*, ch. 33, sec. 1, *Statutes at Large of the United States of America* 1:271–274 (1846).

58. Sawicki, *Infantry Regiments*, pp. 355–57.

59. Doubler, *I Am the Guard*, pp. 87–90, chronicles the demise of the concept of the enrolled militia nationwide after the War of 1812. See also Russell F. Weigley, *History of United States Army* (New York: Macmillan, 1967), pp. 156–57.

60. Sumner's pamphlet, *An Inquiry into the Importance of the Militia to a Free Commonwealth*, quoted in Cunliffe, *Soldiers and Civilians*, pp.199–200. See also John K. Mahon's "A Board of Officers Considers the Condition of the Militia in 1826{in} *Military Affairs* 15 (Summer 1951); 85–94 for more on the role Sumner played in retaining the general militia and opposing attempts to classify the militia by age or other discriminator.

61. W. Eaton, *History of Wakefield*, p. 45.

62. Cunliffe, *Soldiers and Civilians*, p. 207.

63. L. Eaton, *Genealogical History*, p. 275.

64. Lena London, "The Militia Fine 1830–1860," *Military Affairs* 15 (Fall 1951), 133–44.

65. Clarence Danhof, *Change in Agriculture: The Northern United States, 1820–1870* (Cambridge: Harvard University Press, 1969).

66. W. Eaton, *History of Wakefield*, p. 163. For an in-depth study of the rise of the shoe industry in eastern Massachusetts between 1815 and the Civil War, see Alan Dawley, *The Industrial Revolution in Lynn* (Cambridge: Harvard University Press, 1976). Mary H. Blewett's "Work, Gender and the Artisan Tradition in New England Shoemaking, 1780–1860," *Journal of Social History* 17 (Winter 1983): 221–48, explores the sexual division of labor in the New England shoe industry, as well as the transformation of shoe making from a home-based craft system to centralized factory work.

67. L. Eaton, *Genealogical History*, pp. 347–48.

68. For more information see L. Murray Young's *Iron Men and Iron Machines: Wakefield Fire Department, Wakefield, Mass.* (Magnolia, MA: Dick Weir, 1976).

69. L. Eaton, *Genealogical History*, p. 423.

70. From a photo of the sign taken during the Fourth of July celebration in 1889. In the possession of the Wakefield Historical Society.

71. L. Eaton, *Genealogical History*, pp. 671–76.

72. Ibid., 671–73.

73. Cunliffe, *Soldiers and Civilians*, p. 209

74. Secretary of the Commonwealth of Massachusetts, *Acts and Resolves Passed by the Legislature of Massachusetts, in the Year 1840* (Boston: Dutton and Wentworth, Printers to the Commonwealth, 1840), ch. 92. (p. 233).

75. Ibid.

76. Ibid.

77. Doubler, *I Am the Guard*, pp. 88–89.

78. George H. Devereux, Adjutant General, *Senate ... No. 7 Annual Report of 1849*, p. 13.

79. Two militia caps from the period in the possession of the Wakefield Historical Society feature a metal plate on the front that served as visor and plate. On the plate section, the letters "S R R" are stamped.

80. George N. Briggs, "Senate ... No. 15. Annual Report of the Adjutant General, for 1847, submitted January, 1848," p. 25.

81. Ibid., p. 20.

82. Ibid. The report mentions the official date of disbandment of the Washington Rifle Grays, officially Company H, Fourth Brigade, Second Division, as 8 September 1847.

Chapter II

1. The Middlesex Canal, which opened in 1803, ran about five miles west of South Reading and had minimal economic impact on the town. The canal closed in 1851, mainly because it could not compete with the railroad. For more on the fundamental changes on the American economy wrought by canals, turnpikes, and railroads, and especially on their role in the replacement of subsistence farming and self-sustaining communities by the market economy, the classic on the subject remains George Rogers Taylor, *The Transportation Revolution of 1815–1860* (New York: Rinehart, 1951, rpt. New York: M.E. Sharpe, 1977).

2. Dolores Hayden, *Building Suburbia: Green Fields and Urban Growth, 1820–2000* (New York: Vintage, 2004), chs. 3 and 4, explores this linkage between the development of the commuter rail and the suburb. See also Kenneth T. Jackson, *Crabgrass Frontier: Suburbanization of the United States* (New York: Oxford University Press, 1985), pp. 87–102 on the impact of commuter rails. Robert Fishman in *Bourgeois Utopias: The Rise and Fall of Suburbia* (New York: Basic Books, 1987), pp.

134–154, shows a similar development in the Philadelphia area.

3. Henry C. Binford's *The First Suburbs: Residential Communities on the Boston Periphery, 1815–1860* (Chicago: University of Chicago, 1988) fully explores this phenomenon in the nineteenth century.

4. Doubler, *I Am the Guard*, pp. 90–96.

5. Jerry Cooper, *Rise of the National Guard*, pp. 17–18.

6. Secretary of the Commonwealth, "Chapt. 106," *Acts and Resolves Passed by the Legislature of Massachusetts, in the Year 1841* (Boston: Dutton and Wentworth, Printers to the Commonwealth, 1841), p. 382.

7. Ibid.

8. The Secretary of the Commonwealth of Massachusetts, *Acts and Resolves Passed by the General Court of Massachusetts, in the Year 1851: Together with the Messages* (Boston: White & Potter, Printers to the Commonwealth, 1852), ch. 104 (p. 70).

9. George N. Briggs, "Senate ... No. 15. Annual Report of the Adjutant General," for 1847, submitted January 1848, p. 12.

10. Ibid.

11. Fifty years after the event, William Eaton, in his *History of the RLG, 1851–1901*, p. 6, places the event on a "pleasant afternoon in late September, 1851." But as the original petition was submitted to the governor in June 1851, the September date is incorrect.

12. C. Eaton and W. Eaton, *Ancient Redding*, p. 76.

13. Letter 228 (13 June 1851), in "Petitions for New Companies Mass. Militia, 1795–1851," at the Massachusetts National Guard Museum and Archives (hereafter cited as "Mass. NG Museum"), Worcester, Massachusetts.

14. W. Eaton, *History of the RLG, 1851–1901*, p. 7.

15. George N. Briggs, "Senate ... No. 15. Annual Report of the Adjutant General," for 1847, submitted January 1848, p. 71.

16. The town itself left no record of any resident taking part in the Mexican American War, but Alfred S. Roe's *The Fifth Regiment Massachusetts Volunteer Infantry in Its Three Tours of Duty, 1861, 1862–'63, 1864* (Worcester: Blanchard Press, 1911), p. 329, mentions that Charles H. Shepard of South Reading had earlier served in the Mexican American War. He would have been about 18 years old during that war.

17. For more on the impact of the Mexican War on the United States, see Robert W. Johannssen's *To the Halls of the Montezumas: The Mexican War in American Imagination* (New York: Oxford University Press, 1985).

18. Secretary of the Commonwealth, "Chapt. 11," *Acts and Resolves Passed by the General Court of Massachusetts, in the Year 1852: Together with the Messages* (Boston: White & Potter, Printers to the Commonwealth, 1852), p. 70.

19. Letter 229 (dated 4 October 1851), in "Petitions for New Companies Mass. Militia, 1795–1851," at the Mass. NG Museum.

20. Special Order No. 28, in "Special Orders 1846–1856" (bound volume), at the Mass. NG Museum.

21. W. Eaton, *History of RLG, 1851–1901*, p. 12.

22. Edgar W. Martin, *The Standard of Living in 1860: American Consumption Levels on the Eve of the Civil War* (Chicago: University of Chicago Press, 1942), p. 413.

23. Cooper, *Rise of the National Guard*, p. 73.

24. W. Eaton, *History of RLG, 1851–1901*, p. 12.

25. Martin, *The Standard of Living in 1860*, p. 422.

26. Other Massachusetts companies of the period bore names such as the *Worcester City Guards*, the *Cambridge City Guards*, the *Wamesit Light Guard*, and the *Worcester Light Infantry*. See Doubler, *I Am the Guard*, pp. 93–4, on the propensity of volunteer militia companies nationwide to use the terms "light" and "guard" in their names. Units with these terms in their names include the Richmond Light Infantry Blues in Virginia, the Washington Light Infantry in South Carolina, the Hannibal Guards in New York, the Rock City Guards of Tennessee, Swatara Guards of Pennsylvania, the Wallace Guards of California, and the Boston City Guard. The use of a family name in the full name of a militia company was less common.

27. W. Eaton, *History of the RLG, 1851–1901*, p. 14, and L. Eaton, *Genealogical History*, pp. viii-xiii.

28. Ebenezer W. Stone, *Senate ... No. 10 Annual Report of the Adjutant General of the Commonwealth of Massachusetts for the Year ending December 31, 1851* (Boston: Dutton & Wentworth, State Printers, 1852), p. 38.

29. Ibid., p. 13.

30. W. Eaton, *History of RLG, 1851–1901*, p. 17. Although no examples of the RLG's original uniforms are known to exist, similar styles of early and mid-nineteenth-century New England militia uniforms can be seen in John Obed Curtis and William H. Guthman's *New England Militia Uniforms and Accoutrements* (Meriden, CT: The Meriden Gravure Company, 1971). The town of Wakefield maintains in its historical collection two examples of similar hats, from the earlier militia company. The hats have a metal plate on the front bearing the letters "SRR," which probably stood for "South Reading Rifles."

31. W. Eaton, *History of RLG, 1851–1901*, pp. 18–19.

32. Susan G. Davis, in *Parades and Power Street Theatre in Nineteenth-Century Philadelphia* (Berkeley: University of California Press, 1988), pp. 49–72, showed that in Philadelphia, public militia activities, especially parade and commemorative days, were often the only public spectacles celebrating the secular national culture, bringing to bear resources that towns and states lacked. See also R. Claire Synder's *Citizen-Soldiers and Manly Warriors* (New York: Rowman & Littlefield, 1999) on the role of militia in the role militia played in binding together men of a community.

33. The flag of Massachusetts bears the motto "Ense Petit Placidam Sub Libertate Quietem," generally translated as "By the sword we seek peace, but peace only under liberty."

34. W. Eaton, *History of the RLG, 1851–1901*, pp. 19–21.

35. Lajos Kossuth, *Kossuth in New England: A Full Account of the Hungarian Governor's Visit to Massachusetts with his Speeches and the Addresses that Were Made to Him, Carefully Revised and Corrected* (Boston: John P. Jewett, 1853), p. 69. In the list of militia companies that turned out for the event, the South Reading company is erroneously listed as the "Richardson Light Guards."

36. Ann S. Lainhart, comp., "1855 & 1865 State Census" (1990), p. 3. At the Massachusetts State Archives, Boston, Massachusetts.

37. In 1860, Massachusetts was 99.4 percent white, according to the *Eighth Census: Population*, pp. 593–94; *U.S. Statistical Abstracts, LVII* (1935), p. 3, Table 5.

38. *Ibid.*, p. 45.

39. W. Eaton, *History of Wakefield*, p. 125.

40. For more on the declining population of agricultural areas of New England during the second half of the nineteenth century, see Hal S. Barron, *Those Who Stayed Behind: Rural Society in Nineteenth Century New England* (New York: Cambridge University Press, 1984).

41. Ebenezer W. Stone, *Senate ... No. 3 Annual Report of the Adjutant General of the Commonwealth of Massachusetts for the Year ending December 31, 1853* (Boston: William White, Printer to the State, 1854), p. 6.

42. As with much of rural New England, North Reading lost population throughout the latter half of the nineteenth century. The nineteenth century population of the more agriculturally dependant North Reading peaked in 1860 at 1,203 after which it declined to 835 by the end of the century.

43. Cunliffe, *Soldiers and Civilians*, pp. 227, 294.

44. George H. Devereux, Adjutant General, *Senate ... No. 9 Annual Report of the Adjutant General* (1850), p. 16.

45. The town of Reading would create a nine-month Volunteer company during the Civil War, Co. D of the 50th Massachusetts Volunteers, but this unit was formed specifically for war service and disbanded as soon as its term of federal service ended. During the opening days of World War II, the town apparently created the "Reading Home Defense Corps," a home guard-type organization, although little information on it exists.

46. This displacement of local, town-based elites in the decades after the Civil War is explored in depth by Robert H. Wiebe in his *The Search for Order 1877–1920* (New York: Hill & Wang, 1967).

47. Jeremy Adamson, *American Wicker: Woven Furniture from 1850–1930* (Washington, D.C.: National Museum of American Art, 1993), p. 33.

48. Massachusetts Adjutant-General's Report for 1857 (Boston; 1858), 23–26.

49. W. Eaton, *History of the RLG, 1851–1901*, pp. 25–26.

50. *Ibid.*

51. W. Eaton, *History of Wakefield*, p. 178.

Chapter III

1. C. Eaton and W. Eaton, *Ancient Town of Redding*, p. 258.

2. Paul Faler, "Presidential Elections in Wakefield, 1832–1992," in Bertrand, ed., *Wakefield: 350 Years by the Lake*, p. 221.

3. W. Eaton, *History of RLG, 1851–1901*, p. 38.

4. *Ibid.*, pp. 36–39.

5. The Richardson Light Guard, from South Reading, has often been confused with a Volunteer company from Lowell, Massachusetts, called the "Richardson('s) Light Infantry." The Richardson Light Infantry of Lowell formed on 19 April 1861 specifically for service in the Rebellion. It took its name from the Hon. George F. Richardson, who "was foremost in securing enlistments and the necessary equipment for the company." See Greenleaf C. Brock's "Military Organizations," in *The Lowell Book* (Boston: George H. Ellis, Printer, 1899), pp. 30–32. The similarities of the names of these two military companies has long caused confusion. The book *Synonyms of Volunteer Organizations of the United States* (Washington, D.C.: War Department, Office of the Adjutant General, 1885) lists two companies with the name "Richardson Light Guard," the South Reading company and the Lowell company. However, in the copy of the book at the Massachusetts National Guard Museum and Archives, the entry for the Lowell company has the word "Guard" penciled through and "Infantry Lowell" written in. Additionally, the website www.letterscivilwar.com, includes a selection of letters written by members of the Richardson Light Infantry during the war. One letter, however, written by one "A. W.," on 2 May 1861, mentions that "Capt. Locke came in this moment and brought a package from the *South Reading Gazette*..." indicating that the letter has been erroneously credited to a member of the Lowell company. Books and websites for uniform button collectors of the Civil War era often show the "R.L.G." button worn by members of the RLG credited to the Lowell company. The Lowell company later became the Seventh Battery Light Artillery.

6. W. Eaton, *History of the RLG, 1851–1901*, pp. 38–97.

7. *Ibid.*, pp. 42–43.

8. Roe, *The Fifth Regiment*, pp. 24–25.

9. *Ibid.*, pp. 12–22.

10. Roe, *The Fifth Massachusetts*, pp. 24–25.

11. *Ibid.*, p. 27.

12. W. Eaton, *History of the RLG, 1851–1901*, p. 46.

13. "G. W. A." (probably 4th Sergeant George W. Aborn), letter of 27 April 1861, reprinted in the *Woburn Journal*, 18 May 1861.

14. *Ibid.*

15. W. Eaton, *History of the RLG, 1851–1901*, p. 40.

16. The MVM abolished those ranks by General Order on 10 April 1862.

17. "A. W.," letter of 2 May 1861, reprinted in the *Woburn Journal*, 18 May 1861.

18. *Ibid.*

19. "G. W. A.," letter of 27 April 1861, reprinted in the *Woburn Journal*, 18 May 1861.

20. *Ibid.*

21. Roe, *The Fifth Massachusetts*, p. 38.

22. *Ibid.*, p. 40.

23. Letter of 8 May 1861, reprinted in the *Woburn Journal*, 18 May 1861.

24. Roe, *The Fifth Massachusetts*, p. 42.

25. W. Eaton, *History of the RLG, 1851–1901*, p. 45.

26. "G. W. A.," letter of 8 May 1861, reprinted in the *Woburn Journal*, 18 May 1861.

27. *Middlesex Journal*, 1 June 1861.

28. *Ibid.*, 15 June 1861.

29. W. Eaton, *History of the RLG, 1851–1901*, p. 47.

30. Roe, *The Fifth Massachusetts*, p. 64.

31. W. Eaton, *History of the RLG, 1851–1901*, p. 50.

32. Roe, *The Fifth Massachusetts*, p. 70.

33. *Ibid.*, p. 76.

34. Zouaves were militia units influenced by the efforts of Elmer Ellsworth, a militia enthusiast from Chicago. Ellsworth was elected to command a militia company in Chicago in 1859, which he soon reformed as the United States Zouave Cadets of Chicago. Copying the red caps, sashes, and baggy pants, as well as the

intricate drill of the Zouave regiments of the French Army, the Cadets attracted much popular attention. After a twenty city tour in 1860, dozens of similar attired companies were created. Most Zouave companies discarded their fancy uniforms and adopted the standard army uniform after a few months of campaigning during the Civil War.

35. Roe, *The Fifth Massachusetts*, p. 73.
36. *Middlesex Journal*, 27 July 1861.
37. *Ibid*.
38. Roe, *The Fifth Massachusetts*, p. 91.
39. *Middlesex Journal*, 3 August 1861.
40. W. Eaton, in *History of the RLG, 1851–1901*, p. 3, claims the Grays disbanded in 1846, while Massachusetts Adjutant General's report for 1848 says it disbanded that year.
41. *Annual Report of the Adjutant General of the Commonwealth of Massachusetts, with Reports from the Quartermaster-General, Surgeon-General, and Master of Ordnance for the Year Ending December 31, 1864* (Boston: Wright & Potter, State Printers, 1865), pp. 504–06.
42. William Schouler, Adjutant General, *Annual Report of the Adjutant General of the Commonwealth of Massachusetts, with Reports from the Quartermaster-General, Surgeon-General, Commissary-General, and Master of Ordnance, for the Year Ending December 31, 1861* (Boston: William White, Printer to the State, 1861), p. 41.
43. W. Eaton, *History of the RLG, 1851–1901*, p. 57.
44. William Schouler, Adjutant General, *Annual Report of the Adjutant General of the Commonwealth of Massachusetts, with Reports from the Quartermaster-General, Surgeon-General, Commissary-General, and Master of Ordnance, for the Year Ending December 31, 1862* (Boston: Wright & Potter, State Printers, 1862), p. 61.
45. Schouler, *Annual Report of the Adjutant General of the Commonwealth of Massachusetts, for the Year Ending December 31, 1861*, p. 41.
46. Roe, *The Fifth Massachusetts*, pp. 328–34, 371.
47. The term "Evacuation of Boston" normally refers to an event that occurred on 26 March 1776, when, during the War for Independence, the British Army and Navy sailed away from Boston Harbor. The presence of the British ships in the harbor had become untenable after the Patriots emplaced on Dorchester Heights cannons brought from Fort Ticonderoga. The event is still celebrated as a holiday in parts of eastern Massachusetts.
48. W. Eaton, *History of the RLG, 1851–1901*, pp. 61–64.
49. *Ibid.*, pp. 65–68.
50. L. Eaton, *Genealogical History*, p. 466.
51. William B. Stevens, *History of the Fiftieth Regiment of Infantry Massachusetts Volunteer Militia in the Late War of the Rebellion* (Boston: Griffith-Stillings Press, 1907), p. 342.
52. L. Eaton, *Genealogical History*, pp. 597, 652–53.
53. Stevens, *History of the Fiftieth Regiment*, p. 4.
54. Schouler, *Annual Report of the Adjutant General of the Commonwealth of Massachusetts for the Year Ending December 31, 1862*, p. 40.
55. *Ibid.*, p. 61. By the end of the war, North Reading counted some 140 men who served.
56. *Ibid.*, p. 26.
57. Stevens, *History of the Fiftieth Regiment*, p. 5.
58. *Ibid.*, pp. 6–7.
59. For a more detailed, although biased, history of the overall Port Hudson campaign, see Edward Cunningham's *The Port Hudson Campaign, 1862–1863* (Baton Rouge: Louisiana State University Press, 1963).
60. Stevens, *History of the Fiftieth Regiment*, p. 9.
61. *Ibid.*, p. 43.
62. W. Eaton, *History of the RLG, 1851–1901*, p. 72.
63. *Ibid.*, p. 75.
64. Stevens, *History of the Fiftieth Regiment*, p. 51.
65. *Ibid.*, p. 55.
66. Henry A. Willis, *Fitchburg in the War of the Rebellion* (Fitchburg, MA: Stephen Shepley, 1866), p. 77.
67. W. Eaton, *History of RLG, 1851–1901*, p. 76.
68. *Ibid.*, p. 77.
69. Stevens, *History of the Fiftieth Regiment*, p. 89.
70. Diary of Private William C. Eustis, quoted in Stevens, *History of the Fiftieth Regiment*, p. 99.
71. *Ibid.*, pp. 107–08.
72. Technically, a mule and an ass are not the same thing, a mule being a cross between a male ass and a female horse.
73. Diary of Private William C. Eustis, quoted in Stevens, *History of the Fiftieth Regiment*, pp. 135–36.
74. Cunningham, *The Port Hudson Campaign*, pp. 82–93.
75. *Ibid.*, p. 111.
76. Stevens, *History of the Fiftieth Regiment*, p. 187.
77. *Ibid.*, p. 112.
78. On 30 December 1776, during the campaign against the British at Trenton during the War of Independence, General George Washington had to halt his movement to ask his men to extend their service for six weeks to allow him to continue the campaign after the new year. About half of the men did, while those not swayed by Washington's pleas were allowed to leave. See Don Higginbotham, *The American War of Independence: Military Attitudes, Policies, and Practices, 1763–1789* (New York: Macmillan, 1971), p. 169. During the Mexican American War, General Winfield Scott allowed his twelve-month volunteers, seven regiments in all, to leave the army in early June 1847, on the eve of the final drive to take Mexico City. See K. Jack Bauer, *The Mexican War, 1846–1848* (New York: Macmillan, 1974), pp. 269–27.
79. Henry T. Johns, *Life With the Forty-ninth Massachusetts Volunteers* (Washington: Ramsey and Bisbee, Printers, 1890), p. 301.
80. Stevens, *History of the Fiftieth Regiment*, p. 195.
81. W. Eaton, *History of the RLG, 1851–1901*, pp. 89–90.
82. L. Eaton, *Genealogical History*, pp. 466–67.
83. W. Eaton, *History of RLG, 1851–1901*, p. 97.
84. Robert B. Roberts, *Encyclopedia of Historic Forts: The Military, Pioneer, and Trading Posts of the United States* (New York: Macmillan, 1988), p. 405. This book, like many sources, erroneously places Camp Meigs in Reading, Massachusetts. Readville was not a separate town, but a village within the bounds of the town of Hyde Park, and thus finding it can be problematic. In 1911, Hyde Park was absorbed by Boston. Today the former village of Readville forms the southernmost portion of the corporate bounds of the City of Boston.
85. L. Eaton, *Genealogical History*, p. 632.

86. William Schouler, Adjutant General, *Annual Report of the Adjutant General of the Commonwealth of Massachusetts, with Reports from the Quartermaster-General, Surgeon-General, Commissary-General, and Master of Ordnance, for the Year Ending December 31, 1865* (Boston: Wright & Potter, State Printers, 1866), pp. 38–42.

87. *Massachusetts in the Army and Navy, 1861–65* (Boston: Wright & Potter and Co., 1895), p. 1019. Only one of the eight did not go into the Signal Corps.

Chapter IV

1. Mark C. Carnes, *Secret Ritual and Manhood in Victorian America* (New Haven: Yale University Press, 1989), describes this phenomenon of new fraternal organizations in the decades after the Civil War. Carnes sees the rise of fraternal organizations a reaction to the increase role of women in mainstream Protestantism, leading men to carve out masculine spheres elsewhere.

2. W. Eaton, *History of the RLG, 1851–1901*, pp. 98–99.

3. Ibid.

4. Public Document Number 7, *Annual Report of the Adjutant General of Massachusetts with the Report of the Quarter Master General for the Year Ending December 31, 1868* (Boston: Wright & Potter, 1869), p. 23.

5. Ibid., p. 101. See also *Memorial Volume of Ancient Redding*, pp. 48–9.

6. *Annual Report of the Adjutant General of Massachusetts for the Year Ending December 31, 1868*, p. 17.

7. L. Murray Young, *Iron Men and Iron Machines: Wakefield Fire Department, Wakefield, Mass.*, (Magnolia, MA: Dick Weir, 1976), p. 15.

8. *Annual Report of the Adjutant General of Massachusetts for the Year Ending December 31, 1868*, p. 24.

9. See Donald M. Douglas, "Social Soldiers: The Winona Company and the Beginnings of the Minnesota National Guard," *Minnesota History* 45 (Winter 1976): 130–140, and Martin K. Gordon, "The Milwaukee Infantry Militia, 1865–1892," *Historical Messenger of the Milwaukee County Historical Society* 24 (Mar. 1968): 2–15, on units in other states beginning the long transformation from social to military organization.

10. W. Eaton, *History of the RLG 1851–1901*, pp. 111–12.

11. Ibid., p. 109.

12. Ibid., p. 116.

13. W. Eaton, *History of Wakefield, Mass.*, p. 72.

14. Envelope, "Richardson Light Guard Fine Members Association," at the *Wakefield Daily Item* office, Wakefield, Massachusetts. Hereafter cited as "Daily Item office."

15. Cadet companies in high school grew from many of the same impulses that fueled growth in the organized militia, such as fear of the softening of males. See Julia Grant's "A 'Real Boy' and not a Sissy: Gender, Childhood, and Masculinity, 1890–1940," *Journal of Social History* 37 (2004): 829–51.

16. For more on the evolution of the New England Puritan meeting house/church, see Edmund W. Sinnott, *Meeting House & Church in Early New England: The Puritan Tradition as Reflected in their Architecture, History, Builders, and Ministers* (New York: McGraw-Hill, 1963).

17. L. Eaton, *Genealogical History*, pp. 677–83.

18. *Inaugural Exercises in Wakefield, Mass., including the Historical Address and Poem Delivered on the Occasion of the Assumption of Its New Name by the Town Formerly Known as South Reading on Saturday, July 4th, 1868; also, the Exercises at the Dedication of Wakefield Hall, Wednesday, February 22D, 1871* (Boston: Warren Richardson, 1872), pp. 6–9.

19. Howard, *Ancient Redding in Massachusetts Bay Colony*, pp. 1–3.

20. Although the Third Parish was established after the Second Parish, it soon outstripped the Second Parish in population. In 1853, the Second Parish, then called the North Parish, petitioned the Massachusetts General Court to become a separate town. The petition was granted and the former Second Parish became the town of North Reading. Although the main village in what became North Reading predated the main village of the former Third Parish, North Reading had no general movement to change its name. See Samuel M. Page, *A History of North Reading* (Wakefield: Wakefield Item Press, 1944).

21. Douglas Edward Leach, *Flintlock and Tomahawk: New England in King Philip's War* (New York: W.W. Norton, 1958), pp. 157, 158, 162.

22. *Inaugural Exercises in Wakefield, Mass.*, pp. 6–9.

23. Jeremy Adamson, *American Wicker: Woven Furniture from 1850 to 1930* (New York: Rizzoli International, 1993), p. 34.

24. *Inaugural Exercises In Wakefield Mass.*, p. 67.

25. L. Eaton, *Genealogical History*, p. 675.

26. Ibid.

27. W. Eaton, *History of the RLG, 1851–1901*, pp. 105–07. Mount Auburn, established in 1831, claims to be the first large landscaped public cemetery. It has since been designated as a National Historic Landmark by the U.S. Department of the Interior.

28. Ibid.

29. Adamson, *American Wicker*, p. 36.

30. Ibid., p. 73.

31. Russell Stanley Gilmore, "Crackshots and Patriots: The NRA and America's Military-Sporting Tradition 1871–1929," Ph.D. dissertation, University of Wisconsin, 1974.

32. W. Eaton, *History of the RLG, 1851–1901*, p. 178.

33. Ibid., p. 115.

34. W. Eaton, *History of Wakefield*, p. 244.

35. Folder, "Cold War Activities Report," at the Mass. NG Museum.

36. From box "Camp Curtis Guild," folder "Cold War Activities Report," at the Mass. NG Museum.

37. Carnes, *Secret Ritual and Manhood in Victorian America*, 7–8.

38. W. Eaton, *History of Wakefield*, p. 125.

39. *The First Fifty Years*, p. 19.

40. Ibid., p. 20.

41. W. Eaton, *History of Wakefield*, pp. 230–31.

42. Bertrand, *Images of America: Wakefield*, p. 84.

43. *Iron Men and Iron Machines*, pp. 18, 23.

44. W. Eaton, *History of Wakefield*, p. 165.

45. A son, Jasper, died in 1883, two weeks short of his fifth birthday.

46. http:www.wakefield.org/wicker/cyrus-wakefield-ii.htm (accessed 22 October 2010). Mrs. Wakefield was buried next to her husband in Lakeside Cemetery.

47. Bertrand, *Images of America: Wakefield*, p. 83.
48. A photo in the possession of the Wakefield Historical Society shows the mansion so decorated. Two signs bearing the dates "1868{in} and "1893{in} make clear the reason for the patriotic display on the mansion.
49. Bertrand, *Images of America: Wakefield*, p. 85. For more on the borrowing of stereotypical "Indian" culture in the United States, and even the adoption of these images by Native peoples, see Sally J. Southwick's *Building on a Borrowed Past: Place and Identity in Pipestone, Minnesota* (Athens: Ohio University Press, 2005).
50. *The Eighty-Third Annual Report of the Town Officers of Wakefield, Mass. for the Financial Year Ending January 31, 1895* (Wakefield: Citizen and Banner, 1895), p. 34. The committee contained Charles Woodward, Alestead W. Brownell, George H. Taylor, Edward J. Gihon, Silas W. Flint, and William B. Daniel.
51. *Ibid.*
52. *History of the Richardson Light Guard of Wakefield, Mass., Covering the Third Quarter-Century Period, 1901–1926* (Wakefield: Wakefield Item Press, 1926), pp. 61–63.
53. Robert M. Fogelson, *America's Armories: Architecture, Society, and Public Order* (Cambridge: Harvard University Press, 1989).
54. Sam Bass Warner, Jr., *Streetcar Suburbs: The Process of Growth in Boston, 1870–1900*, 2d ed. (Cambridge: Harvard University Press, 1979), for the classic study of the impact of the streetcar network of eastern Massachusetts on drawing surrounding towns closer to Boston. See also Hayden, "Building Suburbia," ch. 5, and Jackson, *Crabgrass Frontier*, pp. 103–15.
55. *Regiments and Armories*, p. 642.
56. *American Wicker*, pp. 67–8.
57. *Wakefield Banner*, 25 March 1898.
58. See Karen L. Cox's *Dixies Daughters: The United Daughters of the Confederacy and the Preservation of Southern Culture* (Gainesville: University Press of Florida, 2003) for more on the involvement of the UDC in establishing almost all Confederate memorials.

Chapter V

1. Leo J. Leamy's "Springfield's Citizen Soldiers in the Spanish American War," *Historical Journal of Massachusetts* 8 (June 1980): 30–45 chronicled the experiences of the three companies from Springfield that served in the Cuban campaign as part of the Second Massachusetts.
2. John H. Sherburne, Jr., *Battery A Field Artillery M.V.M. 1895–1905* (Boston: Sparrell, 1908), pp. 59–61.
3. Charles W. Hall, ed., *Regiments and Armories of Massachusetts: A Historical Narration of the Massachusetts Volunteer Militia, with Portraits and Biographies of Officers Past and Present* (Boston: W.W. Potter, 1899), vol. 1, p. 743. See also Charles Johnson, Jr., *African American Soldiers in the National Guard: Recruitment and Deployment During Peacetime and War* (Santa Barbara: Praeger, 1992), pp. 77–78.
4. Graham A. Cosmas, *An Army for Empire: The United States Army in the Spanish-American War* (Columbia: University of Missouri Press, 1971; rpt. College Station: Texas A&M University Press, 1998), p. 130 (page citations are to the reprint edition). See also William B Gatewood, Jr. "*Smoked Yankees" and the Struggle for Empire: Letters from Negro Soldiers, 1898–1902* (Urbana: University of Illinois Press, 1971), pp. 101–102, on the desire of blacks to see black officers, as well as black enlisted men.
5. Index cards of officers of the MVM, at the Mass. NG Archives.
6. *Wakefield Citizen and Banner*, 1 July 1898.
7. *Ibid.*, 6 May 1898.
8. Hall, ed., *Regiments and Armories of Massachusetts*, vol. 1, pp. 642–43.
9. *Wakefield Banner and Citizen*, 29 April 1898.
10. See Stephen P. Carlson with Thomas W. Harding, *From Boston to the Berkshires: A Pictorial Review of Electric Transportation in Massachusetts* (Boston: Boston Street Railway Association, 1990), pp. 43, 45.
11. *Ibid.*, 6 May 1898.
12. Hall, ed., *Regiments and Armories of Massachusetts*, vol. 1, pp. 642–43.
13. *Wakefield Citizen and Banner*, 1 April 1898.
14. *Ibid.*
15. Edwin P. Conklin, *Middlesex County and Its People: A History* (New York: Lewis Historical, 1927), vol. 5, 63–64.
16. Hall, ed., *Regiments and Armories of Massachusetts*, pp. 533–34.
17. *Massachusetts Soldiers, Sailors, and Marines in the Civil War* (Norwood, MA: Norwood Press, 1932), vol. III, p. 129.
18. *Wakefield Citizen and Banner*, 22 April 1998.
19. *Ibid.*, 29 April 1898.
20. *Ibid.*, 1 April 1898.
21. *Ibid.*, 29 April 1898.
22. *Ibid.*, 15 April 1898.
23. *Ibid.*, 22 April 1898.
24. *Ibid.*, 15 April 1898.
25. *Ibid.*, 29 April 1898.
26. *Ibid.*
27. Cosmas, *An Army for Empire*, p. 166.
28. *Wakefield Citizen and Banner*, 22 April 1898.
29. *Ibid.*
30. *Ibid.*, 29 April 1898.
31. Adjutant General's Office, "General Order #8," 18 May 1898, reproduced in "History Massachusetts State Guard," a monograph produced by the Massachusetts State Guard in 1978.
32. "Massachusetts Soldiers, Sailors, and Marines in the Spanish War, Tabulated by Cities and Towns, Vol. 2, N-Z," Handbound volume at the Mass. NG Archives.
33. See Gerald F. Linderman, *The Mirror of War: American Society and the Spanish American War* (Ann Arbor: University of Michigan Press, 1974), ch. 3, on the role the National Guard and state-raised regiments served in shielding potential war-time soldiers from what they saw as petty Regular Army discipline, and well as the desire to serve with men they knew from civil life.
34. *Wakefield Citizen and Banner*, 29 April 1898.
35. *Ibid.*
36. *Ibid.*
37. *Ibid.*
38. *Ibid.*
39. *Ibid.*, 6 May 1898.
40. W. Eaton, *History of the RLG, 1851–1901*, p. 129.
41. *Wakefield Citizen and Banner*, Fri., 6 May 1898.

42. *Ibid.*, 13 May 1898.
43. *Ibid.*, 20 May 1898.
44. "General Orders No. 8," Section III, relieved from duty men in the units of the MVM that reorganized for federal service, but who did not themselves volunteer for federal service.
45. *Ibid.*, Section IV.
46. Frank E. Edwards, *The '98 Campaign of the 6th Massachusetts U.S.V.* (Boston: Little Brown, 1899.), p. 125.
47. *Wakefield Citizen and Banner*, 29 April 1998. The newspaper erroneously reported the number of days past the signing of a peace treaty that the provisional companies would be allowed to exist as fifty. Also, "General Orders No. 8," Section X.
48. "General Orders No. 8," Section XI.
49. *Wakefield Citizen and Banner*, 17 June 1898.
50. *Ibid.*
51. *Ibid.*, 29 June 1898.
52. *Ibid.*, 1 July 1898.
53. *Ibid.*, 24 June 1898.
54. *Ibid.*, 29 July 1898.
55. *Ibid.*
56. *Ibid.*, 15 July 1898.
57. *Ibid.*, 22 July 1898.
58. *Ibid.*, 29 July 1898.
59. For more on home guard companies in general, see the current author's *The American Home Guard: The State Militia in the Twentieth Century* (College Station: Texas A&M Press, 2002).
60. *Wakefield Citizen and Banner*, 22 July 1898.
61. *Ibid.*, 6 June 1898.
62. The RLG was not part of the Sixth Regiment during the Civil War.
63. *Wakefield Citizen and Banner*, 27 May 1898.
64. *Ibid.*, 6 June 1898.
65. *Ibid.*, 1 July 1898.
66. *Ibid.*, 8 July 1898.
67. W. Eaton, *History of the RLG 1851–1901*, p. 135.
68. *Ibid.*, 10 June 1898.
69. *Ibid.*, 6 June 1898.
70. *Ibid.*, 8 July 1898.
71. *Ibid.*, 6 June 1898.
72. *Ibid.*
73. *Ibid.*
74. *Ibid.*, 10 June 1898.
75. *Ibid.*
76. *Ibid.*, 17 June 1898.
77. *Ibid.*, 1 July 1898.
78. *Ibid.*, 8 July 1898.
79. *Ibid.*, 6 June 1898.
80. *Ibid.*
81. *Ibid.*, 17 June 1898.
82. *Ibid.*, 6 June 1898.
83. *Ibid.*, 17 June 1898.
84. *Ibid.*
85. *Ibid.*, 6 June 1898.
86. *Ibid.*, 1 July 1898.
87. *Ibid.*, 24 June 1898.
88. *Ibid.*, 17 June 1898.
89. Cosmas, *An Army for Empire*, p. 131.
90. *Wakefield Citizen and Banner*, 1 July 1898.
91. *Ibid.*, 1 July 1898.
92. *Ibid.*, 24 June 1898.
93. *Ibid.* At its greatest strength, the Sixth Massachusetts counted forty-one officers and one thousand, two-hundred and thirty enlisted men.
94. *Ibid.*
95. *Ibid.*
96. *Ibid.*
97. *Ibid.*, 8 July 1898.
98. *Ibid.*, 1 July 1898.
99. *Ibid.*, 8 July 1898.
100. *Ibid.*, 22 July 1898.
101. Edwards, *The '98 Campaign*, p. 79.
102. Report of Brigadier General G. A. Garretson, commander of the brigade that included the Sixth Massachusetts, quoted in Hall, ed., *Regiments and Armories of Massachusetts*, p. 442.
103. *Correspondence Relating to the War with Spain* (Washington, D.C.: Center for Military History, 1993), p. 330. See also *Public Document No. 7, Annual Report of the Adjutant General of the Commonwealth of Massachusetts for the Year Ending December 31, 1898* (Boston: Wright & Potter Printing Co., State Printers, 1899), 165. Cosmas, in *An Army for Empire*, p. 234, erroneously states that the Americans suffered no casualties.
104. *Ibid.* However, General Miles, in his dispatch of 4:15 P.M. on 28 July 1898, to the Secretary of War, commenting on the fight, incorrectly refers to "Capt. Edward J. Gibson of Co. A" being wounded in the left hip. See *Correspondence Relating to the War with Spain*, p. 330.
105. W. Eaton, *History of RLG, 1851–1901*, p. 160.
106. *Ibid.*
107. *Ibid.*, pp. 162–64.
108. *Correspondence Relating to the War With Spain* (Washington: Center for Military History, 1993), p. 597.
109. Edwards, *The '98 Campaign of the 6th Massachusetts U.S.V.*, pp. 124–25.
110. *Public Document No. 7, Annual Report of the Adjutant General of the Commonwealth of Massachusetts for the Year Ending December 31, 1899* (Boston: Wright & Potter Printing Co., State Printers, 1900), p. 260.
111. Commonwealth of Massachusetts, Adjutant General's Office, "General Orders Number 5," 15 April 1899, in "History Massachusetts State Guard."

Chapter VI

1. U.S. Volunteers were similar to state Volunteer regiments but were created at the federal rather than the state level. U.S. Volunteers were originally created near the end of the Civil War. The concept was employed again during the Philippine War. The Army recruited men for the U.S. Volunteer regiments from soldiers in state Volunteer regiments pending discharge due to expiration of service.
2. Charles W. Hall, ed., *Regiments and Armories of Massachusetts: A Historical Narration of the Massachusetts Volunteer Militia, with Portraits and Biographies of Officers Past and Present* (Boston: W.W. Potter, 1899), vol. 1, p. 534.
3. W. Eaton, *History of Wakefield*, p. 192. The USWV chapter was named after Corporal Charles F. Parker, the member of the RLG who died on the transport and was buried at sea. Corp. Parker had served in the RLG for nine years.
4. Eaton, *History of Wakefield*, pp. 232–33.

5. *Ibid.*, p. 544.
6. *Richardson Light Guard, 1901–1926*, pp. 13–14.
7. *Ibid.*, p. 16.
8. W. Eaton, *Ancient Town of Redding*, p. 48.
9. *Wakefield Citizen and Banner*, 13 May 1898.
10. *History of the Richardson Light Guard, 1901–1926*, p. 20.
11. W. Eaton, *History of Wakefield*, p. 195. Also, from a photo of the event held by the Wakefield Historical Society.
12. *The One Hundred and Third Annual Report of the Town Officers of Wakefield, Mass. for the Financial Year Ending December 31, 1914, Also the Town Clerk's Record of the Births, Marriages, and Deaths During the Year 1914* (Boston: Falcon Press, 1915), p. 253.
13. Bertrand, *Images of America: Wakefield*, p. 51. The designating of roads as either streets or avenues seems to have been arbitrary. Most of the older roads that parallel Richardson Avenue are designated as "streets." Most likely, the road that bisected the Richardson estate was designated an avenue to avoid confusion with a separate .3-mile road, Richardson Street, which runs east from Main Street a few blocks south of the Richardson estate. Richardson Street had been laid out by the elder Dr. Solon O. Richardson several decades before his heirs created Richardson Avenue.
14. Cooper, *The Rise of the National Guard*, Appendix I.
15. *Ibid.*, pp. 134–35.
16. Martha Derthick, *The National Guard in Politics* (Cambridge: Harvard University Press, 1965), p. 26. See also Jim Dan Hill, *The Minuteman in War and Peace: A History of the National Guard* (Harrisburg: Stackpole, 1964), p. 328.
17. *Militia Act of 1903*, sec. 18. This part of the act stipulated that each unit would hold not less than 24 drills per year. Drill normally consisted of a weekly two-hour period on a weekday evening.
18. *Ibid.*
19. *Ibid.*, sec. 3.
20. George Kearney, ed., *Official Opinions of the Attorneys General of the United States Advising the President and Heads of Departments in Relation to Their Official Duties* (Washington: GPO, 1913), 29:322–329.
21. Section 7 of the *Militia Act of 1903* seems to dispute this. But the *Militia Act of 1908 (An Act to further amend the Act entitled "An Act to Promote the efficiency of the militia, and for other purposes*," ch. 204, sec. 43, *Statutes at Large of the United States of America* 35:399–403 [1909]) supports the notion of federal control only at the governor's consent.
22. Derthick, *The National Guard in Politics*, pp. 28–32.
23. *History of the Richardson Light Guard, 1901–1926*, p. 67.
24. Cooper, *Rise of the National Guard*, p. 134.
25. *Public Document No. 7 Annual Report of the Adjutant General of the Commonwealth of Massachusetts for the Year Ending December 31, 1914* (Boston: Wright & Potter, 1915), p. 8.
26. The photo is in the collection of the Wakefield Historical Society.
27. Property Account Number 35836, Parcel ID 18-222-005, Wakefield Assessor's office.
28. Cooper, *The Rise of the National Guard*, p. 133.

29. *Public Document No. 7 Annual Report of the Adjutant General of the Commonwealth of Massachusetts for the Year Ending December 31, 1911* (Boston: Wright & Potter, 1912), p. 65. ch. 151 of the "Resolves of 1911." The competitive bid to build the new armory was won by Whiton & Haynes of Boston.
30. *The One Hundredth Annual Report of the Town Officers of Wakefield, Mass. for the Financial Year Ending December 31, 1911* (Wakefield: Wakefield Item Press, 1913) p. 29. At the town meeting of 5 September 1911, Colonel Gihon submitted the report on the new armory.
31. From folder "Wakefield" at Mass. NG Museum.
32. *Public Document No. 7 Annual Report of the Adjutant General of the Commonwealth of Massachusetts for the Year Ending December 31, 1912* (Boston: Wright & Potter, 1913), p. 7.
33. *Wakefield Daily Item*, 21 April 1914.
34. *Ibid.*, 4 May 1914.
35. The Metropolitan Police were created in 1893 to enforce law and order in the Metropolitan Park System in eastern Massachusetts. Between 1913 and 1916, the governor used the Metropolitan Police during several strikes. In 1992, the Metropolitan District Commission Police, as they were by then known, were merged into the State Police.
36. *Wakefield Daily Item*, 8 July 1914.
37. Juliane Gerace, "The Rattan Factory Strike," in Bertrand, ed., *Wakefield: 350 Years by the Lake*, pp. 130–133.
38. *Wakefield Daily Item*, 27 June 1914.
39. *History of the Richardson Light Guard, 1901–1926*, pp. 67–69.
40. *Wakefield Daily Item*, 7 July 1914.
41. *Ibid.*, 24 April 1914.
42. *National Defense Act of 1916: An Act for making further and more effectual provisions for the National Defense, and for other purposes*, ch. 134, sec. 2, *Statutes at Large of the United States of America* 39:166–217 (1917).
43. *Ibid.*, sec. 111.
44. U.S. Const. sec. 8., art. 12, and sec. 2, art. 1.
45. *National Defense Act of 1916*, sec. 3.
46. *Ibid.*, sec. 197.
47. This option was followed by the Cleveland Grays in Ohio, which was a more upper middle class institution. See George N. Vourlojianis, *The Cleveland Grays: An Urban Military Company, 1837–1919* (Kent, OH: Kent State University Press, 2002).
48. *National Defense Act of 1916*; "An Act for making further and more effectual provisions for the National Defense, and for other purposes," *Statutes at Large of the United States of America*, vol. 39, part 1, pp. 166–217 (Washington: U.S. GPO, 1917).
49. *History of the Richardson Light Guard, 1901–1926*, p. 82.
50. "Sixth Infantry Headquarters Cos A-D 1895–1917," at Mass. NG Museum.
51. *Ibid.*
52. From photograph, included in *History of the Richardson Light Guard, 1901–1926*, p. 85.
53. *Ibid.*, pp. 87–88.

Chapter VII

1. *History of the RLG*, 1901–1926, p. 108.
2. *Ibid.*, p. 109.

3. Edwin C. Kemp, *Melrose, Massachusetts, 1900–1950* (Wakefield: Murray, 1950), p. 110.

4. *Ibid.*

5. "Sixth Infantry Headquarters Cos A-D 1895–1917," at Mass. NG Museum.

6. *History of the RLG, 1901–1926*, p. 113.

7. From a photo of the event belonging to the Wakefield Historical Society.

8. The term "National Army" referred to those organizations raised as part of the total Army, consisting primarily of drafted soldiers. Under the system created by the U.S. Army during the Great War, the modern division structure came into existence. Divisions numbered 1 to 25 would be from the Regular Army, 26 through 75 would be National Guard, while anything numbered higher than 75 would be National Army. Draftees filled the majority of positions in all wartime divisions.

9. *History of RLG, 1901–1926*, p. 118.

10. Weigley, *History of the United States Army*, pp. 354–56.

11. "General Order Number 95," War Department, 18 July 1917. Devens later served as U.S. Attorney General for the Hays Administration and as Commander in Chief of the Grand Army of the Republic from 1873 until 1875. Devens died in 1891.

12. Eaton, *History of the RLG, 1851–1901*, "Roster of the Richardson Light Guard."

13. From photo of the parade in the possession of the Wakefield Historical Society.

14. J.W.A. Whitehorne, "The Survival of the Duquesne Grays, 1917." *Military Affairs* 50 (October 1986): 179–84

15. *History of the RLG, 1901–1926*, pp. 125–32.

16. *Ibid.*, p. 128.

17. *Ibid.*

18. "Service Records World War I," vol. 29, at Mass. NG Museum.

19. *Wakefield Daily Item*, 14 October 1919.

20. W. Eaton, *History of Wakefield*, p. 92.

21. From box "Camp Curtis Guild," folder "Cold War Activities Report" at Mass. NG Museum.

22. Reserve Officers of Public Affairs Unit 4–1, *The Marine Corps Reserve: A History* (Washington, D.C.: GPO, 1966), p. 6.

23. *Message of 22 March 1917, from Governor McCall to the General Court.* Included in *Public Document No. 49, Fourteenth Annual Report of the Police Commission for the City of Boston: Year Ending November 30, 1919* (Boston: Wright and Potter, 1920).

24. *Ibid.*

25. Chapter 148 of the Acts of 1917, "To Provide for the Organization of a Home Guard in Time of War," approved 5 April 1917.

26. *Ibid.*, Section 1.

27. *Ibid.*, Section 6.

28. Sixty-Fifth Congress, 1st sess., ch. 28, 14 June 1917. In the Army Appropriation Bill for the year ending 30 June 1919, Congress authorized the payment of $2,500,000 to procure the arms and equipment authorized under the Act of 14 June 1917. 65th Congress, 2nd sess., ch. 143.

29. *History of the RLG, 1901–1926*, p. 96.

30. See Christopher Joseph Nicodemus Capozzola, *Uncle Sam Wants You World War I and the Making of the Modern American Citizen* (Oxford: Oxford University Press, 2008) for more on this aspect of the homefront during the war.

31. "Chapter 188 of the Acts of 1918," approved 2 May 1918.

32. Harry A. Benwell, *History of the Yankee Division* (Boston: Cornhill, 1919), p. 245.

33. *History of the RLG, 1901–1926*, p. 188.

34. *Wakefield Daily Item*, 14 October 1919.

35. *Wakefield Daily Item*, 15 October 1919. He was unable to attend, due to the death of his granddaughter in Ohio a few days earlier. Alice Richardson, the nineteen-year-old daughter of Solon III, had been ill for months.

36. From a photograph of the event in the property of the Wakefield Historical Society.

37. *Wakefield Daily Item*, 10 & 14 October 1919.

38. Doubler, *I Am the Guard*, pp. 184–88.

39. "Public Document No. 49," *Fourteenth Annual Report of the Police Commission for the City of Boston: Year Ending November 30, 1919* (Boston: Wright & Potter Printing Co., State Printers, 1920), p. 18.

40. *Ibid.*, p. 5.

41. *Ibid.*, p. 18.

42. Commissioner Curtis to Mayor Peters, September 10, 1919, in the *Report of the Citizen's Committee Appointed By Mayor Peters to Consider the Police Situation*, included in the *Fourteenth Annual Report of the Police Commissioner for the City of Boston: Year Ending November 30, 1919* (Boston: Wright & Potter Printing Co., State Printers, 1920).

43. *Ibid.*, Governor Coolidge's Proclamation, 11 September 1919.

44. *History of the RLG, 1901–1926*, p. 105.

45. *Wakefield Daily Item*, 13 October 1919; also *History of the RLG, 1901–1926*, p. 263.

46. For an example of War Department thinking on the National Guard immediately after the First World War, see the remarks of the Chief of the Militia Bureau, Brigadier General John S. Heavey, who was a Regular Army officer, to the National Guard Association meeting, in *Proceedings of the Convention of the National Guard Association Held at Richmond, Virginia November 14 and 15 1918* (n.p., n.c.).

47. Of course technically, under Massachusetts law, the National Guard was part of the MVM. While the *Wakefield Daily Item*, 13 September 1919, and the *History of the RLG, 1901–1926*, p. 263, referred to the new force as the "National Guard," Eben Putnam, ed., in *Report of the Commission on Massachusetts' Part in the World War History* (Boston: Commonwealth of Massachusetts, 1931), vol. I, p. 164, referred to the force as a new MVM, explaining that the commonwealth was uncertain over the future of the National Guard and was recreating the MVM in case the federal government decided not to recreate the National Guard.

48. Based on comparisons of rosters of the various incarnations of the RLG published in *History of the RLG, 1901–1926*.

49. Russell Stanley Gilmore, "Crackshots and Patriots: The NRA and America's Military-Sporting Tradition 1871–1929," unpublished Ph.D. dissertation, University of Wisconsin, 1974, pp. 167–70.

50. *109th Annual Report of the Town of Wakefield, Mass. for the Year Ending December 31st, 1920* (Wakefield: Abbot Press, 1921), p. 284.

51. Jesse F. Stevens, Adjutant General, *Adjutant Gen-*

eral's Reports, 1926–1927 (Boston: Commonwealth of Massachusetts, 1927), p. 20.

52. Ibid. Curtis Guild, Jr., joined the MVM in 1891 and rose to serve as adjutant general for the commonwealth. During the Spanish American War, he entered the Army as the adjutant of the Sixth Massachusetts Volunteer Regiment. He eventually ro2se to the rank of lieutenant colonel while serving as Inspector General of Havana. After the war he served as Lieutenant Governor from 1903 to 1905, and afterwards as governor for three one-year terms, declining to run for a fourth term. A lifelong Republican, he served as Special Ambassador to Russia from 1911 to 1913. He died in 1915.

53. From box "Camp Curtis Guild," folder "Camp Curtis Guild Map," at Mass. NG Museum. The Boxford Camp was owned by the Second Corp Cadets Camp Association from 1896 until 1930. The Second Corps of Cadets of Salem dated to 1786, and like the RLG, was being integrated into the National Guard. William Eaton, in his *History of Wakefield*, 1944, erroneously states that the name "Camp Curtis Guild" had been given to the Wakefield Rifle Range during the Great War.

54. Robert B. Roberts, *Encyclopedia of Historic Forts: The Military, Pioneer, and Trading Posts of the United States* (New York: Macmillan, 1988), p. 400. While this volume describes the Great War Camp Curtis Guild in Boxford, it makes no mention of the later Camp Curtis Guild in Wakefield.

55. Although the official name of the facility is "Camp Curtis Guild," locals usually refer to it simply as "Camp Curtis."

Chapter VIII

1. Following the tentative reorganization of the Massachusetts National Guard on 12 September 1919, the RLG became Company A of the Sixth Regiment (Provisional). In 1920 it became Company K of the Ninth Regiment (which was later designated as the 101st Regiment) before becoming E of the 182nd on 1 April 1923.

2. See "History of the 182D Infantry Regiment (North Regiment)," prepared by the Historical Services Office, Office of the Adjutant General [of Massachusetts], 2 December 2004.

3. *History of the RLG, 1901–1926*, p. 70.

4. *Historical and Pictorial Review of the National Guard of the Commonwealth of Massachusetts* (Baton Rouge: Army and Navy, 1939), p. 6.

5. Apparently Guardsmen attending the school were only temporarily attached to the school for training, but kept their ranks and assignments with their parent units until selected for commissioning. "The Training School Massachusetts National Guard Year Book 1930{in} (n.p., n.d.), at the National Guard Association Library.

6. *History of the RLG, 1901–1926*, p. 263.

7. *Wakefield Daily Item*, 11 June 1945. Colonel Gihon, childless, and a widower since 1915, lived at the Elks Club, a large home overlooking Lake Quannapowitt, until age finally forced him to move in with a niece near the end of his life.

8. *Ibid.*, p. 289.

9. *Wakefield Citizen and Banner*, 25 March 1898.

10. The statue should not be confused another work, also commonly referred to as the *Hiker* and also commemorating the Spanish American War, by Allen Newman. Newman's statue shows an infantryman in a more relaxed pose.

11. *Boston Daily Globe*, 2 October 1922. Unlike his father, he was buried at Lakeside Cemetery, with a tall gray granite memorial stone, near the grave for the elder Cyrus Wakefield.

12. *Wakefield Daily Item*, 5 October 1922.

13. *History of the RLG, 1901–1926*, p. 293.

14. *Boston Sunday Globe*, 1 November 1931.

15. See Jackson, *Crabgrass Frontier*, pp. 157–171, on the influence of the automobile on suburbs.

16. "Volume N.G. 1920–1941 182 INF 2BN," bound volume of handwritten records at the Mass. NG Museum. For the purposes of this study, members who listed their home as "Greenwood" were combined with those who listed "Wakefield" to get the total number of members from the town of Wakefield. Malden and Melrose also had companies in the 182d, with the Headquarters and Headquarters Company of the 3rd Battalion in Melrose, and Companies K and L in Malden.

17. Envelope, "Richardson Light Guard 1940 #6, Co. E. 182nd Information (News Release 1858)" at the *Wakefield Daily Item* office. The new unit was the Medical Department Detachment, 182nd Infantry. The Medical Detachment served the entire regiment.

18. *Wakefield Daily Item*, 19 September 1930.

19. *Ibid.*

20. *Ibid.*, 22 September 1930.

21. Roberts, *Encyclopedia of Historic Forts*, 397–99. In general, the term "camp" designates either a state-owned military training area, or a temporary, war-time federal mobilization or training area cantonment. "Fort" normally designates a permanent Army installation. However, several exceptions to this rule exist. Additionally, the Marine Corps normally designates its facilities as "camps."

. Diane Shaw, et al., *World War II and the U.S. Army M22obilization Program: A History of 700 and 800 Series Cantonment Structures* (Washington: GPO, 1993), p. 79.

23. U.S. National Guard Bureau, *Annual Report of the Chief of the National Guard Bureau* [hereafter cited as "ARCNGB"] 1934 (Washington, D.C.: GPO, 1934), p. 11.

24. Envelope, "Richardson Light Guard #5 1926–1939" at the *Wakefield Daily Item* office.

25. *Ibid.*

26. *Wakefield Daily Item*, 1 June 1936.

27. Bertrand, *Wakefield 350*, p. 219.

28. Robert Bruegmann, *Sprawl: A Compact History* (Chicago: University of Chicago Press, 2005), pp. 33–41. Bruegmann stresses that sprawl is a pattern as old as the city itself, and while he does not see sprawl itself as a bad thing, he does recognize that it creates problems that have to be addressed.

29. Bertrand, *Wakefield 350*, p. 90.

30. Yanni Tsipis and David Kruh, *Building Route 128* (Charleston, S.C.: Arcadia, 2003), pp. 7–8.

31. *Wakefield Daily Item*, 14 and 17 March 1936.

32. *Ibid.*, 21 March 1936.

33. Richard Dolbeare first entered the RLG on 18 July 1919, when he enlisted in the State Guard version during a recruitment drive following the expiration of the original two year enlistments.

34. *Wakefield Daily Item*, 21 March 1936.
35. *Ibid*.
36. Envelope, "Richardson Light Guard #5 1925–1939," at the *Wakefield Daily Item* office. Also, *Wakefield Daily Item*, 12 March-2 April 1936.
37. *Ibid.*, 31 March 1936.
38. *Ibid.*, 1 June 1936.
39. *Ibid.*
40. *Review of the National Guard of Massachusetts*, pp. 104, 107–09.
41. *Wakefield Daily Item*, 12 November 1936.
42. *Review of the National Guard of Massachusetts*, p. 79.
43. Data Sheet dated 16 June 1938, From folder "Wakefield" at Mass. NG Museum.
44. *ARCNGB*, 1940, p. 14. Despite these additional allotments of drill periods, the 182d Regiment recorded holding only the standard weekly drill.
45. *Ibid.*, p. 15.
46. Executive Order No. 8530.
47. W. Eaton, *History of Wakefield Mass*, p. 240.
48. *Ibid.*
49. Yankee Division Veteran's Association, *The History of the 26th Yankee Division, 1917–1919 1941–1945* (Salem, MA: Deschamps Brothers, 1955), p. 19.
50. This number comes from a comparison of the roster of the men mobilized in January, 1941, with the men listed in the company in the 1939 volume *Review of the National Guard of Massachusetts*. Additionally, Captain Dolbeare, who had long been a member of the RLG/Company E, entered active duty as a member of the regimental staff.
51. *History of the RLG, 1901–1926*, pp. 302, 305.
52. From the extensive records of the men who served in the company during World War II, compiled by James C. Buckle with the aid of David L. Cherry (hereafter cited as "Buckle records.") Mr. Buckle served in Co E, 182d, from 9 September 1940 until his discharge as a staff sergeant in late 1944. He served with Co. E on New Caledonia and in the Guadalcanal and Northern Solomons Campaigns, earning a Silver Star Medal, Bronze Star Medal, Combat Infantry Man Badge, and other decorations. Mr. Buckle's records are currently in his personal collection at his home on Cape Cod, but he plans to donate them to the Wakefield Historical Society.
53. *Reading Chronicle*, 17 January 1941. The same front-page story also listed Reading men in other units, but those in Company E were listed first, and were the largest group.
54. Buckle records.
55. Envelope, "#6 Richardson Light Guard 1940 Co. E. 182nd Information," at the *Wakefield Daily Item*.
56. Cyrille Leblanc and Col. Thomas F. Flynn, *Gardner in World War II* (Gardner: Hatton Press, 1947), p. 25. State Representative Fred A. Blake used this argument to help convince the town of Gardner to create a home guard company.
57. The *Reading Chronicle* regularly carried news of the force throughout 1941 and into 1943, such as the story on 24 January 1941 about its participation in an air defense test. See also Stephen D. Johnson and Gary S. Poppleton's *Cloth Insignia of the U.S. State Guards and State Defense Forces* (Hendersonville, TN: Richard W. Smith, 1993), p. 135, for an example of the unit patch worn by members.

58. As with the Massachusetts State Guard of the Great War, it had no federal standing and was created solely for state missions.
59. "M.S.G. Mo. Rosters 1941–42 23d Inf. 6th-8th Cos," three-ring binder at the Mass. NG Museum.
60. General Order # 7, Section VI, Paragraph 6, quoted in "History Massachusetts State Guard," p. 11.
61. *Ibid.*
62. *Ibid.*, p. 10.
63. *Ibid.*
64. Envelope, "Richardson Light Guard 1940 Co. E 182nd Information," 27 December 1940, at the *Wakefield Daily Item* office.
65. Folder "Cold War Activities Report," at Mass. NG Museum.
66. *The History of the 26th Yankee Division*, p. 19.
67. Buckle records, and Buckle, emails to author, 15 & 16 September 2006.
68. Buckle records.
69. James C. Buckle, telephone interview with author, 14 April 2006.
70. *History of the 26th Yankee Division*, p. 20.
71. *Ibid.*, p. 20. A regimental combat team of the time contained an infantry regiment, an artillery battery, an engineer company, and a medical company. A "battery" is a company-sized unit of artillery.
72. Public Law 213, 77th Congress, approved 18 August 1941, known as the "Service Extension Act."
73. *History of the 26th Yankee Division*, p. 21.
74. James C. Buckle, interview of 14 April 2006.
75. Brown was eventually discharged as a colonel in 1947, while Marchetti was discharged as a lieutenant colonel from the Army Air Forces in 1947. Foster was discharged as a lieutenant colonel in the Quartermaster Corps in 1946.
76. See Christopher R. Gables, *The U.S. Army GHQ Maneuvers of 1941* (Washington, DC: Center of Military History, 1992) for a history of the pre-war series of maneuvers and an analysis of their impact on the U.S. Army during World War II. Chapters 8 and 9 deal specifically with the Carolina Maneuvers.
77. Buckle records. Mr. Buckle published an account of the constant dwindling of the numbers of men from the pre-war company in the *Wakefield Daily Item* under the title "And Then There Were None" in three installments, on 5 April, 30 June, and 3 July 2000.

Chapter IX

1. Shelby L. Stanton, *Order of Battle U.S. Army World War II* (Novato, CA: Presidio Press, 1984), p. 101.
2. Interview with James J. McCarthy by John Coates, 10 July 2001. Part of the Morse Institute Library (Natick, Massachusetts, Community Library), Veterans Oral History Project.
3. Records of James C. Buckle, and email, Buckle to author, 21 September 2006.
4. McCarthy interview.
5. *Wakefield Daily Item*, 9 December 1941.
6. W. Eaton, *History of Wakefield*, p. 96.
7. Email, Buckle to author, 21 September 2006.
8. Dennis A. Connole, *The 26th "Yankee Division on Coast Patrol Duty, 1942–1943* (Jefferson, NC: McFarland, 2008), pp. 113–74.

9. Buckle records.
10. Stanton, *Order of Battle*, p. 101.
11. Buckle records.
12. *Ibid.*
13. Buckle records.
14. *Ibid.* To arrive at the totals for each town, the thirty-four Wakefield men who remained from the company at mobilization were combined with the twenty Wakefield men who remained from the second recruits to arrive at the total. Likewise, the eight Reading men from the original company were combined with the fifteen who remained from the second recruits. Additionally, Captain Dolbeare, who originally entered active federal service as a regimental officer, was a Wakefield resident.
15. Buckle, email to author, 5 June 2006.
16. Francis D. Cronin, *Under the Southern Cross: The Saga of the Americal Division* (Washington: Combat Forces Press, 1951), p. 4.
17. *Ibid.*
18. William K. Wyant, *Sandy Patch: A Biography of Lt. Gen. Alexander M. Patch* (New York: Praeger, 1991), p. 38.
19. Alabama State Military Department Office of the Adjutant General, *Quadrennial Report of the Adjutant General for the Four Year Period Ending September 30, 1942* (Montgomery: State of Alabama, 1942), p. 75.
20. Joseph J. D'Alessandro, interviewed by author, 10 April 2006, Wakefield, Massachusetts. Mr. D'Alessandro served with Co. E, 182d, from January 1941, through late 1943, when he was transferred out for medical reasons. He earned the rank of sergeant, and was awarded the Bronze Star for his actions on Guadalcanal.
21. Cronin, *Under the Southern Cross*, p. 9.
22. Interview with Haskell Saxe by John Coates, 23 October 2001, part of the Morse Institute Library (Natick, Massachusetts, Community Library) Veterans Oral History Project.
23. D'Alessandro interview.
24. McCarthy interview.
25. Peter Dunn, "182nd Infantry Regiment, US Army in Australia During WW2," www.ozatwar.com (April 2003).
26. *Ibid.*
27. D'Alessandro interview.
28. Cronin, *Under the Southern Cross*, p. 10. Also Buckle, "And Then There Were None."
29. Buckle, "And Then There Were None."
30. *Ibid.*
31. D'Alessandro interview.
32. The story of the 164th Regiment during the Second World War is thoroughly examined in Cooper's *Citizens as Soldiers*, pp. 265–332.
33. The Americal Division received the numeral designation as the 23rd Infantry Division when it was recreated after World War II, but was referred to officially throughout World War II as the Americal.
34. Cronin, *Under the Southern Cross*, pp. 28–29.
35. *Ibid.* "General Orders No. 10. Headquarters United Forces in New Caledonia" officially announced the existence of the Americal Division.
36. Buckle, "And Then There Were None."
37. Dunn, "182nd Infantry Regiment, US Army in Australia During WW2." This source indicates that the "Ug Mutton" was fed to them in Australia, but this is probably incorrect. The short period of time the 182d spent in Australia, and that the men were fed by their Australian hosts, argues against that. Cooper, in *Citizens as Soldiers*, p. 278, which in part covers the 164th Infantry Regiment, one of the three regiments in the Americal, also mentions the prevalence of the unpopular mutton in Army chow served on New Caledonia.
38. Wyant, *Sandy Patch*, pp. 43–45.
39. Buckle records.
40. Arthur G. Rees, called "Pop" by the other men in the company, was born 2 November 1897, making him the oldest man in the company. He had served briefly in the National Guard in 1935, but rejoined on 20 July 1940. He normally worked in the kitchen, and eventually was given the rank of Technician 5. He was transferred to Letterman General Hospital in San Francisco, California, on 28 September 1942 and was given a Civilian Disability Discharge on 11 December of that year. Mr. Rees died on 20 August 1963 and was buried in the Soldiers Lot, Forest Glade Cemetery, in Reading.
41. The "Peep Troops" were a battalion-sized unit of volunteers that performed cavalry functions for the division. The name came from their basic vehicle, the quarter-ton truck. Although later more commonly known as a "jeep," the nic-name "peep" was also common during World War II.
42. Unpublished reminisces of James C. Buckle, 19 May 2006.
43. Cronin, *Under the Southern Cross*, pp. 37–38.
44. Reminisces of James C. Buckle.
45. Buckle records. Also Buckle, "And Then There Were None."
46. Buckle, "And Then There Were None."
47. General Order No. 73, Department of the Army, 2 November 1948, reprinted in Cronin *Under the Southern Cross*, appendix "Awards and Decorations."
48. *Ibid.*, p. 111. Although a unit award, the ribbon was worn on above the left pocket on the dress uniforms, as opposed Army unit awards, which were worn above the right pocket. Men who had served in one of the units during the dates for which it received the award could wear permanently the ribbon above their left pocket on their uniforms. Soldiers later assigned to a unit that had won the Citation were to wear the ribbon only while they belonged to that particular unit.
49. Cronin, *Under the Southern Cross*, p. 99.
50. *Ibid.*, p. 109.
51. Buckle records. The men were Richard Schaefer, Francis Nutt, and William Hogquist. All had been in the company in January 1941, when it was originally mobilized, although none appear in the 1939 volume on the Massachusetts National Guard.
52. Buckle, email of 6 June 2006 to author.
53. Cronin, *Under the Southern Cross*, appendix "Order of Battle," p. 420. Major Dolbeare commanded the second battalion of the 182d Infantry from 1 May–10 June 1943, being the last man who had served in the pre-war 182d to command a battalion in the regiment. He had first joined the RLG in summer 1919 when it was a State Guard company. In 1923 he joined the National Guard version of the RLG, and rose to the rank of first sergeant before becoming an officer. He commanded Company E and later Company L, in Stoneham before the war, and Company E again for a while on active duty. After his reassignment stateside, he served as an instruc-

Notes—Chapter IX

tor in the First Service Command, headquartered in Boston. His brief tenure as battalion commander seems to have been a stop-gap measure. He replaced Lieutenant Colonel Bernard B. Twombly, who had served from 27 May 1942 to 30 April 1943, and was in turn replaced by Lieutenant Colonel Dexter Lowry, who served from 11 June 1943 to 20 March 1944.

54. On 10 December 1943, the following officers were assigned to Company E of the 182d: First Lieutenant Melvin E. Carlson as Company Commander, First Lieutenant Gerald B. Lyons as Executive Officer, First Lieutenant George A. Karl, Second Lieutenant Leonard C. Hurley, and Second Lieutenant Elmer C. Nelson as Platoon Leaders. They joined First Lieutenant David J. O'Rourke (assigned 11 September 1943) and First Lieutenant Fred H. Willard (assigned 18 September 1942) as Platoon Leaders.

55. Cronin, *Under the Southern Cross*, appendix "Order of Battle," p. 420. Lieutenant Colonel Francis W. O'Brien, who had been the commander of the Service Company in 1939, commanded the regiment from 15 to 30 June 1945, while the regiment was on the island of Cebu, in the Philippines.

56. Saxe interview.

57. Buckle, email to author, 20 November 2006.

58. John Miller, Jr., *Cartwheel: The Reduction of Rabul* (Washington, D.C.: Office of the Chief of Military History, 1959), pp. 364–65.

59. Miller, *Cartwheel*, p. 366.

60. *Ibid*. The 3rd Battalion of the 13th Infantry, under the command of Colonel Toyoharei Muda.

61. Cronin, *Under the Souther Cross*, pp. 152–53.

62. Buckle records.

63. On page 367 in *Cartwheel*, Miller states that Colonel Long arrived on the south knob at 1715, and infers that he and Colonel Lowery spoke face to face before making the decision for Company E to hold its position for the night. However, part of Miller's account came from notes made by Colonel Long on an earlier draft of the manuscript. Sergeant Buckle claims that he was by the side of Colonel Lowery throughout the battle, and that he never saw Colonel Long on the south knob. Buckle, email of 17 December 2006, to author.

64. The withdrawal was led by Lieutenant O'Rourke. While Cronin, in *Under the Southern Cross*, p. 155, states that the company returned "under the leadership of a wounded commander," Sergeant Buckle says this is erroneous, and that Lieutenant O'Rourke was not wounded when he led the company off the hill. Buckle says that O'Rourke tried to get the story changed for years without success.

65. Buckle, email of 20 November 2006.

66. Morning Reports of Company E, at the Mass. Adjutant General's Office.

67. Cooper, *Citizens as Soldiers*, p. 302.

68. *Ibid*., pp. 301–02.

69. Although Lowry held the rank of lieutenant colonel in the Army of the United States, his Regular Army rank was still captain. After Bougainville, Lowry attended the Command and General Staff College at Fort Leavenworth, Kansas. After the war, he served in occupied Austria. He was eventually promoted to major in the Infantry on 1 July 1948, and to lieutenant colonel on 2 March 1949. He retired in from the Army in 1950 with a disability from his wounds suffered at Bougainville.

70. Buckle, email of 30 September 2006, to author.

71. Long went on to serve in the Rhineland and Central Europe campaigns until the end of World War II. Like Lowry, the rank he held at Bougainville was in the Army of the United States, not in the Regular Army. He was promoted to colonel in the Infantry on 11 March 1948.

72. Buckle, email to author, 30 September 2006.

73. General Orders No. 56, War Department, 12 July 1944, reprinted in Cronin, *Under the Southern Cross*, appendix "Awards and Decorations." The Distinguished Unit Citation (DUC) was the Army's equivalent of the Navy's Presidential Unit Citation (PUC). The two awards, while distinct, were often confused in the records of the soldiers. For example, the Enlisted Report of Separation for First Sergeant Chesley L. Black, from Reading, shows him receiving the DUC but not the PUC. However, since he served on Guadalcanal, but not on Hill 260 on Bougainville, and the date of award is 31 December 1943, it must erroneously refer to the PUC. The Enlisted Record of SSGT James C. Buckle, who was on Guadalcanal and Hill 260, shows him receiving the "Presidential Unit Citation W/Oak Leaf Cluster," indicating a second award of the same decoration. The individual service records of members of Company E are full of these inconsistencies, and the Army itself often seemed unsure of the proper name of its own unit citation. Perhaps in response to the confusion, the Army changed the name of the DUC to the Army PUC in 1966.

74. Buckle records.

75. Buckle records. Thomas J. Burbine received a field commission on 6 May 1945. Although he was discharged in October 1945, he joined the peacetime Regular Army a few years later, and still later served in the Army Reserve, retiring with a total of twenty years service as a Chief Warrant Officer.

76. Buckle, email to author, 16 November 2006. The three were T/Sgt Herbert W. Dillon, SSgt James J. Sullivan, and Pfc Paul E. Muse. Private Muse had been hospitalized on Fiji and did not rejoin the company until after the fight on Hill 260. Three other men who entered active duty with Company E, Captain Blanchard, First Sergeant Burbine, and Major Gromlie, served in other units of the 182d.

77. W. Eaton, *History of Wakefield*, p. 95.

78. From file "Richardson Light Guard 1940 #6, Co. E 182nd Information," (News Release-1858)," 27 October 1944, at the *Wakefield Daily Item* office.

79. "M.S.G. Mo. Rosters 1941–42 23d Inf. 6th-8th Cos," three-ring binder of attendance rosters at Mass. NG Museum.

80. *Wakefield Daily Item*, 12 June 1945.

81. Major Brockbank had first joined the RLG in 1915, served with the 104th Infantry in France during the Great War. While in France he attended Officer Candidate School, after which he was commissioned in the Officers Reserve Corps. He commanded the Wakefield State Guard company for a while during World War II before his promotion.

82. From envelope "Richardson Light Guard 1940 #6 Co. E 182nd Information," at the *Wakefield Daily Item* office.

83. Colonel Gihon was buried near other members of his family, including his father and his wife, in Saint Patrick's cemetery in Stoneham, which was only about

a mile from his niece's home in the western area of Wakefield. The Gihon family members share a single monument, a large gray granite shaft, with the father and mother of Colonel Gihon listed on the front, and several other family members listed on the reverse. The elder Edward J. Gihon's service in the Civil War is noted on the front of the stone, but the nature of Colonel Gihon's service is not.

84. *Wakefield Daily Item*, 12 June 1945.
85. Cronin, *Under the Southern Cross*, p. 345.
86. Ibid., p. 347–48.
87. Ibid., p. 352.
88. Ibid., p. 362.
89. War Department circular number 58, 9 February 1944, first established the point system as a way to establish priority for rotating soldiers back to the United States for discharge.
90. Cronin, *Under the Southern Cross*, pp. 380–81.
91. James A. Sawicki, *Infantry Regiments of the U.S. Army* (Dumfries, VA: Wyvern Publications, 1981), pp. 355–57.
92. Ibid.
93. Cooper, *Citizens as Soldiers*, p. 327.
94. From the Buckle records.
95. From file "Richardson Light Guard 1940 #6 Co. E 182nd Information," at the *Daily Item* office. Clipping from *Wakefield Daily Item*, 30 April 1946.

Chapter X

1. According to the morning reports for the 182d Regiment in 1948, the regiment itself included headquarters company, a service company, a tank company, heavy mortar company, and a medical company, in addition to the three infantry battalions. The third battalion, however, contained only three infantry companies rather than four. The tank company had originally been an anti-tank company, but was converted in 1948.
2. From clipping inside envelope, "Richardson Light Guard 1940 #6," at the *Daily Item* office. The 182nd RCT had been reorganized by the Army on 5 December 1946.
3. The 182d Regimental Combat Team included the 182d Infantry Regiment, the 212 Field Artillery Battalion, and the 379th Engineer Combat Company. *Annual Report Chief National Guard Bureau for the Fiscal Year Ending 30 June 1949* (Washington: GPO, 1950), p. 69.
4. *Daily Item*, 16 December 1946; also Charles L. Doyle, ed., "History of the 182d Infantry" (1980), C-21, 22. The Medical unit was officially the 2d Battalion Section Medical Detachment. In May 1948, all the medical detachments of the 182d were consolidated in Wakefield.
5. Based on a comparison of the roster of Company E aboard the USAT *Santa Elena* in January 1942, with the morning report of Company E for 31 January 1950.
6. *ARCNGB for the Fiscal Year Ending 30 June 1949*, p. 119.
7. *National Guardsman*, July 1945, pp. 4–5.
8. *Annual Report Chief National Guard Bureau* (hereafter cited as "*ARCNGB*") *for the Fiscal Year Ending 30 June 1949* (Washington: GPO, 1950), pp. 90–91, 107, 110.
9. The "Morning Report" of 31 January 1950 lists 72 men who attended drill. Of that number, only 22 appear as residents in the *Wakefield Street Guide 1950*. Most likely, more of the lower ranking men were also residents, but did not appear in the *Street Guide*, which included only residents 20 years of age or older.
10. Renewed interest in the reserve components as the Korean War was in its last stages led the journal *Military Affairs* to dedicate its January 1953 issue to articles about each of the reserve components, but book-length works on the history of each of the reserve components are rare and in serious need of updating. Aside from official histories, most books dealing with the reserve components focus on a specific unit in war, such as Randy Keith Mills and Roxanne Mills's *Unexpected Journey: A Marine Corps Reserve Company in the Korean War* (Annapolis, MD: U.S. Naval Institute Press, 2000). Official histories of the each of the reserve components are limited to Reserve Officers of Public Affairs Unit 4-1 *The Marine Corps Reserve, A History* (Washington: G.P.O., 1966); Gerald T. Cantwell's *Citizen Airmen: A History of the Air Force Reserve, 1946–1994* (Washington: Air Force History and Museums Program, 1997); and Richard B. Curries and James Crossland's *Twice the Citizen: A History of the United States Army Reserve* (Washington: Office of the Army Reserve, 1984). The Air National Guard has fared better, although most works on it are likewise official publications. See Charles Joseph Gross, *Prelude to the Total Force: The Air National Guard 1943–1969* (Washington: Office of Air Force History, 1985) and *The Air National Guard and the American Military Tradition: Militiaman, Volunteer, and Professional* (Washington: NGB Historical Services Division, 1996). See also George Hall, *Air Guard: America's Flying Militia* (Novato, CA: Presidio Press, 1990).
11. Ibid., p. 1, also Doubler, *I Am the Guard*, pp. 219, 228.
12. *ARCNGB for the Fiscal Year Ending 30 June 1957*, p. 28, and *ARCNGB for the Fiscal Year Ending 30 June 1959* (Washington: GPO, 1960), p. 38.
13. *Wakefield Daily Item*, 23 October 1947.
14. Ibid., 24 October 1947.
15. Ibid., 27 October 1947.
16. Mahon, *History of the Militia and National Guard*, p. 209.
17. *Wakefield Daily Item*, 10 October 1951.
18. From envelope "Richardson Light Guard 1940 #6, Co. E 182nd Information," at the *Wakefield Daily Item* office.
19. Ibid., 15 October 1951. Technically this was erroneous. Company E had received the Distinguished Unit Citation for its actions on Hill 260. The company also earned the Navy Presidential Unit Citation for its service on Guadalcanal, and later the Philippine Presidential Unit Citation.
20. The town did, however, financially support programs for veterans, paying $24,250 in 1950. *139th Annual Report of the Town of Wakefield, Mass. Financial Year Ending December Thirty-First, Nineteen Hundred and Fifty* (Wakefield: Wakefield Item Press, 1951), p. 24.
21. Nancy Bertrand, "The 1950{ft}s: Building Wakefield, Tearing Down Wakefield's Hall," in *Wakefield: 350 Years by the Lake*, pp. 171–72.
22. See Bruegmann, *Sprawl*, 41–50, on the pervasiveness of the phenomenon in the post World War II years.
23. "Annual Report of the A.G.O office," p. 2.
24. From folder "Camp Curtis Guild," at the Mass. NG Museum.

Notes—Chapter X

25. In the 1950s, I695 was planned as a true inner loop, about five miles from downtown Boston, but local opposition ended the project in 1971.

26. See Hayden, *Building Suburbia*, chapter 9, on the impact of Route128 on Boston's suburbs.

27. Alison Isenberg. *Downtown America: A History of the Place and the People Who Made It* (Chicago: University of Chicago Press, 2004), pp. 166–202. Isenberg shows that for many downtown areas, replacing the old with the new was an almost constant process, but unlike many of the downtowns in her study, Wakefield Square remained relatively vibrant in the late twentieth century.

28. NG-AROTC 325.4, dated 24 November, effective 1 December 1950, in the morning reports for that year.

29. The morning report for the month of July 1952 (signed 1 August) shows thirty-eight enlisted men and four officers in the tank company. Of that number, only one appears in the town records for the year as a resident, although the town records did not include the names of men younger than 20 years of age.

30. NG-AROTO 325.4, dated 24 November 1950. Quoted in the "Morning Reports" at the Massachusetts National Guard Adjutant General's Office, Milford, Massachusetts.

31. *Ibid.*, 3 February 1955.

32. Clipping of 13 January 1958, inside envelope "Richardson Light Guard 1940 Co. E. 182nd Information (New release -1858) #6," at the *Wakefield Daily Item* office.

33. John B. Wilson, *Armies, Corps, Divisions, and Separate Brigades* (Washington: Center of Military History, 1987), pp. 289–92.

34. Weigley, *History of the United States Army*, p. 537.

35. A few armored cavalry regiments continued to exist as tactical units.

36. General Orders of the Adjutant General of Massachusetts, 1 May 1959. In the volume of Morning Reports of July 1958-April 1959.

37. Doyle, "History of the One-Hundred and Eighty-Second Infantry," C-24.

38. Doubler, *I am the Guard*, pp. 256–61.

39. *Wakefield Daily Item*, 13 October 1976.

40. Doyle, "History of the One-Hundred and Eighty-Second Regiment," C-24–25.

41. From envelope "Richardson Light Guard 1940 #6, Co. E 132nd Information," at the *Wakefield Daily Item* office.

42. *ARCNGB Fiscal Year Ending 30 June 1949*, p. 16–17.

43. *ARCNGB Fiscal Year Ending 30 June 1957*, p. 23, and, *ARCNGB Fiscal Year Ending 30 June 1958*, p. 33. Men between the ages of 26 and 35 who already possessed skills the Army needed were exempt from the requirement to attend active training with the Army.

44. Doubler, *I Am the Guard*, pp. 240–41

45. *ARCNGB Fiscal Year Ending 30 June 1959*, p 33.

46. *Wakefield Daily Item*, 7 September 1961.

47. *Ibid.*, 17 July 1962.

48. General Order Number 7, dated 25 February 1963, effective 1 March 1963, Massachusetts Adjutant General's Office, in volume of Morning Reports "Mar 63 — June 63, 1 BN, 182 INF, HHC thru Co. C." Attachments to the Morning Reports show massive movement of Guardsmen reassigned to other companies.

49. Doyle, "History of the One-Hundred and Eighty-Second Regiment," C-24.

50. Master Sergeant (MA-ARNG, Retired) Donald Tibbetts, interview by author, 10 April 2006, Wakefield. MSG Tibbetts served in the 182d Regiment from 1947 to 1968, mostly in companies L and K.

51. Doyle, "One-Hundred and Eighty-Second Regiment," I-1.

52. *Wakefield Daily Item*, 27 February 1975.

53. *Ibid.*

54. *Ibid.*

55. *Ibid.*, 25 April 1975.

56. *Ibid.*, 27 February 1975.

57. Town Meeting of 16 June 1975, "Article 8," reported in the *Wakefield Daily Item*, 17 June 1975.

58. Doyle, "History of the One-Hundred and Eighty-Second Infantry," C-24.

59. Company B would inherit the Distinguished Unit Citation earned by Company E during World War II. *Ibid.*, F-5. In 2007, the lineage of the RLG remained with Company B, although it was a troop rather than a company and the 182d was a cavalry squadron.

60. *Wakefield Daily Item*, 8 April 1975.

61. *Ibid.*

62. *Ibid.*

63. Bruegmann *Sprawl*, 51–95. Bruegmann, unlike most writers on the growth of large metropolitan areas, sees much in them that makes such areas attractive to many.

64. In the year 2000, the 51st Troop Command, the HHC, 101 Engineer Battalion, the Support Platoon, 101 Engineer Battalion, and the 272 Chemical Company were based on the camp *Massachusetts National Guard Annual Report: There's No Place Like Home and We've Been Defending It for Over 368 Years* (Milford, MA: The Adjutant General's Office, 2000), p. 29.

65. *Ibid.*, p. 8. In the year 2000, the federal government spent $213 million on the Massachusetts National Guard, while the commonwealth contributed $10.3 million.

66. Young, *Iron Men and Iron Machines*, pp. 47–54.

67. The assessor's office lists the building as being built "around 1880{in} which is fifteen years off the mark. See Property Account Number 35836, Parcel ID 18-222-005, at the Wakefield Assessor's office.

68. Other Massachusetts town names of the sort are Ashfield, Brimfield, Brookfield, Chesterfield, Deerfield, Greenfield, Hatfield, Marshfield, Medfield, Middlefield, Northfield, Pittsfield, Plainfield, Sandisfield, Sheffield, Springfield, Topsfield, and Westfield. This list does not include other, fully independent towns with names that are derivatives of the above towns, such as North Brookfield.

69. Wakefield, England, is a city of over 300,000 residents about 160 miles north of London. As the practice of borrowing names of English towns and cities in naming towns in the eastern United States was so prevalent, such an assumption is understandable.

70 Bertrand, *Images of America: Wakefield*, p. 51.

Bibliography

PRIMARY SOURCES

Statutes

Militia Act of 1792. "An Act more effectually to provide for the National Defence by establishing a Uniform Militia throughout the United States." pp. 271–274, also, "An Act to provide for calling forth the Militia to execute the laws of the Union, suppress insurrections and repel invasions." pp. 264–265. From *The Statutes at Large of the United States of America*. Vol. 1. Boston: Charles C. Little and James Brown, 1848.

Militia Act of 1903. "An Act To promote the efficiency of the militia, and for other purposes." *Statutes at Large of the United States of America*. Vol. 32, part 1, pp. 774–780. Washington: U.S. GPO, 1904.

National Defense Act of 1916. "An Act for making further and more effectual provisions for the National Defense, and for other purposes." *Statutes at Large of the United States of America*. Vol. 39, part 1, pp. 166–217. Washington: U.S. GPO, 1917.

Volunteer Act of 1861. "An Act To authorize the employment of volunteers to aid in enforcing the laws and protecting private property." *Statutes at Large of the United States of America*. Vol. 12, pp. 268–269. Boston: Charles C. Little and James Brown, 1863.

Commonwealth of Massachusetts, Military Division, Office of the Adjutant General, Military Records Office Branch, Milford, Massachusetts

"Morning Reports," multiple volumes, 1940–1970.

Massachusetts National Guard Museum and Archives, Worcester, Massachusetts

"Annual Reports of the A.G. Office 1946–48." Three-ring binder of typed reports.
"Camp Curtis Guild." Box of documents.
"Massachusetts Soldiers, Sailors, and Marines in the Spanish War, Tabulated by Cities and Towns." Two volumes, A-M, N-Z.
"M.S.G. Mo. Rosters 1941–42 23d Inf. 6th-8th COs." Three-ring binder.
"Petition for New Companies Mass. Militia, 1795–1851."
"Service Records World War I."
"Special Orders 1846–1856{in} (bound volume).

Government Documents

Annual Report of the Chief National Guard Bureau for the Fiscal Year Ending 30 June 1949. Washington: GPO, 1950.
Annual Report of the Chief National Guard Bureau for the Fiscal Year Ending 30 June 1951. Washington: GPO, 1952.

Annual Report of the Chief National Guard Bureau for the Fiscal Year Ending 30 June 1957. Washington: GPO, 1958.

Annual Report of the Chief National Guard Bureau for the Fiscal Year Ending 30 June 1958. Washington: GPO, 1959.

Annual Report of the Chief National Guard Bureau for the Fiscal Year Ending 30 June 1959. Washington: GPO, 1960.

Graham, George. *Letter from the Secretary of War Transmitting a Plan for the Organization and Discipline of the Militia in Compliance with a Resolution of the House of Representatives, of April 16, 1816.* Washington: William A. Davis, 1816.

Kearney, George, ed. *Official Opinions of the Attorneys General of the United States Advising the President and Heads of Departments in relation to Their Official Duties.* Washington: GPO, 1913.

Office of the Adjutant General, United States Army. *Synonyms of Volunteer Organizations of the United States.* Washington: War Department, 1885.

United States War Department. "Document Number 32: The Organized Militia of the United States," Washington: GPO, 1897.

State Documents

Massachusetts Soldiers, Sailors, and Marines in the Civil War. Norwood, MA: Norwood Press, 1932.

The Militia Law of Massachusetts. General Orders, No. 20 (ch. 604, Acts of 1908, As Amended by Chapters 167, 170, 298 and 300, Acts of 1909). An Act Relative to the Militia. Boston: Adjutant General's Office, 1909.

Pearson, Gardner W., Brigadier General, Adjutant General of Massachusetts. *Records of the Massachusetts Volunteer Militia Called out by the Governor of Massachusetts to Suppress a Threatened Invasion During the War of 1812-'14.* Boston: Wright & Potter Printing Co., State Printers, 1913.

Secretary of the Commonwealth. *Acts and Resolves Passed by the Legislature of Massachusetts, in the Year 1840.* Boston: Dutton and Wentworth, Printers to the Commonwealth, 1840.

_____. *Acts and Resolves Passed by the Legislature of Massachusetts, in the Year 1841.* Boston: Dutton and Wentworth, Printers to the Commonwealth, 1841.

_____. *Acts and Resolves Passed by the Legislature of Massachusetts, in the Year 1852.* Boston: White & Potter, Printers to the Commonwealth, 1852.

Town of Wakefield Documents

Dubley, Dean. *Directory and Register of Wakefield, Stoneham, Reading & C., with Business and Advertising Directories.* Wakefield 1882, 1886, 1888.

. *Directory and Register of Wakefield, Stoneham, Reading & C., with Business and Advertising Directories.* No. 7. Wakefield: Dubley, 1889.

The One Hundred and Third Annual Report of the Town Officers of Wakefield, Mass. for the Financial Year Ending December 31, 1914 also Town Clerk's Record of the Births, Marriages, and Deaths During the Year 1914. Boston: Falcon Press, 1915.

139th Annual Report of the Town of Wakefield, Mass. Financial Year Ending December Thirty-First, Nineteen Hundred and Fifty. Wakefield: Item Press, 1951.

The One Hundredth Report of the Town Officers of Wakefield, Mass. for the Financial Year Ending December 31, 1913 also Town Clerk's Record of the Births, Marriages, and Deaths During the Year 1914. Wakefield: Item Press, 1917.

"Wakefield Street Guide 1950."

Adjutant General Reports

Annual Report of the Adjutant General of the Commonwealth of Massachusetts for the Year Ending December 31, 1917. Boston: Wright & Potter Printing Co., State Printers, 1918.

Annual Report of the Adjutant General of the Commonwealth of Massachusetts for the Year Ending December 31, 1918. Boston: Wright & Potter Printing Co., State Printers, 1919.

Annual Report of the Adjutant General of the Commonwealth of Massachusetts for the Year Ending December 31, 1919. Boston: Wright & Potter Printing Co., State Printers, 1920.

Annual Report of the Adjutant General of the Commonwealth of Massachusetts for the Year Ending December 31, 1920. Boston: Wright & Potter Printing Co., State Printers, [1921].

Massachusetts National Guard Annual Report: There's No Place Like Home and We've Been Defending It for Over 368 Years. Milford, MA: The Adjutant General's Office, 2000.

Public Document No. 7 *Annual Report of the Adjutant General of the Commonwealth of Massachusetts for the Year Ending December 30, 1894.* Boston: Wright & Potter Printing Co., 1895.

Public Document No. 7 *Annual Report of the Adjutant General of the Commonwealth of Massachusetts for the Year Ending December 31, 1911.* Boston: Wright & Potter Printing Co., State Printers, 1912.

Public Document No. 7 *Annual Report of the Adjutant General of the Commonwealth of Massachusetts for the Year Ending December 31, 1912.* Boston: Wright & Potter Printing Co., State Printers, 1913.

Public Document No. 7 *Annual Report of the Adjutant General of the Commonwealth of Massachusetts for the Year Ending December 31, 1914.* Boston: Wright & Potter Printing Co., State Printers, 1915.

Public Document No. 7 Commonwealth of Massachusetts *Annual Report of the Adjutant General for the Year Ending December 31, 1928.*

"The Training School Massachusetts National Guard Year Book 1930." National Guard Association Library.

Monographs

Couris, John G. "History Massachusetts State Guard." 1976?

Doyle, Charles L., ed. *History of the One-Hundred and Eighty-Second Infantry (Fifth Massachusetts Infantry) Massachusetts Army National Guard (The Nation's First).* Melrose, MA: 182nd Infantry, 1977.

Reports

Ford, Ben, and Suzanne Cherau. "Final Technical Report: Intensive (Locational) Archaeological Survey Camp Curtis Guild: Lynnfield, Reading, North Reading, and Wakefield, Massachusetts." Providence, RI: PAL (Public Archeology Laboratory), 2004.

Guzzi, Paul. *Historical Data Relating to Cities and Towns in Massachusetts.* Boston: The Commonwealth of Massachusetts, 1975.

"Public Document Number 49." *Fourteenth Annual Report of the Police Commission for the City of Boston: Year Ending November 30, 1919.* Boston: Wright and Potter Printing Co., State Printers, 1920.

Shaw, Diane, et al. *World War II and the U.S. Army Mobilization Program: A History of 700 and 800 Series Cantonment Construction.* Washington: GPO, 1993.

SECONDARY SOURCES

Books

Adamson, Jeremy Elwell. *American Wicker: Woven Furniture from 1850 to 1930.* New York: Rizzoli/Smithsonian, 1993.

Anderson, Fred. *A People's Army: Massachusetts Soldier and Society in the Seven Years' War.* Chapel Hill: University of North Carolina Press, 1984.

Barron, Hal S. *Those Who Stayed Behind: Rural Society in Nineteenth-Century New England.* New York: Cambridge University Press, 1984.

Barry, Herbert, et al, eds. *Squadron A: History of its First Fifty Years 1889–1939.* New York: Squadron A Association, 1939.

Bauer, K. Jack. *The Mexican War, 1846–1848.* New York: Macmillan, 1974.

Bender, Thomas. *Community and Social Change in America.* Baltimore: Johns Hopkins University Press, 1978.

Bertrand, Nancy. *Images of America: Wakefield.* Charleston, S.C.: Arcadia, 2000.

_____. *Wakefield: 350 Years by the Lake: An Anniversary History.* Wakefield: Item Press, 1994.

Binford, Henry C. *The First Suburbs: Residential Communities on the Boston Periphery, 1815–1860.* Chicago: University of Chicago Press, 1985.

Bodge, George Madison. *Soldiers in King Philip's War being a Critical Account of that War with a Concise History of the Indian Wars of New England from 1620–1677.* Boston: Rockwell & Churchill, 1896.

Breen, T.H. *American Insurgents, American Patriots*. New York: Hill & Wang, 2010.
Bruegmann, Robert. *Sprawl: A Compact History*. Chicago: University of Chicago Press, 2005.
Cantwell, Gerald T. *Citizen Airmen: A History of the Air Force Reserve, 1946–1994*. Washington: Air Force History and Museums Program, 1997.
Carlson, Stephen P., with Thomas W. Harding. *From Boston to the Berkshires: A Pictorial Review of Electric Transportation in Massachusetts*. Boston: Boston Street Railway Association, 1990.
Carnes, Mark C. *Secret Ritual and Manhood in Victorian America*. New Haven: Yale University Press, 1989.
Capozzola, Christopher Joseph Nicodemus. *Uncle Sam Wants You: World War I and the Making of the Modern American Citizen*. Oxford: Oxford University Press, 2008.
A Completed Century, 1826–1926: A History of the Heywood-Wakefield Company. Boston: Printed for the Company, 1926.
Cooper, Jerry M. *The Army and Civil Disorder: Federal Military Intervention in Labor Disputes, 1877–1900*. Westport, CT: Greenwood Press, 1980.
_____. *The Militia and National Guard in America Since Colonial Times: A Research Guide*. Westport, CT: Greenwood Press, 1993.
_____. *The Rise of the National Guard: The Evolution of the American Militia, 1865–1920*. Lincoln: University of Nebraska Press, 1997.
_____, with Glenn Smith. *Citizens as Soldiers: A History of the North Dakota National Guard*. Fargo: North Dakota Institute for Regional Studies, 1986.
Conklin, Edwin P. *Middlesex County and Its People: A History*. New York: Lewis Historical, Inc., 1927.
Connole, Dennis A. *The 26th "Yankee" Division on Coast Patrol Duty, 1942–1943*. Jefferson, N.C.: McFarland, 2008.
Craig, George A. *History: I Massachusetts State Guard, II Massachusetts State Guard Veterans*. Geo. A. Craig, 1931.
Cress, Lawrence Delbert. *Citizens in Arms: The Army and Militia in Americans Society to the War of 1812*. Chapel Hill: University of North Carolina Press, 1982.
Cronin, Francis D. *Under the Southern Cross: The Saga of the Americal Division*. Washington: Combat Forces Press, 1951.
Cumming, William P. *British Maps of Colonial America*. Chicago: University of Chicago Press, 1974.
Cunningham, Edward. *The Port Hudson Campaign, 1862–1863*. Baton Rouge: Louisiana State University Press, 1963.
Curries, Richard B., and James Crossland. *Twice the Citizen: A History of the United States Army Reserve*. Washington: Office of the Army Reserve, 1984.
Curtis, John Obed, and William H. Guthman. *New England Militia Uniforms and Accoutrements*. Meriden, CT: The Meriden Gravure Company, 1971.
Danhof, Clarence H. *Change in Agriculture: The Northern United States, 1820–1870*. Cambridge: Harvard University Press, 1969.
Davis, Susan G. *Parades and Power Street Theatre in Nineteenth-Century Philadelphia*. Berkeley: University of California Press, 1988.
Dawley, Alan. *Class and Community: The Industrial Revolution in Lynn*. Cambridge: Harvard University Press, 1976.
Doubler, Michael D. *I Am the Guard: A History of the Army National Guard, 1636–2000*. Washington: GPO, 2001
Eames, Steven C. *Rustic Warriors: Warfare and the Provincial Soldier on the New England Frontier, 1689–1748*. New York: New York University Press, 2011.
Eaton, Chester W., Warren E. Eaton, and Will Everett Eaton, eds. *Proceedings of the 250th Anniversary of the Ancient Town of Redding Once Including the Territories Now Comprising the Towns of Reading, Wakefield, and North Reading with Historical Chapters*. Reading: Loring & Twombly, 1896.
Eaton, Lilley. *Genealogical History of the Town of Reading, Mass., Including the Present Towns of Wakefield, Reading, and North Reading, with Chronological and Historical Sketches from 1639 to 1874*. Boston: Alfred Mudge and Son, 1874.
Eaton, William E. *History of the Richardson Light Guard of Wakefield, Mass. 1851–1901*. Wakefield: Citizen and Banner, 1901.
_____. *History of the Richardson Light Guard of Wakefield, Mass.* Wakefield: Wakefield Item Press, 1926.
_____. *History of Wakefield, Massachusetts*. Wakefield: Wakefield Item Press, 1944.
Edwards, Frank E. *The '98 Campaign of the 6th Massachusetts, U.S.V.* Boston: Little, Brown, 1899.
Fiftieth Anniversary Committee. *"The First Fifty Years": History of Golden Rule Lodge, A.F. & A.M. Wakefield, Massachusetts: 1888–1938*. Wakefield: Wakefield Item Press, 1938

Fishman, Robert. *Bourgeois Utopias: The Rise and Fall of Suburbia*. New York: Basic Books, 1987.
Fogelson, Robert M. *America's Armories: Architecture, Society, and Public Order*. Cambridge: Harvard University Press, 1989.
Foster, Stephen. *The Long Argument: English Puritanism and the Shaping of New England Culture 1570–1700*. Chapel Hill: University of North Carolina Press, 1991.
Garreau, Joel. *Edge City: Life on the New Frontier*. New York: Anchor Books, 1992.
Gatewood, William B., Jr. *"Smoked Yankees" and the Struggle for Empire: Letters from Negro Soldiers, 1898–1902*. Urbana: University of Illinois Press, 1971.
Gross, Charles J. *The Air National Guard and the American Military Tradition: Militiaman, Volunteer, and Professional*. Washington: NGB Historical Services Division, 1996.
_____. *Prelude to the Total Force: The Air National Guard, 1943–1969*. Washington: GPO, 1984.
Gross, Robert A. *Minutemen and Their World*. New York: Hill & Wang, 1976.
Hall, Charles Winslow, ed. *Regiments and Armories of Massachusetts: A Historical Narration of the Massachusetts Volunteer Milit8ia with Portraits and Biographies of Officers Past and Present*. Boston: W.W. Potter, 1899.
Hall, George. *Air Guard: America's Flying Militia*. Novato, CA: Presidio Press, 1990.
Hayden, Dolores. *Building Suburbia: Green Fields and Urban Growth, 1820–2000*. New York: Vintage, 2004.
Hill, Jim Dan. *The Minute Man in War and Peace: A History of the National Guard*. Harrisburg, PA: Stackpole, 1964.
Higginbotham, Don. *The American War of Independence: Military Attitudes, Policies, and Practices, 1763–1789*. New York: Macmillan, 1971.
Higginson, Thomas Wentworth. *Massachusetts in the Army and Navy 1861–65*. Boston: Wright & Potter Printing Co., State Printers, 1895.
Historical and Pictorial Review National Guard of Commonwealth of Massachusetts 1939. Baton Rouge: Army and Navy, 1939.
Howard, Loea Parker. *Ancient Redding in Massachusetts Bay Colony: Its Planting as a Puritan Village and Sketches of Its Early Settlers From 1639–1652*. Boston: Thomas Todd, 1944.
Hubbs, G. Ward. *Guarding Greensboro: A Confederate Company in the Making of a Southern Community*. Athens: University of Georgia Press, 2003.
Isenberg, Alison. *Downtown America: A History of the Place and the People Who Made It*. Chicago: University of Chicago Press, 2004.
Jackson, Kenneth T. *Crabgrass Frontier: Suburbanization of the United States*. New York: Oxford University Press, 1985.
Johns, Henry T. *Life With the Forty-ninth Massachusetts Volunteers*. Washington: Ramsey and Bisbee, Printers, 1890.
Johnson, Charles Jr. *African American Soldiers in the National Guard: Recruitment and Deployment During Peacetime and War*. Santa Barbara: Praeger, 1992.
Johnson, Stephen D., and Gary S. Poppleton. *Cloth Insignia of the U.S. State Guards and State Defense Forces*. Hendersonville, TN: Richard W. Smith, 1993.
Kemp, Edwin C. *Melrose Massachusetts, 1900–1950: Commemorating the One Hundredth Anniversary of the Founding of the Town of Melrose and the Fiftieth Anniversary of the Incorporation of the City of Melrose*. Wakefield: The Murray Printing Company, 1950.
Kirkland, Edward Chase. *Men, Cities, and Transportation: A Study in New England History, 1820–1900*. Cambridge: Harvard University Press, 1948.
Kossuth, Lajos. *Kossuth in New England: A Full Account of the Hungarian Governor's Visit to Massachusetts with his Speeches and the Addresses that Were Made to Him, Carefully Revised and Corrected*. Boston: John P. Jewett, 1853.
Leach, Douglas Edward. *Flintlock and Tomahawk: New England in King Philip's War*. New York: W.W. Norton, 1958.
Leblanc, Cyrille, and Col. Thomas F. Flynn. *Gardner in World War II*. Gardner: Hatton Press, 1947.
The Lowell Book. Boston: George H. Ellis, Printer, 1899.
Mahon, John K. *The American Militia: Decade of Decision, 1789–1800*. Gainesville: University of Florida Press, 1960.
_____. *History of the Militia and the National Guard*. New York: Macmillan, 1983.
Martin, Edger W. *The Standard of Living in 1860: American Consumption Levels on the Eve of the Civil*. Chicago: University of Chicago Press, 1942.
Miller, John, Jr. *Cartwheel: The Reduction of Rabaul*. Washington: Office of the Chief of Military History, Department of the Army, 1959.

Miller, Perry. *The New England Mind, from Colony to Province*. Cambridge: Harvard University Press, 1983.
Mills, Randy Keith, and Roxanne Mills. *Unexpected Journey: A Marine Corps Reserve Company in the Korean War*. Annapolis, MD: U.S. Naval Institute Press, 2000.
Morans, Bruce N. *A Town that Went to War: A Chronicle of Our Bicentennial*. Reading, MA: Bruce N. Morans, 1975.
Morgan, Edmund S. *Visible Saints: The History of a Puritan Idea*. Ithaca: Cornell University Press, 1963.
Murphy, Emily A. *Merchants, Clerks, Citizens, and Soldiers: The Second Corps of Cadets in Salem, Massachusetts*. Washington: National Park Service, 2005.
National Guard Association. *Proceedings of the Convention of the National Guard Association Held at Richmond, Virginia November 14 and 15 1918*. n.p., n.d.
Nofi, Albert A. *The Spanish-American War, 1898*. Conshohocken, PA: Combined Books, 1996.
Putnam, Eben, ed., *Report of the Commission on Massachusetts' Part in the World War History*. Vol. 1. Boston: The Commonwealth of Massachusetts, 1931.
Reserve Officers of Public Affairs Unit 4–1. *The Marine Corps Reserve: A History*. Washington, D.C.: GPO, 1966.
Ridley, Jasper. *The Freemasons: A History of the World's Most Powerful Secret Society*. New York: Arcade, 1999.
Roberts, Robert B. *Encyclopedia of Historic Forts: The Military, Pioneer, and Trading Posts of the United States*. New York: Macmillan, 1988.
Roe, Alfred S. *The Fifth Regiment Massachusetts Volunteer Infantry in Its Three Tours of Duty, 1861, 1862-'63, 1864*. Worcester: Blanchard Press, 1911.
Rotundo, E. Anthony. *American Manhood: Transformation of Masculinity from the Revolution to the Modern Era*. New York: Basic Books, 1993.
Russell, Francis. *A City in Terror: The 1919 Boston Police Strike*. New York: Viking, 1975.
Sawicki, James A. *Infantry Regiments of the U.S. Army*. Dumfries, VA: Wyvern, 1981.
Schorow, Stephanie. *Boston on Fire: A History of Fires and Firefighting in Boston*. Beverly, MA: Commonwealth Editions, 2003.
Selesky, Harold E. *War and Society in Colonial Connecticut*. New Haven: Yale University Press, 1990.
Sherburne, John H., Jr. *Battery A Field Artillery M.V.M. 1895–1905*. Boston: Sparrell, 1908.
Sligh, Robert Bruce. *The National Guard and National Defense: The Mobilization of the Guard in World War II*. New York: Praeger, 1992.
Snyder, Claire R. *Citizen-Soldiers and Manly Warriors: Military Service and Gender in the Civic Republican Tradition*. New York: Rowman & Littlefield, 1999.
Southwick, Sally Jo. *Building on a Borrowed Part: Place and Identity in Pipestone, Minnesota*. Columbus: Ohio University Press, 2005.
Squadron A: A History of Its First Fifty Years. New York: Ex-Members of Squadron A. 1939.
Stanton, Shelby L. *Order of Battle U.S. Army World War II*. Novato, CA: Presidio Press, 1984.
Stentiford, Barry M. *The American Home Guard: The State Militia in the Twentieth Century*. College Station: Texas A&M University Press, 2002.
Stevens, William B. *History of the Fiftieth Regiment of Infantry Massachusetts Volunteer Militia in the Late War of the Rebellion*. Boston: Griffith-Stillings Press, 1907.
Stewart, George R. *Names on the Land*. New York: Random House, 1945.
Taylor, George Rogers. *The Transportation Revolution of 1815–1860*. New York: Rinehart, 1951.
Town of Wakefield, Massachusetts. *Inaugural Exercises in Wakefield Mass*. Wakefield: Warren Richardson, 1872.
Vaughan, Alden T. *New England Frontier: Puritans and Indians, 1620–1775*, 3d ed. Norman: University of Oklahoma Press, 1995
Vourlojianis, George N. *The Cleveland Grays: An Urban Military Company, 1837–1919*. Kent, OH: Kent State University Press, 2002.
Warner, Sam Bass, Jr. *Streetcar Suburbs: The Process of Growth in Boston, 1870–1900*, 2d ed. Cambridge: Harvard University Press, 1979.
Watson, Bruce. *Bread & Roses: Mills, Migrants, and the Struggle for the American Dream*. New York: Viking Penguin, 2005.
Weaver, Michael E. *Guard Wars: The 28th Infantry Division in World War II*. Bloomington: Indiana University Press, 2010.
Wiebe, Robert H. *The Search for Order 1877–1920*. New York: Hill & Wang, 1967.
Willis, Henry A. *Fitchburg in the War of the Rebellion*. Fitchburg, MA: Stephen Shepley, 1866.
Wilson, John B. *Armies, Corps, Divisions, and Separate Brigades*. Washington: Center of Military History, 1987.

Wyant, William K. *Sandy Patch: A Biography of Lt. Gen. Alexander M. Patch*. New York: Praeger, 1991.
Yankee Division Veterans' Association. *History of the 26th Yankee Division 1917–1919 1941–1945*. Salem, MA: Deschamps Bros., 1955.
Young, L. Murray. *Iron Men and Iron Machines: Wakefield Fire Department Wakefield Mass*. Magnolia, MA: Dick Weir, 1976.
Zelner, Kyle F. *A Rabble in Arms: Massachusetts Towns and Militiamen during King Philip's War*. New York: New York University Press, 2009.

Journal Articles

Adamson, Jeremy Elwell. "The Wakefield Rattan Company." *Antiques* (August 1992): 214–221.
Blewett, Mary H. "Work, Gender, and the Artisan Tradition in New England Shoemaking, 1780–1860." *Journal of Social History* 17 (Winter 1983) 221–48.
Boucher, Ronald L. "The Colonial Militia as a Social Institution: Salem, Massachusetts 1764–1775." *Military Affairs* 37 (December 1973): 125–30.
Cooper, Jerry M. "National Guard Reform, the Army, and the Spanish-American War: The View from Wisconsin." *Military Affairs* 42 (February 1978): 20–23.
Gildrie, Richard P. "Defiance, Division, and the Exercise of Arms: The Several Meanings of Colonial Training Days in Colonial Massachusetts." *Military Affairs* 52 (April 1988): 53–55.
London, Lena. "The Militia Fine 1830–1860." *Military Affairs* 15 (Fall 1951): 133–44.
Lyons, Richard L. "The Boston Police Strike of 1919." *The New England Quarterly* 20 (June 1947): 147–168.
Pushkar, Robert. "Cyrus Wakefield's Smart Idea." *Yankee* (July 1997).
Radabaugh, Jack S. "The Militia of Colonial Massachusetts." *Military Affairs* 18 (January 1954): 1–18.
Whitehorne, J.W.A. "The Survival of the Duquesne Grays, 1917." *Military Affairs* 50 (October 1986): 179–84.

Dissertations

Radabaugh, Jack S. "The Military System of Colonial Massachusetts, 1690–1740." Ph.D. dissertation, University of Southern California, 1965.
Rutman, Darrett Bruce. "A Militant New World, 1607–40: America's First Generation, Its Martial Spirit, Its Tradition of Arms, Its Martial Spirit, Its Wars." Ph.D. dissertation, University of Virginia, 1959.

Newspapers

Middlesex Journal (Woburn)   *Wakefield Banner*
Reading Chronicle   *Wakefield Daily Item*

Interviews

Buckle, James E, (1940–1944). 14 April 2006, and 1 August 2007, with author.
D'alessandro, Joseph J. (E, 1940–1944), 10 April 2006, with author.
McCarthy, James J., (L, 1939–1943), 10 July 2001, with John Coates. Part of the Morse Institute Library (Natick, Massachusetts, Community Library) Veterans Oral History Project.
Saxe, Haskell (L, 1941–1945), 23 October 2001, with Hackell Saxe. Part of the Morse Institute Library (Natick, Massachusetts, Community Library) Veterans Oral History Project.
Tibbetts, Donald, MSG (Ret.), Co.s L & K (Malden) MA-ARNG, 10 April 2006, with author.

Index

Adjuntas, Puerto Rico 97
African Americans 16, 70, 82, 88, 93
Air Force, U.S. 168, 169
Air National Guard 169
Americal Civic Center 175, 177, 178
Americal Division 10, 154–166, 172, 173
American Federation of Labor (AFL) 112
American Legion 124, 131, 142
American Revolution *see* War of Independence
Ancient and Honorable Artillery Company of Massachusetts 34, 115, 134, 142
annual training 42, 64, 87, 90, 113, 131, 137, 143
Arkansas 9, 56
armories 1–3, 11, 32, 35, 41, 44, 49, 53, 54, 60, 64, 67, 77–80, 86, 88, 89, 98, 99, 102, 108–112, 116, 118, 119–122, 126–131, 135, 137, 141–145, 164, 166, 168–175, 177–181
Army Reserve 143, 169, 172, 173, 199n75
Australia 150, 152, 162
automobiles 5, 6, 78, 102, 127, 137, 140, 171

Baltimore, Maryland 46, 59, 60, 89
bands, musical 53, 85, 98, 102, 118, 135, 144, 176
banquets 1, 2, 59, 61, 77, 79, 83, 89, 99, 102, 104, 106, 116, 122, 127, 135, 151, 169, 170
Baptists 3, 19, 25, 43, 65, 177, 180
Bay State Rifle Range 118, 124, 131
Beebe, Lucius 33, 41, 42, 66, 71
Berlin Crisis 173
Blacks *see* African Americans
Boston 2, 4, 10, 13, 18–22, 24, 26–28, 30, 32–34, 36, 37–40, 44–46, 48, 50–53, 55, 59, 60, 63, 68, 69, 74, 79, 80, 82, 83, 87–90, 97–99, 119, 126–129, 133, 137, 140–144, 146, 163, 169, 171, 179
Boston Police Strike 128–129, 133
Bougainville 157–159, 162, 165, 170
Boy Scouts 122, 135
Brown, James G. 129, 141, 143, 146, 147, 149, 197n75
Buckle, James C. 150, 159–160, 162, 197n52
Bull Run, First Battle of 2, 44, 49, 51, 53

Camp Alger 88, 90–93
Camp Andrew 48
Camp Bartlett 123
Camp Curtis Guild 131, 135, 138, 139, 146, 171–173, 177, 196n53–55
Camp Darling 121
Camp Edwards 138, 144–151
Camp Greene 123
Camp Massachusetts 48, 50
Camp Meigs 60–61, 190n84
Camp Plunkett 125
Camp Robert Bancroft 128

Camp Stanton 54
Carolina Maneuvers 147, 149
Carpenter, George O. 31–34, 42, 48, 135
Carter, James H. 2, 48, 73, 75
Catholics, 2, 3, 37, 72, 84, 122, 164
cavalry 16, 17, 20, 22–23, 32, 57, 95, 103, 106, 107, 111, 114, 141, 164, 98
Civil War 1–10, 25, 35, 38, 43–67, 70, 81, 83, 86, 88–89, 102–104, 118, 121, 127, 141, 170
Cold War 5
commissioned officers 23, 45, 89, 90, 96–97, 99, 101, 121, 124, 134, 143–145, 151, 154–155, 166
Concord, Massachusetts 7, 16–19, 31, 34, 45, 90, 141, 176, 177
Congregational Church 3, 15, 20, 22, 25, 32, 43, 44, 65–66, 74–76, 88, 171, 177
Connecticut 6, 7, 25, 30, 46, 55, 79, 107, 115, 142
Connelly, Edward J. 113, 115–117, 120–124, 129, 131–132, 134, 136, 141–142, 145, 170, 180
conscription 11, 17, 108, 117, 121, 126, 128, 143, 145, 146, 156, 163, 172–174; *see also* Selective Service
Constitution, U.S. 21, 23, 28, 45, 108, 114, 121
Continental Army 19, 45, 176
Coolidge, Calvin 126

D'Alessandro, Joseph J 154, 198n20
demobilization 127, 129, 133, 165, 166
Democratic Republicans 21–22, 38
Department of Defense 168, 174
Derthick, Martha 8
Desert Storm 172
Dick, Charles 107
Dolbeare, Richard B. 141, 147, 149, 156, 163, 196, 197, 198n14
draft *see* conscription
drill 2, 11, 18, 21, 22, 24, 35, 41–43, 47–48, 51–54, 57, 59, 62, 88–89, 91–92, 97, 107–108, 111, 116, 118, 120, 123, 126, 129, 137, 143, 145, 163, 168–169, 171, 173, 189n34, 197n17, 197n44
drill pay 113–116, 130, 139, 143

Eaton, John 51
Eaton, Joseph L.R. 31, 34
Eaton, Lilly 33, 42, 102
Eaton, William E. 102, 163
election of officers 1, 8, 42, 53, 83, 113, 134, 180
Emerson, James F. 31
Emerson, Thomas 24, 32–34, 36, 42, 48
enlisted men 18, 28, 45, 48, 53, 57, 70, 87–89, 91, 99, 121, 129, 134, 144–147, 150, 151, 154, 157, 160, 163, 166–167
enlistments 50, 58, 86, 121, 125, 128, 145, 163, 169
enrolled militia 23–24, 28, 38, 134, 187n59

Federalist Party 21
5th Massachusetts Regiment 45–46, 49–50, 52, 134
50th Massachusetts 53–59, 62, 189n45
Fiji Islands Group, 156–157, 160
Fine Members Association 3, 65, 72, 73, 75, 98, 101, 116, 119, 122, 135, 136, 141, 142, 172, 174
fire 25, 49, 64, 69, 70, 75, 76, 77, 109, 110, 140, 169, 170, 172, 178, 179
fire departments 3, 10, 12, 25, 64, 72–73, 76, 87, 109, 169, 170, 181, 127
flood 57, 141, 172
Florida 24–25, 141, 174
Fort Bragg 147, 150
Fort (Camp) Devens 121–123, 127, 137, 146–147
Fort Revere 119–122
4th Pioneer Regiment 123, 124
Fourth Pioneers 123, 125
Framingham, Massachusetts 3, 46, 68, 70, 87–99, 113, 117, 121, 123, 126
French and Indian War 17–19, 34, 45
Fyrd, Anglo-Saxon 185n14

Garde National 8
Gihon, Edward J. 2–3, 83–84, 87, 89, 91, 92, 94–99, 101, 103, 135, 137–138, 142, 164, 192n50, 194n30 196n7 199n83
Grand Army of the Republic (GAR) 3, 7, 65, 72, 87, 92, 98, 101, 102, 103, 121, 122, 127, 135
Gray, Frank 2, 84, 90, 101–102
Gray, William E. 88–89
Great War see World War I
Greenwood, Wakefield, Massachusetts 14, 15, 30, 47, 73, 131–133, 172, 196
Gromlie, George 147, 199
Guadalcanal 154–156, 162–163, 165, 197–200
Guanica, Puerto Rico 94–97

Hartshorn, James 20, 22
Henry H. Miller Piano Company 68, 74, 119
Heywood Brothers and Company 79–80, 110, 112, 130
Heywood Brothers and Wakefield Company 130, 140
high school cadets 3, 33, 65, 67, 87, 130, 135, 139
The Hiker 135–136, 177, 196n10
Hill, Jim Dan 35
Hill 260, Bougainville 157–162, 164, 170
home guard 88–90, 125–126, 134, 145–146, 189n45
horses 16, 24, 69, 72, 76, 117, 135, 156
hurricanes 137, 142, 172

ice 37–38, 139–140
Indians *see* Native Americans
Ipswich River 14–15
Irish 2, 37, 38, 50, 72, 122
Italians 2, 72, 112

Junior Reserve Officers Training Corps (JROTC) 130

Keough, James H. 71
King Philip's War (Metacom's Rebellion) 16–17, 66
Korean War 6, 168–169, 171–172

Lawrence, Massachusetts Strike 111–113
Lincoln, Abraham 43–45, 47, 52–53, 60
Littlefield, Samuel F. 48, 53, 60–62, 72, 99
Locke, John W. 44, 47, 53, 189
Loyalists 19
Lynn, Massachusetts 13–14, 16, 60, 79, 112, 118, 139
Lynnfield, Massachusetts 40, 45, 53, 63, 71, 118, 139, 141, 171, 179

Maine 14, 82, 149–150, 186
Malden, Massachusetts 14, 17, 23, 34, 53, 137, 142, 145, 149, 154, 173, 196n16
Marchetti, Frank 144, 147, 197
Marines 86, 95, 107, 121, 124–125, 131, 135, 154–158, 169
Maryland 36, 46, 59–60
Masons 21, 72
Massachusetts Rifle Association 124
McCall, Samuel 103, 125
McMahon, John H. 91, 101, 103, 109, 110–111, 113, 127, 139, 142–143, 180
medals 71, 124, 127, 135, 154, 155
Melrose, Massachusetts 45, 53, 60, 90, 118, 137, 141, 144, 173, 175, 196
Metropolitan Police 112, 128, 194n35
Mexican American War 1, 29, 32, 45, 81, 188
Mexican border 113, 118
Middlesex Regiment 15–17
Miles, Nelson 93–96, 193
Militia Act of 1792 23, 30, 107
Militia Act of 1903 105, 107, 114–115
Militia Bureau 139
militia tradition 3, 5, 8, 16, 24, 31, 35, 62, 85, 88–89, 106, 114, 116, 163, 175
Miller Piano Company 68, 74, 119
Minute Men 7, 17–19, 45
mobilization 54, 60, 85–86, 88, 108, 113, 127, 128, 134, 141, 143, 146, 155–156, 169, 169; *see also* demobilization
Montrose, Wakefield, Massachusetts 15, 64

National Defense Act of 1916 8, 114–115, 121
National Defense Act of 1920 128, 130
National Guard Association 104, 106–108
National Guard Bureau 139, 143, 173
National Rifle Association 70, 139
Native Americans (Indians) 1, 7, 13, 14–17, 34, 45, 62, 66, 76, 93
naval militia 106, 107, 115, 124
Navy 6, 19, 36, 54, 56, 65, 81, 86, 103, 121, 125, 131, 146, 154, 156, 168–169
Nelson, Harry E. 124, 131
New Caledonia 150–155, 165–166, 197, 198
New England 1, 7, 15
New Hampshire 7, 26, 39, 169
New York 6, 21, 30, 46, 55, 56, 60, 70, 106, 120, 152, 168
North Carolina 53, 82, 123, 147
North Dakota 9, 153
North Reading, Massachusetts 15, 22, 26, 38, 43, 45, 54, 60, 71, 76, 137, 151, 169, 171
North Regiment 16, 30, 134

Officers Candidate School 134, 163, 166, 199n81
Officers' Reserve Corps 139, 145
Officers Training School 120, 134
OHIO ("Over the Hill in October" protests) 147
182d Infantry 134, 137, 141–175, 179
organized militia 5–11, 30, 38, 59–59, 88, 90, 104–107, 112, 114–115, 121, 125, 129, 181

parades 1, 3, 11, 22, 35–36, 50, 62–63, 76, 87, 98, 102–103, 116–118, 122, 126–127, 130, 135, 142, 172, 177
patriotism 6, 32, 51, 80–81, 86–87
Pennsylvania 52, 135, 150, 168
Philippines 89, 91, 97, 99, 101, 113, 135, 162–163, 165
police 102, 122, 128, 133, 135, 141, 179; *see also* Metropolitan Police; State Police

Index

Port Hudson 45, 55–59, 141
posse 16
prisoners of war 17, 19, 53, 57, 59
Puerto Rico 91, 93–97, 99, 101, 113, 135–138, 147
Puritanism 7, 13, 15–16, 65, 122

railroads 3–5, 24, 28–31, 37–41, 44, 46, 49, 60, 93, 112, 122, 140, 163
Reading, Massachusetts (modern town of) 2–3, 13, 21–22, 26, 29, 31, 34, 38–39, 43, 45, 53–54, 60–61, 65–66, 71–72, 76, 79, 82–83, 86–87, 89–91, 99, 115, 118, 131, 137, 141, 143–146, 151, 153, 155, 171, 177
recruiting 32, 34, 51, 53, 90, 92, 113, 117–118, 120, 128–129, 137, 144, 163, 168, 173
Regular Army 1, 6, 8, 50, 61, 65, 70, 82, 90, 96–98, 113–114, 119, 121, 138, 148, 150, 152, 156, 158, 163, 172
reserve components 169, 173–174
Richardson, Solon O. 26–27, 29, 29, 33, 35–37, 39 41, 48, 53, 66, 68–69, 76, 102, 174, 178
Richardson, Solon O., II 2, 26, 36, 54, 64, 68, 71–72, 79, 102–104, 117, 127, 136
Richardson Sherry Wine Bitters 26–27, 79
riots 77, 164, 172
Rockery 75, 135, 176–177, 180
Rogers, Fred H. 109, 110, 114, 118, 121, 124
Roosevelt, Franklin D. 143
Route 128 140, 171

Saugus, Massachusetts 13, 14, 60, 144, 155
Second Amendment 23
Second World War *see* World War II
Selective Service 121, 143–145, 163
shoe industry 24, 26, 33, 54, 83, 87, 140, 170
Sixth Massachusetts Regiment 1–3, 46, 50, 52, 62, 82–101, 110, 113–114, 117–119, 121–123, 129, 137–138
slavery 19, 43, 65
South Carolina 43, 55, 89, 93, 150–15
Spanish American War 4, 10, 62, 65, 84, 86, 88, 92–94, 99–101, 103–104, 108, 113–114, 116, 118, 122, 125–127, 135–137, 143, 164, 177, 180
sports 113, 116, 147
State Guard 129, 133, 143, 145, 149, 163–164, 166, 168, 170
State Police 128, 172, 194n35
street railroads 79, 83, 89, 183
Swedish 2, 65

Total Force 174
26th (Yankee) Division 123–124, 127, 134, 138, 147, 149, 150, 161

uniforms 3, 10, 46, 63, 65, 73, 85, 89–90, 98, 104, 107, 115–116, 122, 127, 130, 141–142, 151–152, 170

Unitarian Church 3, 177
United Spanish War Veterans 101, 120, 122, 135, 136, 142, 193n3
United States Military Academy (West Point) 51, 90, 120, 124, 160
unorganized militia 88, 107
Upton, Emory 6, 8

Vietnam War 6, 8, 11, 32, 172–177
Villa, Pancho 113, 118
Virginia 47–51, 82, 90–91
volunteers 5, 6, 28, 33, 45, 46, 51, 56, 58, 70, 72–73, 81–82, 92–93, 95, 99, 101, 108, 135

Wakefield, Cyrus 33, 39–42, 64, 66–71, 74, 80, 108, 110, 130, 170, 178–179
Wakefield, Cyrus, II 69, 71, 74
Wakefield, Cyrus, III 133
Wakefield Rattan Company 40, 69–70, 74, 79–80, 130, 140
Wakefield Rifle Range 71, 118, 125, 131
Walton, Charles E. 89–90, 93, 101, 116, 122
War Department 47, 114, 129, 137, 143, 149, 153, 154, 156, 168; *see also* Department of Defense
war memorials 80, 104, 132, 133, 136, 180
War of 1812 1, 7, 20, 21, 23, 29, 30, 34, 45
War of Independence (Revolutionary War) 1, 7, 11, 14, 19–23, 106, 141, 175–176, 190n78
Washington, George 21, 67, 170, 178
Washington, D.C. 45–52, 59–61, 81, 89, 90, 116
Washington Rifle Grays (Greens) 1, 22, 23, 28, 29, 32–34, 51, 61, 187n82
wicker furniture 70, 80, 140, 177
Wiley, John, Jr. 29, 31, 34, 42, 51, 52, 54, 83
Wilmington, Massachusetts 17, 30, 31, 54, 126
Wilson, Woodrow 113, 116, 118, 121
women 20, 35, 43, 91, 112, 124, 151, 163, 164
Woodend (Reading Third Parish) 15
Woodward, Charles F. 65, 79, 82–83, 85, 87–89, 92, 94, 96, 99–101, 136
Woodward, James 33, 43
Woodward, Thomas 24
Works Project Administration (WPA) 138, 139, 141
World War I 65, 104, 114–115, 117–134, 136, 138, 139–140, 143–145, 147, 149–150, 171
World War II 6, 10–11, 42, 72, 115, 117, 134, 137, 140, 153–167, 170

Yale, Burrage 25–26

Zouaves 50, 189n34

www.ingramcontent.com/pod-product-compliance
Ingram Content Group UK Ltd.
Pitfield, Milton Keynes, MK11 3LW, UK
UKHW050528150426
5217IPUK00026B/1850